Principles of Food Sanitation

Fourth Edition

Norman G. Marriott, PhD
Extension Food Scientist and Professor
Department of Food Science and Technology
Virginia Polytechnic Institute and State University
Blacksburg, Virginia

A Chapman & Hall Food Science Book

AN ASPEN PUBLICATION®
Aspen Publishers, Inc.
Gaithersburg, Maryland
1999

The author has made every effort to ensure the accuracy of the information herein. However, appropriate information sources should be consulted, especially for new or unfamiliar procedures. It is the responsibility of every practitioner to evaluate the appropriateness of a particular opinion in the context of actual clinical situations and with due considerations to new developments. The author, editors, and the publisher cannot be held responsible for any typographical or other errors found in this book.

Aspen Publishers, Inc., is not affiliated with the American Society of Parenteral and Enteral Nutrition.

Library of Congress Cataloging-in-Publication Data

Marriott, Norman G.
Principles of food sanitation/Norman G. Marriott.—4th ed.
p. cm.
Includes bibliographical references and index.
ISBN 0-8342-1232-3
1. Food industry and trade—Sanitation. 2. Food handling.
I. Title.
TP373.6.M37 1998
664—dc21
98-43373
CIP

Orders: (800) 638-8437
Customer Service: (800) 234-1660

About Aspen Publishers • For more than 35 years, Aspen has been a leading professional publisher in a variety of disciplines. Aspen's vast information resources are available in both print and electronic formats. We are committed to providing the highest quality information available in the most appropriate format for our customers. Visit Aspen's Internet site for more information resources, directories, articles, and a searchable version of Aspen's full catalog, including the most recent publications: **http://www.aspenpublishers.com**
Aspen Publishers, Inc. • The hallmark of quality in publishing
Member of the worldwide Wolters Kluwer group.

Editorial Services: Denise H. Coursey
Library of Congress Catalog Card Number: 98-43373
ISBN: 0-8342-1232-3

Printed in the United States of America

1 2 3 4 5

*To my family and fellow NBA (Noontime Basketball Association) players
who served as an inspiration during the revision of this book.*

Table of Contents

Preface .... **xi**

Chapter 1   **Sanitation and the Food Industry** **1**
Why Sanitation? 1
What Is Sanitation? 2
Sanitation Laws and Regulations 4
Voluntary Sanitation Programs 8
Establishment of Sanitary Practices 8
Summary 9
Study Questions 10

Chapter 2   **The Relationship of Microorganisms to Sanitation** **11**
How Microorganisms Relate to Food Sanitation 11
What Causes Microorganisms To Grow 16
Effects of Microbes on Spoilage 20
Effects of Microorganisms on Foodborne Illness 21
Foodborne Illnesses 22
Microbial Destruction 37
Microbial Growth Control 39
Microbial Load Determination 40
Diagnostic Tests 46
Summary 50
Study Questions 51

Chapter 3   **Food Contamination Sources** **53**
Transfer of Contamination 54
Contamination of Foods 55
Other Contamination Sources 56
Protection against Contamination 57
Summary 58
Study Questions 59

Chapter 4   Personal Hygiene and Sanitary Food Handling **60**
Personal Hygiene 60
Sanitary Food Handling 72
Summary 73
Study Questions 74

Chapter 5   The Role of HACCP in Sanitation **75**
What Is HACCP? 75
Interface with GMPs and SSOPs 81
HACCP Principles 81
Organization, Implementation, and Maintenance 86
Summary 88
Study Questions 89

Chapter 6   Quality Assurance **90**
The Role of Total Quality Management 91
Quality Assurance for Effective Sanitation 91
Organization for Quality Assurance 92
Establishment of a Quality Assurance Program 95
Summary 112
Study Questions 112

Chapter 7   Cleaning Compounds **114**
Soil Characteristics 114
Effects of Surface Characteristics on Soil Deposition 116
Soil Attachment Characteristics 117
Cleaning Compound Characteristics 119
Classification of Cleaning Compounds 121
Cleaning Auxiliaries 128
Scouring Compounds 130
Cleaning Compound Selection 131
Handling and Storage Precautions 131
Summary 137
Study Questions 138

Chapter 8   Sanitizers **139**
Sanitizing Methods 139
Summary 155
Study Questions 156

Chapter 9   Sanitation Equipment **158**
Sanitation Costs 158
Equipment Selection 159
Cleaning Equipment 160

Sanitizing Equipment 182
Lubrication Equipment 184
Summary 184
Study Questions 186

Chapter 10   Waste Product Disposal **187**
Strategy for Waste Disposal 188
Solid Waste Disposal 192
Liquid Waste Disposal 192
Summary 205
Study Questions 206

Chapter 11   Pest Control **207**
Insect Infestation 207
Insect Destruction 211
Rodents 215
Birds 220
Use of Pesticides 220
Integrated Pest Management 222
Summary 224
Study Questions 225

Chapter 12   Sanitary Design and Construction of Food Facilities **226**
Site Selection 226
Site Preparation 226
Building Construction Considerations 227
Processing and Design Considerations 229
Pest Control Design 230
Construction Materials 230
Summary 230
Study Questions 231

Chapter 13   Low-Moisture Food Manufacturing and Storage Sanitation **232**
Sanitary Construction Considerations 232
Receipt and Storage of Raw Materials 236
Cleaning of Low-Moisture Food Manufacturing Plants 240
Summary 241
Study Questions 242

Chapter 14   Dairy Processing Plant Sanitation **243**
Role of Pathogens 244
Sanitary Construction Considerations 246
Soil Characteristics in Dairy Plants 247
Sanitation Principles 248

Cleaning Equipment 251
Summary 256
Study Questions 256

Chapter 15 Meat and Poultry Plant Sanitation **258**
Role of Sanitation 258
Sanitation Principles 264
Cleaning Compounds for Meat and Poultry Plants 266
Sanitizers for Meat and Poultry Plants 267
Sanitation Practices 268
Sanitation Procedures 270
Troubleshooting Tips 281
Summary 281
Study Questions 282

Chapter 16 Seafood Plant Sanitation **283**
Sanitary Construction Considerations 283
Contamination Sources 285
Sanitation Principles 286
Recovery of By-Products 288
Voluntary Inspection To Enhance Sanitation 289
Summary 289
Study Questions 290

Chapter 17 Fruit and Vegetable Processing Plant Sanitation **291**
Contamination Sources 291
Sanitary Construction Considerations 292
Cleaning Considerations 294
Cleaning of Processing Plants 295
Cleaners and Sanitizers 297
Cleaning Procedures 298
Evaluation of Sanitation Effectiveness 300
Summary 300
Study Questions 302

Chapter 18 Beverage Plant Sanitation **303**
Mycology of Beverage Manufacture 303
Sanitation Principles 303
Nonalcoholic Beverage Plant Sanitation 304
Brewery Sanitation 307
Winery Sanitation 312
Distillery Sanitation 318

Summary		319
Study Questions		320

Chapter 19   Food Service Sanitation **321**
Sanitary Design 321
Contamination Reduction 324
Sanitary Procedures for Food Preparation 325
Sanitation Principles 326
Food Service Sanitation Requirements 341
Summary 343
Study Questions 344

Chapter 20   Management and Sanitation **345**
Management Requirements 345
Employee Selection 347
Management of a Sanitation Operation 349
Total Quality Management 351
Summary 352
Study Questions 352

Appendix A—Glossary **353**

Index **357**

Preface

In this era of emphasis on food safety, high-volume food processing and preparation operations have increased the need for improved sanitary practices from processing to consumption. This trend presents a challenge for the food processing and food preparation industry.

Sanitation is an applied science for the attainment of hygienic conditions. Because of increased emphasis on food safety, sanitation is receiving additional attention from those in the food industry. During the past, inexperienced employees with few skills who have received little or no training have been given sanitation tasks. Still, sanitation employees should have knowledge about the attainment of hygienic conditions. In the past, these employees, including sanitation program managers, have had only limited exposure to this subject. Technical information has been limited primarily to a number of training manuals provided by regulatory agencies, industry and association manuals, and recommendations from equipment and cleaning compound firms. Most of this material lacks specific information about the selection of appropriate cleaning methods, equipment, compounds, and sanitizers for maintaining hygienic conditions in food processing and preparation facilities.

The purpose of this text, as with previous editions, is to provide sanitation information needed to ensure hygienic practices. Sanitation is a broad subject; thus, principles related to contamination, cleaning compounds, sanitizers, and cleaning equipment, as well as specific directions for applying these concepts to attain hygienic conditions in food processing or food preparation operations, are discussed.

The discussion starts with the importance of sanitation and also includes information about regulations. To enable the reader to understand more fully the fundamentals of food sanitation, Chapter 2 is devoted to microorganisms and their effects on food products. Updated information is provided on pathogenic microorganisms and rapid microbial determination methods. A discussion of contamination sources and hygiene has been updated (Chapters 3 and 4), including how management can encourage improved sanitation. Chapter 5, a new topic, provides updated information on Hazard Analysis Critical Control Points (HACCP).

Chapter 6 is about quality assurance (QA) and sanitation. Information given here presents specific details on how to organize, implement, and monitor an effective program.

Chapter 7 discusses cleaning compounds and contains updated information on this subject. It examines characteristics of soil deposits and identifies the appropriate generic cleaning compounds for the removal of various soils. Also, it looks at how cleaning compounds function, identifies their chemical and physical properties, and offers information on their appropriate handling. Because of the importance of sanitizing, Chapter 8 discusses sanitizers and their characteristics. Specific generic compounds for various equipment and areas, as well as updated information on such compounds, are discussed.

Chapter 9 provides updated information on cleaning and sanitizing equipment best suited for various applications in the food industry. It provides detailed descriptions, including new illustrations of most cleaning equipment that may be used in food processing and food preparation facilities.

Waste product handling, which is a major challenge for the food industry, is discussed in detail in Chapter 10. This chapter contains updated information about the treatment and monitoring of liquid wastes. Pest control is another problem for the food industry. Chapter 11 discusses common pests found in the food industry; their prevention, including chemical poisoning; Integrated Pest Management (IPM) and biological control; and the potential advantages and limitations of each method. The importance of sanitary design and construction, a new topic, is reviewed in Chapter 12.

Because sanitation is so important in low-moisture food processing, dairy, meat and poultry, seafood, fruit and vegetable, and beverage plants, a chapter is devoted to each of these areas. These chapters (13 through 18) present information on plant construction, cleaning compounds, sanitizers, and cleaning equipment that applies to those segments of the industry. These chapters provide the food industry with valuable guidelines for sanitation operations and specific cleaning procedures.

Chapter 19 is devoted entirely to sanitation in the food service industry. It provides instructions on how to clean specific areas and major equipment found in a food service operation.

Effective management practices can promote improved sanitation, a topic addressed in Chapter 20. The intent is not to provide an extensive discussion of management principles, but to suggest how effective management practices can improve sanitation.

This book is intended to provide a concise discussion about sanitation of low-, intermediate-, and high-moisture foods. It can be used as a text for college students and in continuing education courses about sanitation. It will serve as a reference for food processing courses, industry-sponsored courses, and the food industry itself.

Appreciation is expressed to Mrs. LuAnn Oliver for typing the manuscript. Also, I remember the support of my family during the preparation of this revised edition.

CHAPTER 1

Sanitation and the Food Industry

Because the food industry has become larger and more diversified, sanitary practices have become more complex. Modern and large-volume operations have increased the need for workers to understand sanitary practices and how to attain and maintain hygienic conditions. Workers who comprehend the logic and biologic basis behind these practices will become more effective sanitation workers.

WHY SANITATION?

More processing is conducted now at plants near the area of production, a trend that should continue. Many of these food plants are hygienically designed; nevertheless, foods can be contaminated with spoilage microorganisms or those that cause foodborne illness if proper sanitary practices are not followed. However, hygienic and safe foods can be produced with sanitary practices, even in older plants. Sanitary practices can be as important to the wholesomeness and safety of food as are the characteristics of the physical plant.

Scientific advances in food processing, preparation, and packaging during the past century have contributed to improved food acceptability that is less expensive. However, with increased productivity, convenience foods and other processed foods are affected by problems created through advanced technology. The major problems have been food contamination and waste disposal.

Few programs provide formal training in food sanitation. Only a limited number of institutions offer even one course related to food sanitation. Limited resource materials are available to sanitarians. Only a few manuals are published by trade associations and regulatory agencies.

Gravani (1997) stated that never in recent history have Americans been more concerned about the food supply than they are today. He stated that an estimated 6.5 to 33 million people become ill each year from pathogenic microorganisms in food, and approximately 9,000 die. The national economic impact is estimated to be $6.5 billion to $35 billion per year. This problem is exacerbated by 26% of consumers not washing cutting boards and up to 50% undercooking some foods. Thus, it is necessary for consumers to learn new behavioral patterns through effective education.

Many food processing and food service operators offer excuses for poor sanitation in their establishment(s). Yet, the reasons for not establishing such a program are more compelling, because they relate to the bottom line of a profit and loss statement. A *sanitation program* is "a planned way of practicing sanitation." It results in a number of crucial benefits for both the public and the businesses conducting the program. The word *sanitation* is not a "dirty" word!

Most owners or managers of food facilities want a clean operation. However, unsanitary operations frequently result from a lack of understanding of the principles of sanitation and the benefits that effective sanitation will provide. The following brief discussion of these benefits shows that *sanitation* is not a "dirty" word.

1. Inspection is becoming more stringent because inspectors are relying more on modern methods of microbial and chemical determination to establish compliance. Thus, an effective sanitation program is essential.
2. An effective sanitation program can prevent trouble. Foodborne illness can be controlled when sanitation is properly implemented in food operations. Common problems caused by poor sanitation are food spoilage through off-odor and flavor. Spoiled foods are disliked by consumers and cause reduced sales and increased claims. Off-condition products convey the lack of an effective sanitation program.
3. An effective sanitation program can improve product quality and shelf life because the microbial population can be reduced. Increased labor, trim loss, and packaging costs due to poor sanitation can cause a decrease of 5% to 10% of profit of meat operations in a supermarket. A rigid sanitation program can increase the storage life of food.
4. An effective sanitation program includes regular cleaning and sanitizing of heating, air conditioning, and refrigeration equipment, reducing energy and maintenance costs. Dirty, clogged coils harbor microorganisms; blowers and fans spread these flora throughout the establishment. Coil cleaning and sanitizing lower the risk of airborne contamination. Clean coils are more effective heat exchangers and can reduce costs by up to 20%. Insurance carriers may reduce rates for clean establishments because of fewer slips and falls on greasy floors.
5. Various, less tangible benefits of an effective sanitation program include: (a) improved product acceptability, (b) increased product shelf life, (c) improved customer relations, (d) reduced public health risks, (e) increased trust of compliance agencies and inspectors, (f) decreased product salvaging, and (g) improved employee morale.

WHAT IS SANITATION?

The word *sanitation* is derived from the Latin word *sanitas*, meaning "health." Applied to the food industry, *sanitation* is "the creation and maintenance of hygienic and healthful conditions." It is the application of a science: to provide wholesome food handled in a clean environment by healthy food handlers, to prevent contamination with microorganisms that cause foodborne illness, and to minimize the proliferation of food spoilage microorganisms. Effective sanitation refers to the mechanisms that help accomplish these goals.

Sanitation: An Applied Science

Sanitation is an applied science that incorporates the principles of maintenance, restoration, or improvement of hygienic practices and conditions. Food sanitation is an applied sanitary science related to the processing, preparation, and handling of food. Sanitation applications refer to hygienic practices designed to maintain a clean and wholesome environment for food production, preparation, and storage. Sanitation is equated with more than cleanliness. It is responsible for the improvement of the aesthetic qualities of homes, commercial operations, and public facilities. Also, applied sanitary science can improve waste disposal (see Chapter 10), which results in less pollution and an improved ecologic balance. Therefore, food sanitation and general sanitary practices for our environment should be closely allied.

Sanitation is considered to be an applied science because of its importance to the protection of human health and its affiliation with those segments of the environment that relate to health. Therefore, this applied science relates to the physical, chemical, and biological factors that constitute the environment. Sanitation scientists must thoroughly understand the living organisms that are most likely to affect human health. Microorganisms can be responsible for food spoilage and foodborne illness, but they can also be beneficial in the processing and/or preparation of foods. Thus, sanitation seeks to control microorganisms so that they are beneficial instead of detrimental.

Sanitation: An Application of Food Safety Practices

Proper sanitation practices are important in maintaining food safety. Poor hygienic practices can contribute to outbreaks of foodborne illnesses, as the following examples illustrate.

Chunky

Several years ago, an outbreak of foodborne illness was traced to a brand of chocolate candy containing fruit and nuts. The manufacturer of Chunky candy bars responded by recalling the product and removing from the marketplace those bars still on the shelves. An extended time period was required before the problem was solved, and consumer confidence in the product was restored.

Starlac

Starlac, one of the leading brands of powdered milk during the past, was produced by a large food-manufacturing company. However, several years ago, a foodborne illness outbreak caused by *Salmonella* was traced to this product. Adverse publicity caused the product to be withdrawn from the market, and the brand name was discontinued.

Bon Vivant

This company produced a line of canned gourmet foods until a foodborne illness outbreak related to *Clostridium botulinum* was traced to one of its products. The unfavorable publicity forced the company to go out of business.

A Supermarket

A woman purchased chicken legs from the heated delicatessen case in her supermarket. Later, she vomited and suffered other symptoms of foodborne illness so severe that she was hospitalized. Subsequent examination of the chicken revealed that it contained *Staphylococcus aureus*. The point of contamination was never identified, but the fact that this foodstuff resulted in serious illness points to the importance of hygienic practices.

A U.S. Navy Ship

A foodborne illness outbreak affecting 28 sailors that occurred aboard a U.S. Navy ship was caused by shrimp salad. An inspection of the food preparation facilities failed to reveal any conditions that could have contributed to food infection. However, examination of the personnel revealed that the employee who prepared the salad had a draining sore on his thumb. Microbial analyses of the drainage from the employee's thumb revealed the same strain of *S. aureus* that infected the salad.

A Country Club Buffet

An outbreak of acute gastrointestinal illness followed a buffet served to approximately 855 people at a New Mexico country club. *Staphylococcus aureus* phage type 95 was isolated from the turkey and dressing that was served and from some of the food handlers' nares and stools.

These are examples of the importance of sanitation during food processing or preparation, as well as proper cleaning and sanitizing of food service and food manufacturing facilities and equipment. The consequences of improper sanitation are severe: loss of sales and profits, damaged product acceptability and consumer confidence, adverse publicity, and, sometimes, legal action. Sanitary practices can prevent these problems. Moreover, consumers have the right to expect wholesome and safe food products.

The nutrients in food that nourish humans also provide nutrition for the microorganisms that cause foodborne illness. Foods, especially fresh meats, that are packaged for self-service retail cases have a large surface area exposed for a long period of time. Modern merchandising demands an effective sanitation program so that fresh foods can have the required shelf life and remain attractive. Fresh food products with poor stability will not look good long enough to be sold. If off-condition products are sold to consumers, they become an inferior example for the food industry.

SANITATION LAWS AND REGULATIONS

Because thousands of laws and regulations are currently in effect in the United States to control the food and related industries in the preparation and manufacture of wholesome food, this treatise cannot and should not address all of these rules. Thus, it is not the intent of this chapter or this book to emphasize the details of food handling, preparation, or processing regulations. Only the major agencies involved with food safety and their primary responsibilities will be discussed. The reader should consult regulations available from various jurisdictions to determine specific requirements for the food operation and area where it is located. It is inappropriate to discuss regulatory requirements for cities and countries because they have designated governmental entities with their own criteria related to food safety (Bauman, 1991) that differ from one area to another and can change periodically.

Sanitation requirements developed by legislative bodies and regulatory agencies in response to public demands are spelled out in laws and regulations. They are not static but change in response to sanitation and other production problems brought to public attention.

Laws are passed by legislators and must be signed by the chief executive. After a law has been passed, the agency responsible for its enforcement prepares *regulations* designed to implement the intention of the law, or act. Regulations are developed to cover a wide range of requirements and are more specific and detailed than are laws. Regulations for food provide standards for building design, equipment design, commodities, tolerances for chemical or other food additives, sanitary practices and qualifications, labeling requirements, and training for positions that require certification.

Regulation development is a multistep process. For example, in the federal process, the relevant agency prepares the proposed regulation, which is then published as a proposed rule in the *Federal Register*. Accompanying the proposal is information related to background. Any comments, suggestions, or recommendations are to be directed to the agency, usually within 60 days after proposal publication, although time extensions are frequently provided. The regulation is published in final form after comments on the proposal have been reviewed, with another statement of how the comments were handled and specifying effective dates for compliance. This statement suggests that comments on matters not previously considered in the regulations may be submitted for further review. Amendments may be initiated by any individual, organization, other government office, or by the agency itself. A petition is necessary, with appropriate documents that justify the request.

There are two types of regulations: *substantive* and *advisory*. Substantive regulations are the more important because they have the power of law. Advisory regulations are intended to serve as guidelines. Sanitation regulations are substantive because food must be made safe for the public. In regulations, the use of the word *shall* means a requirement, whereas *should* implies a recommendation. We will look at some regulations important to sanitation by various governmental agencies.

Food and Drug Administration Regulations

The Food and Drug Administration (FDA), responsible for enforcing the Food, Drug, and Cosmetic Act, as well as other statutes, has wide-ranging authority. It is under the jurisdiction of the U.S. Department of Health and Human Services.

The FDA has impacted the food industry, especially in controlling adulterated foods. Under the Food, Drug, and Cosmetic Act, food is considered to be adulterated if it contains any filth or putrid and/or decomposed material or if it is otherwise unfit as food. This act states that food prepared, packed, or held under unsanitary conditions that may cause contamination from filth or that is injurious to health is *adulterated*. The act gives the FDA inspector authority, after proper identification and presentation of a written notice to the person in charge, to enter and inspect any establishment where food is processed, packaged, or held for shipment in interstate commerce or after shipment. Also, the inspector has the authority to enter and inspect vehicles used to transport or hold food in interstate commerce. This official can check all pertinent equipment, finished products, containers, and labeling.

Adulterated or misbranded products that are in interstate commerce are subject to seizure. This is considered a civil action against merchandise that has been examined by the FDA and determined to be in violation of the act. Although the FDA initiates action through the federal district courts, seizure is performed by the U.S. Marshal's office.

Legal action can also be taken against an organization through an injunction. This form of legal action is usually taken when serious violations occur. However, the FDA can prevent interstate shipments of adulterated or misbranded products by requesting a court injunction or restraining order against the involved firm or individual. This order is effective until the FDA is assured that the violations have been corrected. To correct flagrant violations, the FDA has taken legal steps against finished products made from interstate raw materials, even though they were never shipped outside the state.

The FDA does not approve cleaning compounds and sanitizers for food plants by their trade names. However, the FDA regulations indicate approved sanitizing compounds by their chemical names. For example, sodium hydrochlorite is approved for "bleach-type" sanitizers, sodium or potassium salts of isocyanuric acid for "organic chlorine" sanitizers, n-alkyldimethylbenzyl-ammonium chloride for quaternary ammonium products, sodium dodecylbenzenesulfonate as an acid anionic sanitizer component, and oxypolyethoxyethanol–iodine complex for iodophor sanitizers. A statement of maximum allowable-use concentrations for these compounds without a potable water rinse on product contact surfaces after use is also provided.

Good Manufacturing Practices

On April 26, 1969, the FDA published the first Good Manufacturing Practice (GMP) regulations, commonly referred to as the *umbrella* GMPs. These regulations deal primarily with sanitation in manufacturing, processing, packing, or holding food.

The sanitary operations section establishes basic rules for sanitation in a food establishment. General requirements are provided for the maintenance of physical facilities; cleaning and sanitizing of equipment and utensils; storage and handling of clean equipment and utensils; pest control; and the proper use and storage of cleaning compounds, sanitizers, and pesticides. Minimum demands for sanitary facilities are included through requirements for water, plumbing design, sewage disposal, toilet and hand-washing facilities and supplies, and solid waste disposal.

Specific GMPs supplement the umbrella GMPs and emphasize wholesomeness and safety of manufactured products. Each regulation covers a specific industry or a closely related class of foods. The critical steps in the processing operations are addressed in specific detail, including time-and-temperature relationships, storage conditions, use of additives, cleaning and sanitizing, testing procedures, and specialized employee training.

According to Marriott et al. (1991), inspection has been used by regulatory agencies to ensure food safety. However, this approach has limitations because laws that are supposed to be upheld by inspectors are frequently not clearly written, and what constitutes compliance is questionable. Furthermore, it is difficult to distinguish between requirements critical to safety and those related to aesthetics. Laws are not always specific; for example, some GMPs contain phrases such as "clean as frequently as necessary." Because of this vagueness, the umbrella GMPs do not have the force of law.

U.S. Department of Agriculture Regulations

The U.S. Department of Agriculture (USDA) has jurisdiction over three areas of food processing, based on the following laws: the Federal Meat Inspection Act, the Poultry Products Inspection Act, and the Egg Products Inspection Act. The agency that administers the area of inspection is the Food Safety and Inspection Service (FSIS), which was established in 1981.

By design, federal jurisdiction usually involves only interstate commerce. However, the three statutes on meat, poultry, and eggs have extended USDA jurisdiction to the intrastate level if state inspection programs are unable to provide proper enforcement as required by federal law. Products shipped from official USDA-inspected plants into distribution channels and subsequently found to be adulterated or misbranded come under the jurisdiction of the Food, Drug, and Cosmetic Act. The FDA can take legal steps to remove this product from the market. Normally, the product is referred back to the USDA for disposition.

Environmental Regulations

Provisions of numerous statutes related to the environment, many of which affect food processing establishments, are enforced by the Environmental Protection Agency (EPA).

Environmental regulations that affect sanitation of the food facility include the Federal Water Pollution Control Act; Clean Air Act; Federal Insecticide, Fungicide, and Rodenticide Act (FIFRA); and the Resource Conservation and Recovery Act.

The EPA is involved in the registration of sanitizers by both their trade and chemical names. Sanitizing compounds are recognized by federal regulators as pesticides; thus, their uses are derived from the FIFRA. Antimicrobial efficacy, toxicologic profiles, and environmental impact information are required by the EPA. Furthermore, specific label information and technical literature that detail recommended use of applications and directions are required. Disinfectants must be identified by the phrase "It is a violation of federal law to use this product in a manner inconsistent with its labeling."

Federal Water Pollution Control Act

This act is important to the food industry because it provides for an administrative permit procedure for controlling water pollution. The National Pollutant Discharge Elimination System (NPDES), which is under this permit system, requires that industrial, municipal, and other point-source dischargers obtain permits that establish specific limitations on the discharge of pollutants into navigable waters. The purpose of this permit is to effect the gradual reduction of pollutants discharged into streams and lakes. Effluent guidelines and standards have been developed specific to industry groups or product groups. Regulations covering meat products and selected seafood products, grain and cereal products, dairy products, selected fruit and vegetable products, and beet and cane sugar refining have been published by the EPA.

Clean Air Act

This act, devised to reduce air pollution, gives the EPA direct control over polluting sources in the industry, such as emission controls on automobiles. Generally, state and local agencies set pollution standards based on EPA recommendations and are responsible for their enforcement. This statute is of concern to the food operation that may discharge air pollutants through odors, smokestacks, incineration, or other methods.

Federal Insecticide, Fungicide, and Rodenticide Act

The FIFRA gives the EPA control of the manufacture, composition, labeling, classification, and application of pesticides. Through the registration provisions of the act, the EPA must classify each pesticide either for restricted use or for common use, with periodic reclassification and registration as necessary. A pesticide classified for restricted use must be applied only by or under the direct supervision and guidance of a certified applicator. Those who are certified, either by the EPA or by a state, to use or supervise the use of restricted pesticides must meet certain standards, demonstrated through written examination and/or performance testing. Commercial applicators are required to have certain standards of competence in the specific category in which they are certified.

Current EPA regulations permit the use of certain residual insecticides for crack and crevice treatment in food areas of food establishments. The EPA lists residual pesticides that are permitted in crack and crevice treatment during an interim period of 6 months, while registrants apply for label modification.

Resource Conservation and Recovery Act

The Resource Conservation and Recovery Act was provided to develop a national program designed to control solid waste disposal. The act authorizes the EPA to recommend guidelines in cooperation with federal, state, and local agencies for solid waste management. It also authorizes funds for research, construction, disposal, and utilization projects in solid waste management at all regulator levels.

VOLUNTARY SANITATION PROGRAMS

As the size and complexity of the food industry has increased, it has become more expensive to inspect facilities continuously. Thus, it has been essential to investigate avenues to reduce costs by altering the traditional inspection procedures. During the past, the USDA and FDA have proposed voluntary programs to replace continuous inspection with constant supervision.

Hazard Analysis Critical Control Points

Although other voluntary programs have been developed, Hazard Analysis Critical Control Points (HACCP) is the approach that is being emphasized. Since this concept was developed jointly by the Pillsbury Company, the National Aeronautics and Space Administration, and the U.S. Army Natick Laboratories in 1971 for use in the food processing industry, it has been adopted as a voluntary or mandatory program to ensure food safety through the prevention of hazards. A major portion of the hazards in the food supply is microbial-related and is affected by the effectiveness of sanitary measures adopted. Thus, HACCP has been recognized as a voluntary program to enhance sanitation. Although this program was initially voluntary, more recent regulations have been developed that require HACCP in parts of the food industry and have changed the status of this program from voluntary to required. Because of the importance of HACCP to effective sanitation, this subject will be discussed in detail in Chapter 5.

ESTABLISHMENT OF SANITARY PRACTICES

The employer is responsible for establishing and maintaining sanitary practices to protect the public health and maintain a positive image. The problem of establishing as well as applying and maintaining sanitary practices within the food industry is the challenge of the sanitarian or the food technologist in charge of sanitation. The sanitarian must ensure that the practices adopted are essential to public health and are economical. The sanitarian is both the guardian of public health and the counselor to management in quality control as influenced by sanitary practices.

A large food-processing company should have a separate sanitation department in the general office, on the same level as production or research, that is in charge of all operating plants. A sanitation department should exist in each plant on a level with other plant departments. In a large organization, sanitation maintenance should be separated from production and mechanical maintenance, an arrangement that will enable the sanitation department to exercise company-wide surveillance of sanitary practices and, thus, to maintain a high level of sanitation. Production practice, quality control, and sanitary practice do not always appear compatible when subjected to the administration of a single department or individual; however, all of these functions are complementary and are best performed when properly coordinated and synchronized. Although production deserves major consideration, the proper application of sanitary practices is essential to ensure efficient and effective production. More discussion about the subject will be included in Chapter 6.

Ideally, an organization should have a full-time sanitarian with assistants, but this is not always practical. Instead, a trained individual who was originally employed as a quality-control technician, a production foreman, a superintendent, or some other individual experienced in production can be charged with the responsibility of the sanitation operation. This situation is fairly common and usually effective. However, unless the sanitarian has an assistant to take care of some of the routine tasks and is given sufficient time for proper attention to sanitary details, the program may not succeed.

A one-person quality assurance department with a full schedule of control work will be generally inadequate to assume the tasks of a sanitarian. However, with proper assistance, quality assurance and sanitation supervision can be successfully conducted by a single qualified individual who knows how to divide his or her effort between sanitation and quality assurance. It is beneficial for this person to have the advice and service of an outside agency, such as a university, trade association, or private consultant, to avoid becoming submerged in the conflicting interests of different departments. The extra expense can be a worthy investment.

A planned sanitation maintenance program is essential to meet legal requirements; to protect brand and product reputation; and to ensure product safety, quality, and freedom from contamination. All phases of food production and plant sanitation should be included in the program to supplement the cleaning and sanitizing procedures for equipment and floors. A sanitation program should start with surveying and monitoring the raw materials that enter the facility because these items are potential contamination sources.

The sanitation survey should be comprehensive and critical. As each item is considered, the ideal solution should be noted, irrespective of cost. When the survey is completed, all items should be reevaluated and more practical and/or economic solutions determined. Aesthetic sanitary practices should not be adopted without clear evidence of their ability to pay dividends in increased sales or because they are necessary to meet competitive sales pressure.

SUMMARY

Large-volume food-processing and/or preparation operations have increased the need for sanitary practices and hygienic conditions in the food industry. Even in hygienically designed plants, foods can be contaminated with spoilage microorganisms or those causing foodborne illness if proper sanitary practices are not followed.

Sanitation is the creation and maintenance of hygienic and healthful conditions. It is an applied science that incorporates principles regarding the maintenance of hygienic practices and conditions. It is also considered to be an application of food safety practices.

The Food, Drug, and Cosmetic Act covers all food commodities—except meat and poultry products—in interstate commerce, from harvest through processing and distribution channels. Meat and poultry products are under the jurisdiction of the USDA. GMP regulations are specific requirements developed to establish criteria for sanitation practices. A number of statutes related to pollution control of the air, water, and other resources is enforced by the EPA.

The progressive food processing or preparation firm should take responsibility for establishing and maintaining sanitary practices. However, new regulations have changed the status of HACCP from voluntary to required by parts of the industry. HACCP is the current trend and may be replaced by total quality management in the future. A planned sanitation program is essential to meet regulatory requirements, protect brand and product reputation, and ensure product safety, quality, and freedom from contamination.

STUDY QUESTIONS

1. What is sanitation?
2. What is a law?
3. What is a regulation?
4. What is an advisory regulation?
5. What is a substantive regulation?

REFERENCES

Bauman, H.E. 1991. Safety and regulatory aspects. In *Food product development,* ed. E. Graf and I.S. Saguy, 133. New York: Van Nostrand Reinhold.

Gravani, R.B. 1997. Coordinated approach to food safety education is needed. *FoodTech* 51, no. 7: 160.

Marriott, N.G. et al. 1991. *Quality assurance manual for the food industry.* Virginia Cooperative Extension, Virginia Polytechnic Institute and State University, Publication No. 458-013.

SUGGESTED READINGS

Collins, W.F. 1979. Why food safety? In *Sanitation notebook for the seafood industry*, eds. G.J. Flick Jr., et al. Blacksburg, VA: Virginia Polytechnic Institute and State University.

Guthrie, R.K. 1988. *Food sanitation*. 3rd ed. New York: Van Nostrand Reinhold.

Katsuyama, A.M. 1980. *Principles of food processing sanitation*. Washington, DC: The Food Processors Institute.

National Restaurant Association Educational Foundation. 1992. *Applied foodservice sanitation*. 4th ed. New York: John Wiley & Sons, Inc. In cooperation with the Education Foundation of the National Restaurant Association, Chicago.

CHAPTER 2

The Relationship of Microorganisms to Sanitation

To understand the principles of food sanitation, knowledge of the role of microorganisms in food spoilage and foodborne illness is needed. Microorganisms (also called *microbes* and *microbial flora*) are found throughout the natural environment. They cause food spoilage through color and flavor degradation and foodborne infection when food that contains microorganisms of public health concern is ingested. Microorganisms are also important in connection with food sanitation because certain disease-producing microbes may be transmitted through food. Sanitation practices are needed to combat the proliferation and activity of food spoilage and food-poisoning microorganisms.

HOW MICROORGANISMS RELATE TO FOOD SANITATION

Microbiology is the science of microscopic forms of life known as *microorganisms*. Knowledge of microorganisms is important to the sanitation specialist because their control is part of a sanitation program.

What Are Microorganisms?

A microorganism is a microscopic form of life found on all nonsterilized matter that can be decomposed. The word is of Greek origin and means "small" and "living beings." These organisms metabolize in a manner similar to humans. They intake nourishment, discharge waste products, and reproduce. It is believed that these tiny organisms were discovered in 1693 by a Dutch merchant, Anton van Leeuwenhoek, through lenses that he developed. Approximately 100 years ago, Louis Pasteur and his associates concluded that microorganisms cause disease and discovered that heat could destroy these pathogens. Joseph Lister, an English physician, further supported the theory that disease and infection are caused by invading microorganisms through his attempt to prevent infection in surgical patients by the use of antiseptic techniques. The work of these pioneers encouraged further study of microorganisms and resulted in our subsequent knowledge of how to control them.

Most foods are highly perishable because they contain nutrients required for microbial growth. To reduce food spoilage and to eliminate foodborne illness, this microbial proliferation must be controlled. Food deterioration should be minimized to prolong the time during which an acceptable level of flavor and wholesomeness can be maintained. If proper sanitation practices are not followed during food processing, preparation, and serving, the rate and extent of the deteriorative changes that lead to spoilage will increase.

Microorganisms Common to Food

A major challenge for the sanitarian is to protect the production area and other involved locations against microorganisms that can reduce the wholesomeness of food. Microorganisms are omnipresent. They can infect and affect food, with dangerous consequences to consumers. The microorganisms most common to food are *bacteria* and *fungi*. The fungi, which are less common than bacteria, consist of two major microorganisms: molds (which are multicellular) and yeasts (which are usually unicellular). Bacteria, which usually grow at the expense of fungi, are unicellular. Viruses, although transmitted more from person to person than via food, should also be mentioned because they can present a problem among unhealthy employees.

Molds

Molds are multicellular microorganisms (eukaryotic cells) with mycelial (filamentous) morphology. They consist of tubular cells, ranging from 30 to 100 μm in diameter, called *lyphae,* which form a macroscopic mass called a *mycelium*. Molds are characterized by their display of a variety of colors and are generally recognized by their mildewy or fuzzy, cottonlike appearance. They can develop numerous tiny spores that are found in the air and can be spread by air currents. These can produce new mold growth if they are transferred to a location that has conditions conducive to germination. Molds generally withstand greater variations in pH than do bacteria and yeasts and can frequently tolerate greater temperature variations. Although molds thrive best at or near a pH of 7.0, a range from 2.0 to 8.0 can be tolerated, though an acid-to-neutral pH is preferred. Molds thrive better at ambient temperature than in a colder environment, even though growth can occur below 0°C. Although most molds prefer a minimum water activity (A_w) of approximately 0.90, growth of a few osmiophilic molds can and do occur at as low as 0.60. (Water activity is explained later in this chapter.) At an A_w of 0.90 or higher, bacteria and yeasts grow more effectively and normally utilize available nutrients for growth at the expense of molds. When the A_w goes below 0.90, molds are more likely to grow. Foods such as pastries, cheeses, and nuts that are low in moisture content are more likely to spoil from mold growth.

Molds have been considered very beneficial and very troublesome, ubiquitous microorganisms. They often work in combination with yeasts and bacteria to produce numerous indigenous fermented foods and are involved in industrial processes to produce organic acids and enzymes. However, molds are a major contributor to annual food product recalls and can complicate food processing and storage. Most molds do not cause health hazards, but some produce mycotoxins under certain conditions that are toxic to humans.

Almost any food can be invaded by mold growth. Grains, nuts, vegetables, and fruits are susceptible to mold contamination prior to harvesting and during storage. This mold may spread throughout the food processing chain. Molds are airborne easily, but this is a possible mode of contamination during processing. Molds cause various degrees of visible deterioration and decomposition of foods. Their growth is identifiable through rot spots, scabs, slime, cottony mycelium, or colored sporulating mold. Molds may produce abnormal flavors and odors due to fermentative, lipolytic, and proteolytic changes caused by enzymatic reactions with carbohydrates, fats, and proteins in foods.

Molds have an absolute requirement for oxygen and are inhibited by high levels of carbon dioxide (5% to 8%). Their diversity is evident through the ability of some molds to function as oxygen scavengers and to grow at very low levels of oxygen and even in vacuum packages. Some halophilic molds can tolerate salt concentration at over 20%.

Because molds are difficult to control, food processors have encountered spoilage problems. In the past, 6,000 cases of ready-to-eat pudding were recalled because of mold contamination (FDA, 1996a). During 1996, two manufacturers of fruit juice issued recalls on products contaminated with mold (FDA, 1996b).

Yeasts

Yeasts are generally unicellular. They differ from bacteria in their larger cell sizes and morphology, and because they produce buds during the process of reproduction by fission. The generation time of yeasts is slower than that of bacteria, with a typical time of 2 to 3 hours in foods, leading from an original contamination of one yeast/g of food to spoilage in approximately 40 to 60 hours. Like molds, yeasts can be spread through the air or by other means and can alight on the surface of foodstuffs. Yeast colonies are generally moist or slimy in appearance and creamy white. Yeasts prefer an A_w of 0.90 to 0.94, but can grow below 0.90. In fact, some osmiophilic yeasts can grow at an A_w as low as 0.60. These microorganisms grow best in the intermediate acid range, a pH of from 4.0 to 4.5. Yeasts are more likely to grow on foods with lower pH and on those that are vacuum-packaged. Food that is highly contaminated with yeasts will frequently have a slightly fruity odor.

Bacteria

Bacteria are unicellular microorganisms (prokaryotic cells) that are approximately 1 μm in diameter, with morphology variation from short and elongated rods (bacilli) to spherical or ovoid forms. Cocci (meaning "berry") are spherically shaped bacteria. Individual bacteria closely combine in various forms, according to genera. Some sphere-shaped bacteria occur in clusters similar to a bunch of grapes (e.g., staphylococci). Other bacteria (rod-shaped or sphere-shaped) are linked together to form chains (e.g., streptococci). Also, certain genera of sphere-shaped bacteria are formed together in pairs (diploid formation), such as pneumococci. Microorganisms such as sarcinia form as a group of four (tetrad formation). Other genera appear as an individual bacterium. Some bacteria possess flagella and are motile.

Bacteria produce pigments ranging from variations of yellow to dark shades, such as brown or black. Certain bacteria have pigmentation of intermediate colors—red, pink, orange, blue, green, or purple. These bacteria cause food discoloration, especially among foods with unstable color pigments, such as meat. Some bacteria also cause discoloration by slime formation.

Some species of bacteria also produce spores, some of which are resistant to heat, chemicals, and other environmental conditions. Many of these spore-forming bacteria are thermophilic microorganisms that produce a toxin that will cause foodborne illness.

Viruses

Viruses are infective microorganisms with dimensions that range from 20 to 300 nm, or about 1/100 to 1/10 the size of a bacterium. Most viruses can be seen only with an electron microscope. A virus particle consists of a single molecule of DNA or RNA, surrounded by a coat made from protein. Viruses cannot reproduce outside of another organism and are obligate parasites of all living organisms, such as bacteria, fungi, algae, protozoa, higher plants, and invertebrate and vertebrate animals. When a protein cell becomes attached to the surface of the appropriate host cell, either the host cell engulfs the virus particle or the nucleic acid is injected from the virus particle into the host cell, as with bacteriophages active against bacteria.

In animals, some infected host cells may die, but others survive infection with the virus and resume their normal function. It is not necessary for the host cells to die for the host organism—in the case of humans—to become ill (Shapton and Shapton, 1991). Viruses are transmitted to food by workers who are carriers. An infected food handler can excrete the organism through the feces and respiratory tract infection. Transmission occurs through coughing, sneezing, touching a runny nose, and from not washing the hands after using the toilet. The inability of host cells to perform their normal function causes illness. After the normal function is reestablished, recovery from illness occurs. The inability of viruses to reproduce themselves outside the host and their small size complicates their isolation from foods suspected of being the cause of illness in humans. There is no evidence of the human immunodeficiency virus (HIV) (acquired immune deficiency syndrome [AIDS]) being transmitted by foods. Viruses can be destroyed by sanitizers, such as the iodophors (see Chapter 8).

A virus that has caused a major increase in outbreaks in restaurants during the past 10 years is hepatitis. Apparently, intravenous drug use is one factor that accounts for some of this rise. Infectious hepatitis A can be transmitted through food that has not been handled in a sanitary manner. Symptoms include nausea, cramps, and, sometimes, jaundice, which can last from several weeks to several months. A major source of hepatitis is raw shellfish from polluted waters. The most likely foods to transmit viral illnesses are those handled frequently and those that receive no heating after handling,

such as sandwiches, salads, and desserts. Because this disease is highly contagious, it is mandatory that employees handling food practice thorough hand-washing after using the toilet, before handling food and eating utensils, and after diapering, nursing, or feeding infants. Viruses also cause diseases such as influenza and the common cold.

Microbial Growth Kinetics

With minor exceptions, multiplication of microbial cells by binary fission occurs in a growth pattern of various phases, according to the typical microbial growth curve illustrated in Figure 2–1.

Lag Phase

After contamination occurs, the period of adjustment (or adaptation) to the environment, with a slight decrease in microbial load due to stress (Figure 2–1), followed by limited growth in the number of microbes, is called the *lag phase* of microbial growth. The lag phase can be extended with less microbial proliferation through reduced temperature. This increases the "generation interval" of microorganisms. Microbial proliferation can also be retarded by decreasing the number of microbes that contaminate food, equipment, or buildings. When initial counts of microorganisms are lowered through improved sanitation and hygienic practices, initial contamination will be reduced; the lag phase may be extended and entry into the next growth phase deferred. Figure 2–2 illustrates how differences in temperature and initial contamination load can affect microbial proliferation.

Logarithmic Growth Phase

Bacteria multiply by binary fission, characterized by the duplication of components within each cell, followed by prompt separa-

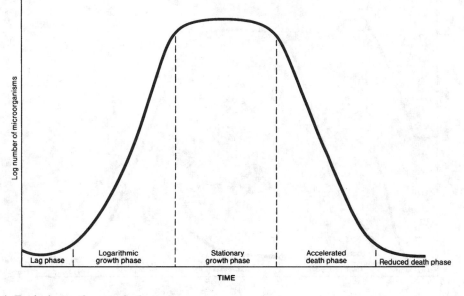

Figure 2–1 Typical growth curve for bacteria.

tion to form two daughter cells. During this phase, the number of microorganisms increases to the point that, when cells divide, the increase in number of microbes occurs at an exponential rate until some environmental factor becomes limiting. The length of this phase may vary from 2 to several hours. The number of microorganisms and environmental factors, such as nutrient availability and temperature, affect the logarithmic growth rate of the number of microorganisms. Effective sanitation to reduce the microbial load can limit the number of microbes that can contribute to microbial proliferation during this growth phase.

Stationary Growth Phase

When environmental factors such as nutrient availability, temperature, and competition from another microbial population become limiting, the growth rate slows and reaches an equilibrium point. Growth becomes relatively constant, resulting in the stationary phase. During this phase, the number of microorganisms is frequently large enough that their metabolic by-products and competition for space and nourishment reduce proliferation to the point that it is nearly stopped, is stopped, or a slight decrease in the microbial proliferation occurs. The length of this phase usually ranges from 24 hours to more than 30 days but depends on both the availability of energy sources for the maintenance of cell viability and the degree of pollution in (hostility of) the environment.

Accelerated Death Phase

Lack of nutrients, effect of metabolic waste products, and competition from other microbial populations contribute to the rapid death of microbial cells at an exponential rate. The accelerated death rate is similar to the logarithmic growth rate and may range from 24 hours to 30 days but depends on temperature, nutrient supply, microbial genus and species, age of the microorganisms, application of sanitation tech-

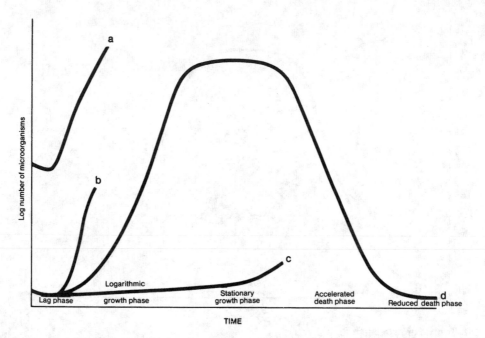

Figure 2–2 The effect of amount of initial contamination and length of lag phase on the growth curve of microorganisms. *a,* high initial contamination and poor temperature control (short lag phase). *b,* low initial contamination and poor temperature control (short lag phase). *c,* low initial contamination and rigid temperature control (long lag phase). *d,* typical growth curve.

niques and sanitizers, and competition from other microbial flora.

Reduced Death Phase

This phase is nearly the opposite of the lag phase. It is caused by a sustained accelerated death phase, so that the microbial population number is decreased to the extent that the death rate decelerates. After this phase, the organism has been degraded, sterilization has occurred, or another microbial population continues decomposition.

WHAT CAUSES MICROORGANISMS TO GROW

Factors that affect the rate of proliferation of microorganisms can be categorized as *extrinsic* and *intrinsic*.

Extrinsic Factors

Extrinsic factors relate to the environmental factors that affect the growth rate of microorganisms.

Temperature

Microbes have an optimum, minimum, and maximum temperature for growth. Therefore, the environmental temperature determines not only the proliferation rate but also the genera of microorganisms that will thrive and the extent of microbial activity that occurs. For example, a change of only a few degrees in temperature may favor the growth of entirely different organisms and result in a different type of food spoilage and foodborne illness. These characteristics have been responsible for the use of temperature as a method of controlling microbial activity.

The optimal temperature for the proliferation of most microorganisms is from 14°C to 40°C, although some microbes will grow below 0°C, and other genera will grow at temperatures up to and exceeding 100°C.

Microbes classified according to temperature of optimal growth include:

1. Thermophiles (high-temperature-loving microorganisms), with growth optima at temperatures above 45°C. Examples are *Bacillus stearothermophilus, Bacillus coagulans,* and *Lactobacillus thermophilus*.
2. Mesophiles (medium-temperature-loving microorganisms), with growth optima between 20°C and 45°C. Examples are most lactobacilli and staphylococci.
3. Psychrotrophs (cold-temperature-tolerant microorganisms), which tolerate and thrive at temperatures below 20°C. Examples are *Pseudomonas* and *Moraxella-Acinetobacter*).

Bacteria, molds, and yeasts each have some genera with temperature optimal in the range characteristic of thermophiles, mesophiles, and psychrotrophs. Molds and yeasts tend to be less thermophilic than do bacteria. As the temperature approaches 0°C, fewer microorganisms can thrive, and their proliferation is slower. As the temperature falls below approximately 5°C, proliferation of spoilage microorganisms is retarded, and growth of nearly all pathogens ceases.

Oxygen Availability

As with temperature, availability of oxygen determines which microorganisms will be active. Some microorganisms have an absolute requirement for oxygen. Others grow in the total absence of oxygen, and others grow either with or without available oxygen. Microorganisms that require free oxygen are called *aerobic* microorganisms (*Pseudomonas* species is an example). Those that thrive in the absence of oxygen are called *anaerobic* microorganisms (i.e., *Clostridium* species). Microorganisms that grow with or without the presence of free oxygen are called *facultative* microorganisms (e.g., *Lactobacillus* species).

Relative Humidity

This extrinsic factor affects microbial growth and can be influenced by temperature. All microorganisms have high requirements for water to support their growth and activity. A high relative humidity can cause moisture condensation on food, equipment, walls, and ceilings. Condensation causes moist surfaces, which are conducive to microbial growth and spoilage. Also, microbial growth is inhibited by a low relative humidity.

Bacteria require a higher humidity than do yeasts and molds. Optimal relative humidity for bacteria is 92% or higher, whereas yeasts prefer 90% or higher. Molds thrive more if the relative humidity is 85% to 90%.

Intrinsic Factors

Intrinsic factors that affect the rate or proliferation relate more to the characteristics of the substrates (foodstuff or debris) that support or affect growth of microorganisms.

Water Activity

Because microorganisms require water, a reduction of water availability will reduce microbial proliferation. It is important to recognize that it is not the total amount of moisture present that determines the limit of microbial growth but the amount of moisture that is readily available for metabolic activity. The unit of measurement for water requirement of microorganisms is usually expressed as water activity (A_w), defined as the vapor pressure of the subject solution divided by the vapor pressure of the pure solvent: $A_w = p \div p_o$, where p is the vapor pressure of the solution and p_o is the vapor pressure of pure water. The approximate optimal A_w for the growth of many microorganisms is 0.99, and most microbes require an A_w higher than 0.91 for growth. The approximate relationship between fractional equilibrium relative humidity (RH) and A_w is RH=$A_w \times 100$. Therefore an A_w of 0.95 is approximately equivalent to an RH of 95% in the atmosphere above the solution. Most natural food products have an A_w of approximately 0.99. Generally, bacteria have the highest water activity requirements of the microorganisms. Molds normally have the lowest A_w requirement, and yeasts are intermediate. Most spoilage bacteria do not grow at an A_w below 0.91, but molds and yeasts can grow at an A_w of 0.80 or lower. Molds and yeasts are more likely to grow in partially dehydrated surfaces (including food), whereas bacterial growth is retarded.

pH

pH is a measurement of $\log_{10}$ of the reciprocal of the hydrogen ion concentration (g/L) and is represented as pH=$\log_{10}$[H$^+$]. The pH for optimal growth of most microorganisms is near neutrality (7.0). Yeasts can grow in an acid environment and thrive best in an intermediate acid (4.0 to 4.5) range. Molds tolerate a wider range of pH (2.0 to 8.0), although their growth is generally greater with an acid pH. Molds can thrive in a medium that is too acid for either bacteria or yeasts. Bacterial growth is usually favored by near-neutral pH values. However, acidophilic (acid-loving) bacteria will grow on food or debris down to a pH of approximately 5.2. Below 5.2, micro-

bial growth is dramatically reduced from that in the normal pH range.

Oxidation-Reduction Potential

The oxidation-reduction potential is an indication of the oxidizing and reducing power of the substrate. To attain optimal growth, some microorganisms require reduced conditions; others need oxidized conditions. Thus, the importance of the oxidation-reduction potential is apparent. All saprophytic microorganisms that are able to transfer hydrogen as H$^+$ and E$^-$ (electrons) to molecular oxygen are called *aerobes*. Aerobic microorganisms grow more rapidly under a high oxidation-reduction potential (oxidizing reactivity). A low potential (reducing reactivity) favors the growth of anaerobes. Facultative microorganisms are capable of growth under either condition. Microorganisms can alter the oxidation-reduction potential of food to the extent that the activity of other microorganisms is restricted. For example, anaerobes can decrease the oxidation-reduction potential to such a low level that the growth of aerobes can be inhibited.

Nutrient Requirements

In addition to water and oxygen (except for anaerobes), microorganisms have other nutrient requirements. Most microbes need external sources of nitrogen, energy (carbohydrates, proteins, or lipids), minerals, and vitamins to support their growth. Nitrogen is normally obtained from amino acids and nonprotein nitrogen sources. However, some microorganisms utilize peptides and proteins. Molds are the most effective in the utilization of proteins, complex carbohydrates, and lipids because they contain enzymes capable of hydrolyzing these molecules into less complex components. Many bacteria have a similar capability, but most yeasts require the simple forms of these compounds. All microorganisms need minerals, but requirements for vitamins vary. Molds and some bacteria can synthesize enough B vitamins for their needs, whereas other microorganisms require a ready-made supply.

Inhibitory Substances

Microbial proliferation can be affected by the presence or absence of inhibitory substances. Substances or agents that inhibit microbial activity are called *bacteriostats*. Those that destroy microorganisms are called *bactericides*. Some bacteriostatic substances, such as nitrites, are added during food processing. Most bactericides are utilized as a method of decontaminating foodstuffs or as a sanitizer for cleaned equipment, utensils, and rooms. (Sanitizers are discussed in detail in Chapter 8.)

Interaction between Growth Factors

The effects that factors such as temperature, oxygen, pH, and A_w have on microbial activity may be dependent on each other. Microorganisms generally become more sensitive to oxygen availability, pH, and A_w at temperatures near growth minima or maxima. For example, bacteria may require a higher pH, A_w and minimum temperature for growth under anaerobic conditions than when aerobic conditions prevail. Microorganisms that grow at lower temperatures are usually aerobic and generally have a high A_w requirement. Lowering A_w by adding salt or excluding oxygen from foods (such as meat) that have been held at a refrigerated temperature dramatically reduces the rate of microbial spoilage. Normally, some microbial growth

occurs when any one of the factors that controls the growth rate is at a limiting level. If more than one factor becomes limiting, microbial growth is drastically curtailed or completely stopped.

Role of Biofilms

Biofilms, which were discovered in the mid 1970s, are microcolonies of bacteria closely associated with an inert surface attached by a matrix of complex polysaccharide-like material in which other debris, including nutrients and microorganisms, may be trapped. A biofilm is a unique environment that microorganisms generate for themselves, enabling the establishment of a "beachhead" on a surface resistant to intense assaults by sanitizing agents. When a microorganism lands on a surface, it attaches itself with the aid of filaments or tendrils. The organism produces a polysaccharide-like material, a sticky substance that will cement in a matter of hours the bacteria's position on the surface and act as a glue to which nutrient material will adhere with other bacteria and, sometimes, viruses. The bacteria become entrenched on the surface, clinging to it with the aid of numerous appendages.

A biofilm builds upon itself, adding several layers of the polysaccharide material populated with microorganisms, such as *Salmonella, Listeria, Pseudomonas,* and others common to the specific environment. Increased time of organism contact with the surface contributes to the size of the microcolonies formed, amount of attachment, and difficulty of removal. The biofilm will eventually become a tough plastic that often can be removed only by scraping. Although cleaned surfaces may be sanitized, a firmly established biofilm has layers of organisms that may be protected from the sanitizer. Biofilm buildup can be responsible for portions of it being sheared off by the action of food or liquid passing over the surface. Because the shear force is greater than the adherence force in the topmost layers of the biofilm, chunks of the polysaccharide cement, with the accompanying microbial population, will be transferred to the product, with subsequent contamination.

There has been additional interest in biofilms since the mid 1980s because it has been demonstrated that *Listeria monocytogenes* will adhere to stainless steel and form a biofilm. Biofilms form in two stages. First, an electrostatic attraction occurs between the surface and the microbe. The process is reversible at this state. The next phase occurs when the microorganism exudes an extracellular polysaccharide, which firmly attaches the cell to the surface. The cells continue to grow, forming microcolonies and, ultimately, the biofilm.

These films are very difficult to remove during the cleaning operation. Microorganisms that appear to be more of a problem to remove because of biofilm protection are *Pseudomonas* and *L. monocytogenes*. Current information suggests that the application of heat appears to be more effective than that of chemical sanitizers, and Teflon appears to be easier to clear of biofilm than does stainless steel.

Biofilms are not effectively penetrated by water-soluble chemicals such as caustics, bleaches, iodophors, phenols, and quaternary ammonium sanitizers. Therefore, the organisms within them may not be destroyed. According to Kramer (1992), there are no procedural specifications or regulations on the removal and disinfection of biofilms. A biocide may require use at 10 to 100 times normal strength to achieve inactivation.

In tests of sanitizers—including hot water at 82°C; chlorine at 20, 50, and 200 ppm; and iodine at 25 ppm—the bacteria on stainless steel chips survived, even after immersion in the sanitizer for 5 minutes. The only true germicide tested was a hydrogen-peroxide-based powder that was found to be effective against biofilms at 3% and 6% solutions (Felix, 1991).

Relationship of Amount of Contamination, Temperature, and Time to Microbial Growth

As temperature decreases, the *generation interval* (time required for one bacterial cell to become two cells) is increased. This is especially true when the temperature goes below 4°C. The effect of temperature on microbial proliferation is illustrated in Figure 2–2. For example, freshly ground beef usually contains approximately 1 million bacteria/g. When the number of this microbial population reaches approximately 300 million/g, abnormal odor and some slime development, with resultant spoilage, can occur. This trend does not apply to all genera and species of bacteria. However, it can be determined from these data that initial contamination and storage temperature dramatically affect the shelf life of food. The storage life of ground beef that contains 1 million bacteria/g is approximately 28 hours at 15.5°C. At normal refrigerated storage temperature of approximately –1°C to 3°C, the storage life exceeds 96 hours.

EFFECTS OF MICROBES ON SPOILAGE

Food is considered spoiled when it becomes unfit for human consumption. Spoilage is usually equated with the decomposition and putrefaction that results from microorganisms.

Physical Changes

The physical changes caused by microorganisms usually are more apparent than the chemical changes. Microbial spoilage usually results in an obvious change in physical characteristics such as color, body, thickening, odor, and flavor degradation. Food spoilage is normally classified as being either aerobic or anaerobic, depending on the spoilage conditions, including whether the principal microorganisms causing the spoilage were bacteria, molds, or yeasts.

Aerobic spoilage of foods from molds is normally limited to the food surface, where oxygen is available. Molded surfaces of foods such as meats and cheeses can be trimmed off, and the remainder is generally acceptable for consumption. This is especially true for aged meats and cheeses. When these surface molds are trimmed, surfaces underneath usually have limited microbial growth. If extensive bacterial growth occurs on the surface, penetration inside the food surface usually follows, and toxins may be present.

Anaerobic spoilage occurs within the interior of food products or in sealed containers, where oxygen is either absent or present in limited quantities. Spoilage is caused by facultative and anaerobic bacteria, and is expressed through souring, putrefaction, or taint. Souring occurs from the accumulation of organic acids during the bacterial enzymatic degradation of complex molecules. Also, proteolysis without putrefaction may contribute to souring. Souring can be accompanied by the production of various gases. Examples of souring are milk, round sour or ham sour, and bone sour in meat. These meat sours, or taints, are caused by anaerobic bacteria that may originally have been present in lymph nodes or bone joints, or that might have gained entrance along the bones during storage and processing.

Chemical Changes

Through the activity of endogenous hydrolytic enzymes that are present in foodstuffs (and the action of enzymes that microorganisms produce), proteins, lipids, carbohydrates, and other complex molecules are degraded into smaller and simpler compounds. Initially, the endogenous enzymes are responsible for the degradation of complex molecules. As microbial load and activity increase, degradation subsequently occurs. These enzymes hydrolyze the complex molecules into simpler compounds, which are subsequently utilized as nutrient sources for supporting microbial growth and activity. Oxygen availability determines the end products of microbial action. Availability of oxygen permits hydrolysis of proteins into end products such as simple peptides and amino acids. Under anaerobic conditions, proteins may be degraded to a variety of sulfur-containing compounds, which are odorous and generally obnoxious. The end products of nonprotein nitrogenous compounds usually include ammonia.

Other chemical changes include action of lipases secreted by microorganisms that hydrolyze triglycerides and phospholipids into glycerol and fatty acids. Phospholipids are hydrolyzed into nitrogenous bases and phosphorus. Lipid oxidation is also accelerated by extensive lipolysis.

Most microorganisms prefer carbohydrates to other compounds as an energy source. When available, carbohydrates are more readily utilized for energy. Utilization of carbohydrates by microorganisms results in a variety of end products, such as alcohols and organic acids. In many foods, such as sausage products and cultured dairy products, microbial fermentation of sugar that has been added yields organic acids (such as lactic acid), which contribute to their distinct and unique flavors.

EFFECTS OF MICROORGANISMS ON FOODBORNE ILLNESS

The United States has the safest food supply of all nations. However, it is estimated that 25 million foodborne illness cases and 16,000 deaths occur each year. Snyder (1992) estimated that the annual cost of foodborne illness and death in the United States averages $3,000 per individual, or approximately $75 billion. The cost of each death related to foodborne illness, including insurance and other expenses, is estimated to be $42,300, or approximately $676 billion per year.

The development of gastrointestinal disturbances following the ingestion of food can result from any one of several plausible causes. Although the sanitarian is most interested in those related to microbial origin, other causes are chemical contaminants, toxic plants, animal parasites, allergies, and overeating. Although each of these conditions is recognized as a potential source of illness in humans, subsequent discussions will be confined to those illnesses caused by microorganisms.

Foodborne Disease

A foodborne disease is considered to be any illness associated with or in which the causative agent is obtained by the ingestion of food. A *foodborne disease outbreak* is defined as "two or more persons experiencing a similar illness, usually gastrointestinal, after eating a common food, if analysis identifies the food as the source of illness" (Gillespie, 1981). Approximately 66% of all foodborne illness outbreaks are caused by bacterial

pathogens. The incidence of foodborne diseases is unknown. An estimated 10 to 30 million Americans become ill each year from microorganisms in their food, and 200,000 to 1 million cases of salmonellosis have been suspected to occur annually in the United States. Approximately 9,000 people die each year from foodborne illness, and about 60% of the 200 foodborne outbreaks reported each year are of undetermined etiology. Unidentified causes may be from the *Salmonella* and *Campylobacter* species, *Staphylococcus aureus, Clostridium perfringens, Clostridium botulinum, Listeria monocytogenes,* and *Yersinia enterocolitica*, which are transmitted through foods. A wide variety of home-cooked and commercially prepared foods have been implicated in outbreaks, but they are most frequently related to foods of animal origin, such as poultry, eggs, red meat, seafood, and dairy products.

FOODBORNE ILLNESSES

Food poisoning is considered to be an illness caused by the consumption of food containing microbial toxins or chemical poisons. Food poisoning caused by bacterial toxins is called *food intoxication;* whereas, that caused by chemicals that have gotten into food is referred to as *chemical poisoning*. Illnesses caused by microorganisms exceed those of chemical origin. Illnesses that are not caused by bacterial by-products, such as toxins, but through ingestion of infectious microorganisms, such as bacteria, rickettsia, viruses, or parasites, are referred to as *food infections*. Foodborne illnesses caused from a combination of food intoxication and food infection are called *food toxicoinfections*. In this foodborne disease, pathogenic bacteria grow in the food. Large numbers are then ingested with the food by the host and, when in the gut, pathogen proliferation continues, with resultant toxin production, which subsequently causes illness symptoms. Illness caused by the mind, due to one witnessing another human sick or to the sight of a foreign object, such as an insect or rodent, in a food product, is termed *psychosomatic food illness*.

To provide protection against foodborne illness, it is necessary to have up-to-date knowledge of production, harvesting, and storage techniques to accurately evaluate the quality and safety of raw materials. Thorough knowledge of design, construction, and operation of food equipment is essential to exercise control over processing, preservation, preparation, and packaging of food products. An understanding of the vulnerability of food products to contamination will help establish safeguards against food poisoning.

Staphylococcal Food Poisoning

This poisoning is caused by ingesting the enterotoxin produced by *S. aureus*. This facultative, sphere shaped, gram-positive non–spore-forming microorganism produces an enterotoxin that causes an inflammation of the stomach and intestines, known as *gastroenteritis*. Although mortality seldom occurs from staphylococcal food poisoning, the central nervous system can be affected. If death does occur, it is usually due to added stress among people with other illnesses. The microorganisms causing staphylococcal food poisoning are widely distributed and can be present among healthy individuals. It appears that the handling of food by infected individuals is one of the greatest sources of contamination. The most common foods that may cause staphylococcal food poisoning are potato salad, custard-filled pastries, dairy products (including cream), poultry, cooked ham, and tongue. With ideal temperature and high contamination levels, staphylococci can multiply enough to cause food poisoning without noticeable changes in color, flavor, or odor. *Staphylococcus aureus* organisms

are destroyed by heating at 66°C for 12 minutes, but the toxin requires heating for 30 minutes at 131°C. Therefore, the normal cooking time and temperature for most foods will not destroy the enterotoxin.

Salmonellosis

Salmonellosis is considered a food infection because it results from the ingestion of any one of numerous species of living *Salmonella* organisms. These microorganisms grow and produce an endotoxin (a toxin retained within the bacterial cell) that causes the illness. The usual symptoms of salmonellosis are nausea, vomiting, and diarrhea, which appear to result from the irritation of the intestinal wall by the endotoxins. About 1 million of these microorganisms must be ingested for an infection to occur. The time lapse between ingestion and appearance of symptoms of salmonellosis is generally longer than that of staphylococcal food poisoning symptoms. Mortality from salmonellosis is generally low. Most deaths that occur are among infants, the aged, or those already debilitated from other illnesses. Celum et al. (1987) reported that salmonellosis may be especially harmful to persons with AIDS. Archer (1988) stated that AIDS patients are quite susceptible to this foodborne illness.

Salmonellae are gram-negative non–spore-forming, ova-shaped facultative bacteria that primarily originate from the intestinal tract. These bacteria may be present in the intestinal tract and other tissues of poultry and red-meat animals without producing any apparent symptoms of infection in the animal. This microorganism has been an enduring problem for fresh poultry and has been found on up to 70% broiler carcasses. The epidemic of *Salmonella enteritidis* in the northeastern United States during 1988 was partially attributable to poultry and shell eggs. A five-fold increase of this serotype has occurred since the late 1970s. This contaminant appears to have entered the egg through fecal soiling of shell eggs, hairline cracks in the shell, and ovarian infection in the hen. During 1987, eight food handlers in a grocery store in McLean, Virginia, were found to be positive for *S. enteritidis,* which was attributed to their using cracked eggs for food preparation in gourmet foods.

Although *Salmonella* organisms can be present in skeletal tissues, the major source of the infection results from the contamination of food by the handlers during processing, through recontamination or cross-contamination. *Salmonellae* transferred by the fingertips are capable of surviving for several hours and still infecting food. Thermal processing conditions for the destruction of *S. aureus* will destroy most species of *Salmonella*. Because of the origin of these bacteria and their sensitivity to cold temperature, salmonellosis can usually be blamed on poor sanitation and temperature abuse.

Clostridium perfringens Foodborne Illness

Clostridium perfringens is an anaerobic, gram-positive, rod-shaped, spore-former that produces a variety of toxins as well as gas during growth. These microorganisms and their spores have been isolated in many foods—especially among red meats, poultry, and seafood. Numbers of these microbes tend to be higher among meat items that have been cooked, allowed to cool slowly, and subsequently held for an extended period of time before serving. As with *Salmonella* microorganisms, large numbers of active bacteria must be ingested for this type of foodborne illness to occur.

The spores from various strains of this microorganism have differing resistances to heat. Some spores are killed in a few minutes at 100°C, whereas others require from 1 to 4 hours at this temperature for complete destruction. *Clostridium perfringens* can be controlled most

effectively by rapidly cooling cooked and heat-processed foods. An outbreak of foodborne illness by *C. perfringens* can usually be prevented through proper sanitation and refrigeration of foods at all times, especially of leftovers. Leftover foods should be reheated to 60°C to destroy living microorganisms.

Botulism

Botulism is a true food poisoning that results from the ingestion of a toxin produced by *C. botulinum* during its growth in food. This microbe is an anaerobic, gram-positive, rod-shaped, spore-forming, gas-forming bacterium that is found primarily in the soil. There are currently eight different botulin toxins recognized and serologically classified (see Table 2–1). The extremely potent toxin (the second most powerful biological poison known to humans) produced by this microorganism affects the peripheral nervous system of the victim. Death occurs in approximately 60% of the cases from respiratory failure. The characteristics, including symptoms, incubation time, foods involved, and preventive measures, of botulism and other common food poisonings are presented in Table 2–2.

Because *C. botulinum* may occur in the soil, it is also present in water. Therefore seafoods are a more viable source of botulism than are other muscle foods. However, the largest potential sources of botulism are home-canned vegetables and fruits with a low to medium acid content. Because this bacterium is anaerobic, canned and vacuum-packaged foods are also viable sources for botulism. Canned foods with a swell should not be eaten because the swelling results from the gas produced by the organism. Smoked fish should be heated to at least 83°C for 30 minutes during processing to provide additional protection.

To prevent botulism, proper sanitation, proper refrigeration, and thorough cooking are essential. This toxin is relatively heat-la-

Table 2–1 Types of Botulin Toxin

Type	Characteristics
A	Toxin is poisonous to humans; the most common cause of botulism in the United States
B	Toxin is poisonous to humans; found more often than Type A in most soils of the world
C_1	Toxin is poisonous to waterfowl, turkeys, and several mammals, but not to humans
C_2	Toxin is poisonous to waterfowl, turkeys, and several mammals, but not to humans
D	Toxin is responsible for forage poisoning of cattle, but rarely poisonous to humans
E	Toxin is poisonous to humans; usually associated with fish and fish products
F	Toxin is poisonous to humans; only recently isolated and extremely rare
G	Toxin is poisonous but rarely found.

Source: Adapted from Zottola, 1987.

bile, but the bacterial spores are very heat-resistant, and severe heat treatment is required to destroy them. Thermal processing at 85°C for 15 minutes inactivates the toxin. The combinations of temperatures and times given in Table 2–3 are required to destroy the spores completely.

Legionellosis

Legionella pneumophila is a vibrant bacterium that causes Legionnaires' disease. This facultative gram-negative microbe is found in contaminated waters in most of the environment and is becoming a widespread concern. This bacterium is able to multiply intracellularly within a variety of cells. The dominant extracellular enzyme produced by *L. pneumophila* is a zinc metalloprotease, also called a *tissue-destructive protease*, *cytolysin*, or *major secretory* protein. This protease is toxic to different types of cells and causes tissue destruction and pulmonary damage, which suggests its involvement in the pathogenesis of Legionnaires' disease.

This microorganism causes 1% to 5% of community-acquired pneumonia in adults, with most cases occurring sporadically. The Centers for Disease Control and Prevention receives 1,000 to 3,000 reports of cases of Legionnaires' disease each year. Most of the outbreaks have been shown to be caused by aerosol-producing devices, such as cooling towers, evaporating condensers, whirlpool spas, humidifiers, decorative fountains, shower heads, and tap water faucets.

Water is the major reservoir for *Legionella* organisms; however, this microorganism is found in other sources, such as potting soil. Amoebae and biofilms, which are ubiquitous within plumbing systems, have a critical role in the amplification process of supporting the bacterial growth.

Legionellosis is usually transmitted through the inhalation of *Legionella* organisms as liquid that has been aerosolized to respirable size (1 to 5 μm). Occasional transmission occurs through other routes, such as inoculation of surgical wounds with contaminated water during the placement of surgical dressings.

Campylobacteriosis

Campylobacter is commonly found as commensals of the gastrointestinal tract of wild and domesticated animals. This fastidious, facultative (microaerophilic), gram-negative, spiral-curve-shaped microorganism, which is motile by means of flagella, is now the greatest cause of foodborne illness in the United States. It has been identified as the causative agent of veteri-

Table 2–2 Characteristics of Foodborne Illnesses

Illness	Causative Agent	Symptoms	Average Time before Onset of Symptoms	Foods Usually Involved	Preventive Measures
Bacillus	*Bacillus cereus*	Nausea, vomiting, abdominal pain	1–16 hours	Cooked products, pasta, fried rice, and dried milk	Sanitary handling and rigid temperature control
Botulism	Toxins produced by *Clostridium botulinum*	Impaired swallowing, speaking, respiration, and coordination, dizziness and double vision	12–48 hours	Canned low-acid foods, including canned meat and seafood, smoked and processed fish	Proper canning, smoking and processing procedures. Cooking to destroy toxins, proper refrigeration and sanitation.
Staphylococcal (foodborne illness)	Enterotoxin produced by *Staphylococcus aureus*	Nausea, vomiting, abdominal cramps due to gastroenteritis (inflammation of the lining of the stomach and intestines)	3–6 hours	Custard and cream-filled pastries, potato salad, dairy products, ham, tongue, and poultry	Pasteurization of susceptible foods, proper refrigeration and sanitation
Clostridium perfringens (foodborne illness)	Toxin produced by *Clostridium perfringens* (infection?)	Nausea, occasional vomiting, diarrhea, and abdominal pain	8–12 hours	Cooked meat, poultry, and fish held at nonrefrigerated temperatures for long periods of time	Prompt refrigeration of unconsumed cooked meat, poultry, or fish; maintain proper refrigeration and sanitation
Salmonellosis (food infection)	Infection produced by ingestion of any of over 1,200 species of *Salmonella* that can grow in the gastrointestinal tract of the consumer	Nausea, vomiting, diarrhea, fever, abdominal pain, may be proceeded by chills and headache	6–24 hours	Insufficiently cooked or warmed-over meat, poultry, eggs, and dairy products; these are especially susceptible when kept refrigerated for a long time	Cleanliness and sanitation of food handlers and equipment, pasteurization, proper refrigeration and packaging
Shigella infection (bacillary dysentery)	*Shigella* species	Nausea, vomiting, water diarrhea, fever, abdominal pain, chills, and headache	1–7 days	Foods handled by unsanitary workers	Hygienic practices of food handlers

continues

Table 2–2 continued

Illness	Causative Agent	Symptoms	Average Time before Onset of Symptoms	Foods Usually Involved	Preventive Measures
Trichinosis (infection)	*Trichinella spiralis* (nematode worm) found in pork	Nausea, vomiting, diarrhea, profuse sweating, fever, and muscle soreness	2–28 days	Insufficiently cooked pork and products containing pork	Thorough cooking of pork to an internal temperature of 59°C to 77°C or higher with microwave cooking; frozen storage of uncooked pork at −15°C or lower, for a minimum of 20 days; avoid feeding pigs raw garbage
Aeromonas (foodborne illness)	*Aeromonas hydrophila*	Gastroenteritis	—	Water, poultry, red meats	Sanitary handling, processing, preparation and storage of foods; store foods below 2°C
Campylobacter (foodborne illness)	*Campylobacter* species	Diarrhea, abdominal pain, cramping, fever, prostration, bloody stools, headache, muscle pain, dizziness, and rarely death	1–7 days	Poultry and red meats	Sanitary handling, processing, preparation and storage of muscle foods
Listerosis	*Listeria monocytogenes*	Meningitis or miningoencephalitis, listerial septicemia (blood poisoning), fever, intense headache, nausea, vomiting, lesions after contact, collapse, shock, coma, mimics influenza, interrupted pregnancy, stillbirth, 30% fatality rate in infants and immunocompromised children and adults	4 days to several weeks	Milk, cole slaw, cheese, ice cream, poultry, red meats	Avoid consumption of raw foods with contact with infected animals; store foods below 2°C

continues

Table 2–2 continued

Illness	Causative Agent	Symptoms	Average Time before Onset of Symptoms	Foods Usually Involved	Preventive Measures
Yersiniosis	*Yersinia enterocolitica*	Abdominal pain, fever, diarrhea, vomiting, skin rashes for 2–3 days and rarely death	1–3 days	Dairy products, raw meats, seafoods, fresh vegetables	Sanitary handling, processing, preparation, and storage of foods
Escherichia coli O157:H7 (infection)	Enterohemorrhagic *Escherichia coli* O157:H7	Hemorrhagic colitis, hemolytic uremic syndrome with 5–10% acute mortality rate, abdominal pain, vomiting, anemia, thrombosicytopenia, acute renal injury with bloody urine, seizures, pancreatitis	—	Ground beef, dairy products, raw beef, water, apple cider, mayonnaise	Sanitary handling, irradiation, cooking to 65°C or higher
Hepatitis	Infectious Hepatitis A	Fever, abdominal pain, nausea, cramps, jaundice	1–7 weeks approx. 25 days	Raw shellfish from polluted waters, sandwiches, salads, desserts	Thorough hand washing, sanitary food handling, cooking to 70°C

nary diseases in poultry, cattle, and sheep, and is quite common on raw poultry. As detection and isolation of this microorganism have been improved, it has been incriminated in foodborne disease outbreaks. This microbe is now recognized as one of the most frequent causes of bacterial diarrhea and other illnesses, and there is a mounting body of evidence that it causes ulcers. *Campylobacter* has become a major concern because it is transmitted by food, especially inadequately cooked foods and through cross-contamination.

The infective dose of *Campylobacter* is 400 to 500 bacteria, depending on individual resistance. The pathogenic mechanisms of this pathogen allow it to produce a heat-liable toxin that may cause diarrhea.

The symptoms of foodborne illness from *Campylobacter* vary. Humans with a mild case may reflect no visible signs of illness but excrete this microorganism in their feces. Symptoms of those with a severe case may include muscle pain, dizziness, headache, vomiting, cramping, abdominal pain, diarrhea, fever, prostration, and delirium. Diarrhea usually occurs at the beginning of the illness or after fever is apparent. Blood is frequently present in the stool after 1 to 3 days of diarrhea. The length of illness normally varies from 2 to 7 days. Although death is rare, it can occur. *Campylobacter* can be controlled most effectively through sanitary handling and proper cooking of foods from animal origin.

Campylobacteriosis can occur at least twice as frequently as salmonellosis. Approximately 4 million cases in the United States at a cost that may exceed $2 billion are

Table 2–3 Temperatures and Times Required To Destroy Completely *Clostridium botulinum* Spores

Temperature (°C)	Time (min)
100	360
105	120
110	36
115	12
120	4

attributable to foodborne illnesses from *Campylobacter* each year.

Campylobacteriosis is found in the intestinal tract of cattle, sheep, swine, chickens, ducks, and turkeys. Because this microorganism is found in fecal material, muscle foods can be contaminated during the slaughtering process if sanitary precautions are not observed. *Campylobacter jejuni* has also been detected in milk, eggs, and water that have been in contact with animal feces. Limited studies have shown that the incidence of *C. jejuni* on retail cuts of red meat is lower than on retail poultry cuts. Symptoms and signs of *C. jejuni* infection lack special distinctive features and cannot be differentiated from illnesses caused by other enteric pathogens. Isolation of this pathogen is difficult because it is usually present in low numbers.

Normal levels of oxygen in the air will inhibit the growth of this microorganism. Survival in raw foods is predicated on the strain of *C. jejuni*, initial contamination load, and environmental conditions, especially storage temperature. This microbe is easily destroyed by heating contaminated foods to 60°C internal temperature and holding at this temperature for several minutes for beef and approximately 10 minutes for poultry. Infection with this pathogen can be reduced through thorough hand-washing with soap and hot running water for at least 18 seconds before food preparation and between handling of raw and prepared foods.

Campylobacter outbreaks have occurred most frequently in children over 10 years old and in young adults, although all age groups have been affected. This infection causes both the large and small intestines to produce a diarrheal illness. Although symptoms may occur between 1 and 7 days after eating contaminated food, illness usually develops 3 to 5 days after ingestion of this microbe.

The total elimination of this pathogen is unlikely. The web of causation (see Chapter 3) of campylobacteriosis is so diverse that complete elimination of *Campylobacter* species from domestic animals is not currently feasible.

Listeriosis

Listeria monocytogenes is especially dangerous because it can survive at refrigerated temperatures. Previously, listeriosis has been considered rare in humans. However, foodborne outbreaks in the 1980s have increased public health concern over this pathogen. Busch (1971) suggested that this disease has been incorrectly diagnosed during the past. The rate of infection is estimated to be 7 cases per 1 million persons, with individuals in certain high-risk groups more likely to acquire listeriosis. *Listeria monocytogenes* is an opportunistic pathogen, as it does not cause disease in healthy individuals with strong immune systems (Russell, 1997). The Centers for Disease Control and Prevention has estimated that 2,000 cases of listeriosis occur annually in the United States, and a survey during 1992 suggested that approximately 425 deaths from this illness occur each year.

A facultative gram-positive, rod-shaped, non–spore-forming microaerophilic bacterium, *L. monocytogenes* is found in the intestinal tracts of over 50 domestic and wild species of birds and animals, including sheep, cattle, chickens, and swine, as well as in soil and decaying vegetation. Other potential sources of this microorganism are stream water, sewage, mud, trout, crustaceans, houseflies, ticks, and the intestinal tracts of symptomatic human carriers. This pathogen has been found in most foods, from chocolate and garlic bread to diary products and meat and poultry. Elimination of *Listeria* is impractical and may be impossible. The critical issue is how to control its survival.

The optimal temperature for the proliferation of this microbe is 37°C; however, growth can occur at a temperature range of 0°C to 45°C. This microorganism is considered to be a psychrotrophic pathogen, which grows well in damp environments. It will grow twice as fast at 10°C as at 4°C, will survive freezing temperatures, and is usually destroyed at processing temperatures above 61.5°C. Although *L. monocytogenes* is most frequently found in milk, cheese, and other dairy products, it can be present in vegetables that have been fertilized with the manure of infected animals. This microorganism thrives in substrates of neutral to alkaline pH but not in highly acidic environments. Growth can occur in a pH range from 5.0 to 9.6, depending on the substrate and temperature. *L. monocytogenes* operates through intracellular growth in mononuclear phagocytes. Once the bacterium enters the hosts, monocytes, microphages, or polymorphonucleus leukocytes, it is bloodborne and can grow.

Listeriosis primarily affects pregnant women, infants, people over 50 years old, those debilitated by a disease, and other individuals who are in an immunocompromised state of health. Gravani (1987) reported that meningitis or meningoencephalitis are the most common manifestations of this disease in adults. This disease may occur as a mild illness with influenza-like symptoms, septicemia, endocarditis, abscesses, osteomyelitis, encephalitis, local lesions, or minigranulomas (in the spleen, gall bladder, skin, and lymph nodes) and fever. Fetuses of pregnant women with this disease may also be infected. These women might suffer an interrupted pregnancy or give birth to a stillborn child. Infants who survive birth may be born with septicemia or develop meningitis in the neonatal period. The fatality rate is approximately 30% in newborn infants and nearly 50% when the infection occurs in the first 4 days after birth.

Mascola et al. (1988) reported that listeriosis may be particularly dangerous to persons with AIDS. Because AIDS severely damages the immune system, those with the disease are more susceptible to a foodborne illness such as listeriosis (Archer, 1988). Mascola et al. (1988) reported that AIDS-diagnosed males are more than 300 times as susceptible to listeriosis as those of the same age who were AIDS-negative. The infectious dose for *L. monocytogenes* has not been established because of the presence of unknown factors in persons with normal immune systems that make them as susceptible to the bacteria as immunocompromised persons. The infectious dose depends on both the strains of *Listeria* and on the individual. However, it appears that thousands or even millions of cells may be required to infect healthy animals, whereas 1 to 100 cells may infect those who are immunocompromised. Human listeriosis usually does not occur in the absence of a predisposing infection.

Listeria monocytogenes can adhere to food contact surfaces by producing attachment fibrils, with the subsequent formation of a biofilm, which impedes removal during cleaning. The attachment of *Listeria* to solid surfaces involves two phases. They are primary attraction of the cells to the surface and firm attachment following an incubation period. A primary acidic polysaccharide is responsible for initial bacterial adhesion. This microbe adheres by producing a mass of tangled polysaccharide fibers that extend from the bacterial surface to form a "glycocalyx," which surrounds the cell of the colony and functions to channel nutrients into the cell and to release enzymes and toxins. These microbes are also potential contaminants of raw materials utilized in plants, which contribute to constant reintroduction of this organism into the plant environment. Utilization of Hazard Analysis Critical Control Points (HACCP) and other process control practices is the most effective method of controlling this pathogen in the processing environment. The HACCP approach has helped to identify critical points and to evaluate the effectiveness of control systems through verification procedures.

This pathogen is most effectively transmitted through the consumption of contaminated food, but it can also occur from person-to-person contact or by inhalation of this microorganism. For example, a person who has had direct contact with infected materials, such as animals, soil, or feces, may develop lesions on the hands and arms. This pathogen is likely to be found in home refrigerators, suggesting the need for regular cleaning and sanitizing of this equipment.

A study reported by the Centers for Disease Control and Prevention (Felix, 1992) found *Listeria* species present in 64% of 123 home refrigerators that were checked. The most effective prevention against listeriosis is to avoid the consumption of raw milk, raw meat, and foods made from contaminated ingredients. It is important for pregnant women, especially, to avoid contact with infected animals. One novel suggestion (Anon., 1988) is the use of a naturally occurring lysozyme in certain products to destroy *L. monocytogenes* during manufacture because sanitizers are effective against this pathogen. However, fail-safe procedures for the production of *Listeria*-free products have not been developed. Thus, food processors must rely on a rigid environmental sanitation pro-

gram and HACCP principles to establish a controlled process. The most critical areas for the prevention of contamination are plant design and functional layout, equipment design, process control operational practices, sanitation practices, and verification of *L. monocytogenes* control.

Various studies have demonstrated that *L. monocytogenes* is resistant to the effects of sanitizers. This pathogen has resistance to the effects of trisodium phosphate (TSP), and exposure to a high (8%) level of TSP for 10 minutes at room temperature is required to reduce bacterial numbers by 1 log after a colony has grown on the surface and a biofilm has formed. Furthermore, washing skin with 0.5% sodium hydroxide has a minimal effect on the proliferation of *L. monocytogenes*. This microorganism is more resistant to the cooking process than are other pathogens, and cooking may not be a definitive means of eliminating the organism from foods. Although *L. monocytogenes* is susceptible to irradiation, it is not the final solution with regard to eliminating this pathogen from fresh meat and poultry.

Russell (1997) has recognized that although a minimal number of listeriosis cases are reported in the United States each year, a significant number of those affected die from the disease. He has identified this microorganism as a "super bacterium" that can survive environmental extremes that will eliminate other pathogenic bacteria. Thus, food processors and food service operators should focus on reducing the presence of this microorganism in products, even though it is nearly impossible to completely eliminate this pathogen from the food supply.

Foodborne Illness from *Arcobacter butzleri*

Ongoing research is being conducted on this pathogen that is related to the foodborne pathogen, the *Campylobacter* species. This microorganism, which is found in beef, poultry, pork, and nonchlorinated drinking water, occurs in up to 81% of poultry carcasses. It is more resistant to irradiation and more tolerant of oxygen than is *C. jejuni* and will grow at refrigerated temperatures in atmospheric oxygen.

Foodborne Illness from *Helicobacter pylori*

Research results suggest that this pathogen, which is related to *Campylobacter*, may cause gastroenteritis and is a causative agent for gastritis, stomach and intestine ulcers, and stomach cancer in humans. It is suspected that this microorganism, which is the most common chronic bacterial infection in humans, can swim and resist muscle contractions that empty the stomach during contraction. This bacterium is found in the digestive tract of animals, especially pigs. It is present in 95% of duodenal and in up to 80% of human gastric ulcer cases, in addition to clinically healthy individuals, including family members of patients. Sewage-contaminated water is a source of transmission of this microorganism (Wesley, 1997).

Yersiniosis

Yersinia enterocolitica, a psychrotrophic pathogen, is found in the intestinal tracts and feces of wild and domestic animals. Other sources are raw foods of animal origin and nonchlorinated water from wells, streams, lakes, and rivers. Also, this microorganism appears to be transmitted from person to person. Fortunately, most strains isolated from food and animals are avirulent.

Y. enterocolitica will multiply at refrigerated temperatures, but at a slower rate than at room temperature. This microorganism is heat-sensitive and is destroyed at temperatures over 60°C. The presence of this microbe in processed foods suggests postheat treatment contamination. *Y. enterocolitica* has been isolated from raw or rare red meats; the tonsils of swine and poultry; dairy products such as milk, ice cream, cream, eggnog, and cheese curd; most seafoods; and fresh vegetables.

Not all types of *Y. enterocolitica* cause illness in humans. Yersiniosis can occur in adults but most frequently appears in children and teenagers. The symptoms, which normally occur from 1 to 3 days after the contaminated food is ingested, include fever, abdominal pain, and diarrhea. Vomiting and skin rashes can also occur. Abdominal pain associated with Yersiniosis closely resembles appendicitis. In food-related outbreaks in the past, some children have had appendectomies because of an incorrect diagnosis.

The illness from Yersiniosis normally lasts 2 to 3 days, although mild diarrhea and abdominal pain may persist 1 to 2 weeks. Death is rare but can occur due to complications. The most effective prevention measure against Yersiniosis is proper sanitation in food processing, handling, storage, and preparation.

Escherichia coli O157:H7 Foodborne Illness

Outbreaks of hemorrhagic colitis and hemolytic uremic syndrome caused by *Escherichia coli* O157:H7, a facultative, gram-negative, rod-shaped bacterium, have elevated this pathogen to a higher echelon of concern. It is uncertain how this microorganism mutated from *E. coli*, but some scientists speculate that it picked up genes from *Shigella,* which causes similar symptoms. This microorganism is shed in the feces of cattle and can contaminate meat during processing. It is important to establish intervention procedures during slaughter and meat processing operations to control the proliferation of this pathogen. Until approval of an absolute critical control point, such as irradiation, beef should be cooked to 70°C to ensure sufficient heat treatment to destroy this pathogen. A rigid sanitation program is essential to reduce foodborne illness outbreaks from this microorganism.

E. coli O157:H7, which is designated by its somatic O and flagellar H antigens, was discovered as a human pathogen following two hemorrhagic colitis outbreaks in 1982. Six classes of diarrheagenic *E. coli* are recognized. They are enterohemorrhagic, enterotoxigenic, enteroinvasive, enteroaggregative, enteropathogenic, and diffusely adherent. All enterohemorrhagic strains produce Shiga toxin 1 and/or Shiga toxin 2, also referred to as *veratoxin 1* and *veratoxin 2*. The ability to produce Shiga toxin was acquired from a bacteriophage, presumably directly or indirectly from *Shigella* (Buchanan and Doyle, 1997).

According to Buchanan and Doyle (1997), the initial symptoms of hemorrhagic colitis generally occur 1 to 2 days after eating contaminated food, although periods of 3 to 5 days have been reported. This bacterium attaches itself to the walls of the intestine, producing a toxin that attacks the intestinal lining. Symptoms start with mild, nonbloody diarrhea that may be followed by a period of abdominal pain and short-lived fever. During the next 24 to 48 hours, the diarrhea increases in intensity to a 4- to 10-day phase of overtly bloody diarrhea, severe abdominal pain, and moderate dehydration.

A life-threatening complication that may occur in hemorrhagic colitis patients is hemolytic uremic syndrome, which may occur a week after the onset of gastrointestinal symptoms. Characteristics of this condition include edema and acute renal failure. It occurs most frequently in children less than 10 years old. Approximately 50% of these patients require dialysis, and the mortality rate is 3% to 5%. Other associated complications may include seizures, coma, stroke, hypertension, pancreatitis, and hypertension. Approximately 15% of these cases lead to early development of chronic kidney failure and/or insulin-dependent diabetes, and a small number of cases may recur (Siegler et al., 1993).

Thrombotic thrombocytopenic purpurea is another illness associated with *E. coli* O157:H7. It resembles hemorrhagic uremic syndrome, except that it normally causes renal damage, has significant neurologic involvement (i.e., seizures, strokes, and central nervous system deterioration), and is restricted primarily to adults.

Ground beef has been the food most often associated with outbreaks in the United States. Dry-cured salami was associated with an *E. coli* O157:H7 outbreak in the western United States, revealing that low levels of this pathogen can survive in acidic fermented meats and cause illness. Another food associated with this pathogen is unpasteurized apple juice and cider. The largest reported *E. coli* outbreak, which caused thousands of illnesses, occurred in Japan in 1996. This outbreak and one that followed a year later were associated with radish sprouts. Alfalfa sprouts have been implicated in an outbreak in the United States. Drinking water and recreational waters have been vehicles of several *E. coli* O157:H7 outbreaks (Doyle et al., 1997).

The infectious dose associated with foodborne illness outbreaks from this pathogen has been low (2,000 cells or less), due to the organism's acid tolerance. However, Fratamico et al. (1993) reported that the inability of *E. coli* O157:H7 to ferment sorbitol is not associated with its virulence.

Zhao et al. (1995b) found that 3.2% of dairy calves and 1.6% of feed-lot cattle tested were found to be positive for *E. coli* O157:H7.

Deer have been found to be a source of this pathogen, and the transmission of this microorganism may occur between deer and cattle. Fecal shedding of this pathogen has been found to be transient and seasonal (Kudva et al., 1995).

It appears that *E. coli* O157:H7 can grow at 8°C to 45°C. Growth rates are similar at pH values between 5.5 and 7.5 but decline rapidly under more acidic conditions. Buchanan and Dagi (1994) reported that the minimum pH for *E. coli* O157:H7 is 4.0 to 4.5. The survival of *E. coli* O157:H7 in acidic foods is important, as several outbreaks have been associated with low levels surviving in acidic foods, such as fermented sausages, apple cider, and apple juice. This pathogen has been shown experimentally to survive for several weeks in a variety of acidic foods, such as mayonnaise, sausages, and apple cider. Survival in these foods is extended when stored at a refrigerated temperature (Zhao et al., 1995a).

The destruction of *E. coli* O157:H7 can be accomplished by cooking ground beef to 72°C or higher, or incorporating a procedure that kills this pathogen in the manufacture of fermented sausages or the pasteurization of apple cider. According to Buchanan and Doyle (1997), the HACCP system continues to be the most effective means for systematically developing food safety protocols that can reduce infection from this pathogen. The low incidence of this pathogen limits the utility of direct microbial testing as a means of verifying the effectiveness of HACCP.

Why Psychrotrophic and Other Pathogens Have Emerged

In addition to improved detection methods for emerging pathogens, other reasons exist for the emergence of these microbes. Cox (1989) suggested seven basic reasons, which are interrelated.

1. *Changes in eating habits.* Some "organically grown" products perceived to be healthy can be unsafe. An outbreak of listeriosis was linked to coleslaw in Canada that was made from cabbage fertilized with sheep manure.

2. *Changes in perception and awareness of what constitutes hazards, risks, and hygiene.* Advances in epidemiology, especially the collection of data by computer, have contributed to the recognition of foodborne listeriosis.

3. *Demographic changes.* Ill people are kept alive much longer, increasing the probability of new infections. Tourism and immigration may affect the emergence of certain disease. More immunocompromised people are living longer.

4. *Changes in food production.* Large-scale production of raw materials increases the possibility of creating ecologic niches where microorganisms may grow and from which they may be spread. Fruits and vegetables grown in countries with less rigid hygienic practices have introduced additional contamination.

5. *Changes in food processing.* The use of vacuum packaging and chill storage could affect the survival of facultative microorganisms.

6. *Changes in food handling and preparation.* Longer storage life of foods such as vegetables, salads, soft cheeses, and muscle foods can give rise to psychrotrophic pathogens, such as *Listeria monocytogenes.*

7. *Changes in the behavior of microorganisms.* Many of the factors responsible for pathogenicity are determined by plasmids that can be transferred from one species to another. The emergence of foodborne diseases is the result of complex mutual interaction of

many factors. New microbial hazards can be the result of a change in behavior of microorganisms not previously recognized as pathogens and the occurrence of conditions allowing the expression of these changes.

Mycotoxins

Mycotoxins are compounds or metabolites that are toxic or have other adverse biological effects on humans and animals (Table 2–4). They are produced by a wide range of fungi. The acute diseases caused by mycotoxins are called *mycotoxicoses*. Mycotoxicoses are not common in humans. However, epidemiologic evidence suggests an association between primary liver cancer and aflatoxin, one type of mycotoxin, in the diet. Bullerman (1979) reported that in large doses, aflatoxins are acutely toxic, causing gross liver damage with intestinal and peritoneal hemorrhaging, resulting in death. Mycotoxins may enter the food supply by direct contamination, resulting from mold growth on the food. Also, entry can occur by indirect contamination through the use of contaminated ingredients in processed foods or from the consumption of foods containing mycotoxin residues.

Molds that are capable of producing mycotoxins are frequent contaminants of food commodities. Those that are important in the food industry because of potential mycotoxin production include members of the genera *Aspergillus, Penicillium, Fusarium, Cladosporium, Alternaria, Trichothecium, Byssochlamys,* and *Sclerotinia.* Most foods are susceptible to invasion by these or other fungi during some stage of production, processing, distribution, storage, or merchandising. If there is mold growth, mycotoxins may be produced. The existence of mold in a food product, however, does not necessarily signify the presence of mycotoxins. Furthermore, the absence of mold growth on a commodity does not indicate that it is free of mycotoxins, because a toxin can exist after the mold has disappeared.

Of the mycotoxins, aflatoxin is considered to pose the greatest potential hazard to human

Table 2–4 Mycotoxins of Significance to the Food Industry

Mycotoxin	Major* Producing Microorganism	Potential Foods Involved
Aflatoxin	*Aspergillus flavus* *Aspergillus parasiticus*	Cereal, grains, flour, bread, corn meal, popcorn, peanut butter
Patulin	*Penicillium cyclopium* *Penicillium expansanum*	Apples and apple products
Penicillic acid	*Aspergillus* species	Moldy supermarket foods
Ochratoxin	*Aspergillus ocharaceus* *Penicillium vitidicatum*	Cereal grains, green coffee beans
Sterigmatocystin	*Aspergillus versicolor*	Cereal grains, cheese, dried meats, refrigerated and frozen pastries

*Other genera and species may produce these mycotoxins.

health. It is produced by *Aspergillus flavus* and *Aspergillus parasiticus,* which are nearly ubiquitous with spores that are widely disseminated by air currents. These molds are frequently found among cereal grains, almonds, pecans, walnuts, peanuts, cottonseed, and sorghum. The microorganisms will normally not proliferate unless these commodities are damaged by insects or not dried quickly and stored in a dry environment. Growth can occur by the invasion of the kernels with mold mycelium and subsequent aflatoxin production on the surface and/or between cotyledons.

The clinical signs of acute aflatoxicosis include lack of appetite, listlessness, weight loss, neurological abnormalities, jaundice of mucous membranes, and convulsions. Death may occur. Other evidence of this condition is gross liver damage through pale color, other discoloration, necrosis, and fat accumulation. Edema in the body cavity and hemorrhaging of the kidneys and intestinal tract may also occur.

Control of mycotoxin production is complex and difficult. Insufficient information exists regarding toxicity, carcinogenicity, and teratogenicity to humans, stability of mycotoxins in foods, and extent of contamination. Such knowledge is required to establish guidelines and tolerances. The best approach to eliminating mycotoxins from foods is to prevent mold growth at all levels of production, harvesting, transporting, processing, storage, and marketing. Prevention of insect damage and mechanical damage throughout the entire process—from production to consumption—as well as moisture control, is essential. Mycotoxins are not known to be produced at an A_w level below 0.83, or approximately 8% to 12% kernel moisture, depending on the type of grain. Therefore, rapid and thorough drying and storage in a dry environment is necessary. Photoelectric eyes that examine and pneumatically remove

discolored kernels that may contain aflatoxins are used in the peanut industry to aid in control and to avoid the difficult, tedious, and costly process of hand sorting.

Other Bacterial Infections

Other bacterial infections that occur in humans will cause illnesses with symptoms similar to food poisoning. The most common of these infections is caused by the *Streptococcus faecalis* bacterium. Although this microorganism is not a proven pathogen, products manufactured from muscle foods and dairy products have been implicated in some cases of this illness. Similar effects have been reported from infections caused by *E. coli.* Enterotoxigenic *E. coli* is the most common cause of "traveler's diarrhea," an illness frequently acquired by individuals from developed countries during visits to developing nations where hygienic practices may be substandard. Evisceration and cold storage of chickens at 3°C may permit an increase in *Aeromonas hydrophila.* Chill waters and the evisceration process itself appear to be probable sources of contamination in the typical broiler processing operation and may contribute to the high efficiency of occurrence of this microorganism at the retail level.

A microorganism that is widely distributed and can produce foodborne illness is *Bacillus cereus,* a gram-positive, spore-forming obligate aerobe. One type of *B. cereus* foodborne illness is called the *diarrhetic type*. It is characterized by relatively mild symptoms, such as diarrhea and abdominal pain that occur 8 to 12 hours after infection and may last for approximately 12 hours. In the emetic form of *B. cereus* illness, the symptom is primarily vomiting, which occurs within 1 to 5 hours after infection, although diarrhea may occur also. The *B. cereus* emetic toxin is performed in the food and, like *S. faecalis,* it is heat stable. The emetic form, which is more severe than the diarrhetic type, is caused by the production of an enterotoxin within the gut. Outbreaks have occurred as a result of consuming rice or fried rice served in restaurants or from warmed-over mashed potatoes. This foodborne illness is best controlled by proper sanitation in restaurants and by holding starchy cooked foods above 50°C or refrigerating at below 4°C within 2 hours after cooking to prevent growth and toxin production.

MICROBIAL DESTRUCTION

Microorganisms are considered dead when they cannot multiply, even after being in a suitable growth medium under favorable environmental conditions. Death differs from dormancy, especially among bacterial spores, because dormant microbes have not lost the ability to reproduce, as evidenced by eventual multiplication after prolonged incubation, transfer to a different growth medium, or some form of activation.

Regardless of the cause of death, microorganisms follow a logarithmic rate of death, as in the accelerated death phase of Figure 2–1. This pattern suggests that the population of microbial cells is dying at a relatively constant rate. Deviations from this death rate can occur due to accelerated effects from a lethal agent, effects due to a population mixture of sensitive and resistant cells, or with chain- or clump-forming microbial flora with uniform resistance to the environment.

Heat

Historically, application of heat has been the most widely used method of killing spoilage and pathogenic bacteria in foods. Heat processing has been considered a way to cook food products and destroy spoilage and pathogenic microorganisms. Therefore, extensive studies have been conducted to determine optimal heat treatment to destroy microorganisms. A measurement of time required to sterilize completely a suspension of bacterial cells or spores at a given temperature is the *thermal death time* (TDT). The value of TDT will depend on the nature of the microorganisms, its number of cells, and factors related to the nature of the growth medium.

Another measurement of microbial destruction is *decimal reduction time* (D value). This value is the time in minutes required to destroy 90% of the cells at a given temperature. The value depends on the nature of the microorganism, characteristics of the medium, and the calculation method for determining the D value. This value is calculated for a period of exponential death of microbial cells (following the logarithmic order of death). The D value can be determined through an experimental survivor curve.

A thermal resistance curve (phantom thermal death time curve) may be plotted from D values or TDT values at different heating temperatures. The log of D (in minutes) is plotted against the heating temperatures (Figure 2–3). The slope of this curve is the Z value, defined as the number of degrees that the temperature must be increased to cause a 1 log (90%) reduction in D. The ordinate of this thermal death time curve may also be TDT values (Figure 2–3 uses D values). Therefore, the destruction rate of a specific strain of bacteria (or its spores) in a foodstuff is best described by its D and Z values.

Increased concern about pathogens of fecal origin, such as *E. coli* O157:H7, has been responsible for the investigation and implementation of hot-water spray-washing of beef carcasses immediately after slaughter and dressing as a method of cleaning and decontamination. Smith (1994) identified the best combination (and sequence) of interventions reducing microbial load to be: use of 74°C water in the first wash and 20 kg/cm² pressure and spray-wash with hydrogen peroxide or ozone in the second wash (especially if 74°C

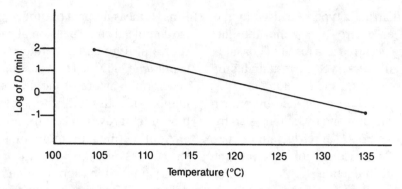

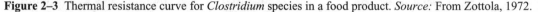

Figure 2–3 Thermal resistance curve for *Clostridium* species in a food product. *Source:* From Zottola, 1972.

water temperature is not incorporated in the first wash). Additional investigations are evaluating the efficacy of steam pasteurization/steam-vacuum as a technique for microbial reduction of beef carcasses.

Chemicals

Many chemical compounds that destroy microorganisms are not appropriate for killing bacteria in or on a foodstuff. Those that can be used are applied as sanitizing agents for equipment and utensils that can contaminate food. As the cost of energy for thermal sanitizing has increased, the use of chemical sanitizers has grown. It is hypothesized that chlorine disinfection may result from slow penetration into the cell or the necessity of inactivating multiple sites within the cell before death results. (Additional discussion related to this subject is presented in Chapter 8.) Chlorine, acids, and phosphates are potential decontaminants for microbial load on red meat and poultry carcasses.

Radiation

When microorganisms in foods are irradiated with high-speed electrons (beta rays) or with X-rays (or gamma rays), the log of the number of survivors is directly proportional to the radiation dose. The relative sensitivity of a specific strain of microorganisms subjected to specific conditions is normally expressed as the slope of the survivor curve. The $\log_{10}$ of survivors from radiation is plotted against the radiation dosage, and the radiation D or D_{10} value, which is comparable with the thermal D value, is obtained. The D_{10} value is defined as the amount of radiation in rads (ergs of energy per 100 g of material) required to reduce the microbial population by 1 log (90%).

The destructive mechanism of radiation is not fully understood. It appears that death is caused by inactivation of cellular components through energy absorbed within the cell. A cell inactivated by radiation cannot divide and produce visible outgrowth. (Additional information related to radiation as a sanitizer is presented in Chapter 8.)

Electronic Pasteurization

Electron-beam accelerators can be used for electron pasteurization of food products by impacting the products directly with electrons or optimizing the conversion of electron energy to X-rays and treating the product with these X-rays. For electron treatment, 10 million electron volt meV kinetic energy is the maximum allowed by international agreement.

Accelerators provide X-rays or electrons for treatment of food. An accelerator provides energy to electrons by providing an electric field (potential energy) to accelerate the electrons. Electrons are atomic particles, rather than electromagnetic waves, and their depth of penetration in the product is smaller. Therefore, the direct use of electrons is limited to packages less than 10 cm thick (Prestwich et al., 1994).

Pulsed Light

A potential method of microbial reduction on both packaging and food surfaces (including red meats and poultry carcasses) is the utilization of intense pulses of light. Pulsed light is energy released as short, high-intensity pulses of broad-spectrum "white" light that can sterilize packaging materials and decrease microbial populations on food surfaces. Microorganisms exposed to pulse light are destroyed. Reductions of more than 8 logs of vegetative cells and 6 logs of spores on packaging materials and in transmissive fluids and beverages, and 1 to 3 logs on complex or rough surfaces, such as meat, may be achieved.

Pulsed-light flashes are created by compressing electrical energy into short pulses and using these pulses to energize an inert gas lamp. The lamp emits an intense flash of light with a duration of a few hundred microseconds. Because this lamp can be flashed many times per second, only a few flashes are required to produce a high level of microbial kill. Thus, an on-line procedure for food processing can be very rapid (Pruett and Dunn, 1994).

Pruett and Dunn (1994) reported that the incorporation of an acetic acid spray before pulsed-light treatment led to higher levels of pathogen kill. Further analysis of the multi-hurdle concept is planned using a hot-water spray in combination with pulsed light, but the results are unavailable at the time of this writing.

Past investigations have revealed no nutritional or sensory changes attributable to pulsed light. Anticipated treatment costs are approximately $0.01/m^2 of treated surface (Pruett and Dunn, 1994).

MICROBIAL GROWTH CONTROL

Most methods used to kill microorganisms may be applied in a milder treatment to inhibit microbial growth. Sublethal heating, irradiation, or treatment with toxic chemicals frequently causes injury to microorganisms and impaired growth without death. Injury is reflected through an increased lag phase, less resistance to environmental conditions, and greater sensitivity to other inhibitory conditions. Synergistic combinations of inhibitory agents, such as irradiation plus heat and heat plus chemicals, can increase microbial sensitivity to inhibitory conditions. Injured cells appear to require synthesis of some essential cell materials (that is, ribonucleic acid or enzymes) before recovery is accomplished. Microbial growth is inhibited through maintenance of hygienic conditions to reduce debris available to support bacterial proliferation.

Refrigeration

The effect of temperature on microbial proliferation has been discussed. Freezing and subsequent thawing will kill some of the microbes. Those that survive freezing will not proliferate during frozen storage. Yet, this method of reducing the microbial load is not practical. Also, microorganisms that survive frozen storage will grow on thawed foods at a rate similar to those that have not been frozen. Refrigerated storage can be used with other methods of inhibition—preservatives, heat, and irradiation.

Chemicals

Chemicals that increase osmotic pressure with reduced A_w below the level that permits growth of most bacteria can be used as bacteriostats. Examples include salt and sugar.

Dehydration

Reduction of microbial growth by dehydration is another method of reducing the A_w to a level that prevents microbial proliferation. Some dehydration techniques restrict the types of microorganisms that may multiply and cause spoilage. Dehydration is most effective when combined with other methods of controlling microbial growth, such as salting and refrigeration.

Fermentation

In addition to producing desirable flavors, fermentation can control microbial growth. It functions through anaerobic metabolism of sugars by acid-producing bacteria that lower the pH of the substrate, the foodstuff. A pH below 5.0 restricts growth of spoilage microorganisms. Acid products that result from fermentation contribute to a lower pH and reduced action of microorganisms. Foods that are acidified and heated may be packed in hermetically sealed containers to prevent spoilage by aerobic growth of yeasts and molds.

MICROBIAL LOAD DETERMINATION

Various methods are available for determining microbial growth and activity in foods. The choice of method depends on the information required, food product being tested, and the characteristics of the microbe(s). One of the most important factors in obtaining accurate and precise results is the collection of representative samples. Because of the large numbers and variability of microorganisms present, microbial analyses are less accurate and precise and, therefore, less objective than are chemical methods of analysis. Therefore, the results of microbial analysis must be interpreted. Technical knowledge and experience related to microbiology and food products are essential for selection of the most appropriate method and the ultimate application of results.

Although microbial analysis may not provide precise results, it can indicate the degree of hygiene reflected through equipment, utensils, other portions of the environment, and food products. In addition to reflecting sanitary conditions, product contamination, and potential spoilage problems, microbial analysis can indicate anticipated shelf life. We will look at some potential methods of assessment of microbial load here. (Readers interested in more information should consult a book on microbiology.)

Aerobic Plate Count Technique

This technique is among the most reproducible methods used to determine the total population of microorganisms present on equipment or food products. The total plate count method can be used to assess the amount of contamination from the air, water, equipment surfaces, facilities, and food products. In this technique, the equipment, walls, or food products to be analyzed are swabbed. The swab is diluted in a dilutant such as peptone water or phosphate buffer, according to the anticipated amount of contamination, and subsequently applied to a growth medium containing agar in a sterile, covered plate (Petri dish). The diluted material is transferred to a culture medium (such as standard methods agar or plate count agar) that nonselectively supports the growth of microorganisms. Also, a diluted sample of food product can be added to the medium.

The number of colonies that grow on the growth medium in the sterile, covered plate during an incubation period of 2 to 20 days (depending on incubation temperature and potential microorganisms) at an incubation temperature consistent with the environment of the product being tested reflects the number of microorganisms contained by the sample. This technique provides limited information related to the specific genera and species of the sample, although physical characteristics of the colonies can provide a clue. Special methods that permit the selective growth of specific microorganisms are available to determine their presence and quantity.

Although this method is reliable, it is a slow and laborious means of testing for microbial contamination. Furthermore, the need for a faster response to a high-volume production environment has encouraged the investigation of more rapid methods. Slowness of end-product testing can retard production and does not provide an actual total count. Its use continues because of reliability and wide acceptance of this method.

Surface Contact Technique

This method of assessment, which has also been called the *contact plate technique,* is similar to the plate count technique except that swabbing is not done. A covered dish or rehydrated Petrifilm is opened, and the growth medium (agar) is pressed against the area to be sampled. The incubation process is the same as for the total plate count method. This method is easy to conduct, and less chance for error (including contamination)

exists. The greatest limitation of this technique is that it can be used only for surfaces that are lightly contaminated because dilution is not possible. Press plates can be used to monitor the effectiveness of a sanitation program. The amount of growth on the media suggests the amount of contamination.

Indicator and Dye Reduction Tests

Various microorganisms secrete enzymes as a normal metabolic function of their growth. These enzymes are capable of inducing reduction reactions. Some indicator substances (such as dyes) are used as the basis of these tests. The rate of their reduction, which is indicated by a color change, is proportional to the number of microorganisms present. The length of time required for the complete reduction of a standard amount of the indicator is a measure of the microbial load of the sample. In a modification of these methods, a dye-impregnated filter paper is applied directly onto a food sample or piece of equipment. The time required for the filter paper to change color is used to determine the microbial load. This method can lack utility because of biofilms and not all microorganisms being picked up, and because of material cost.

The major limitation of this technique is that it does not quantify the extent of contamination. However, it is quicker and easier to conduct than the plate count technique and, therefore, has become and acceptable tool for evaluation of the effectiveness of a sanitation program.

Direct Microscopic Count

An aliquot of known volume is dried and fixed to a microscope slide and stained; then a number of fields (frequently 50) are counted. Because viable and nonviable bacteria are not distinguished by most staining techniques, this method estimates the total number of microorganisms present in a sample. Sophisticated digital cameras may be attached to microscopes, and images may be captured to a computer using image analysis software. These images may be analyzed to differentiate bacteria based on size and to enumerate organisms/field, thus eliminating human error. Although this method provides morphologic or specific staining information and the slides may be retained for future reference, it is not used much because analyst fatigue can produce errors, and only a small quantity is examined.

Most Probable Number

This estimate of bacterial populations is obtained by placing various dilutions of a sample in replicate tubes containing a liquid medium. The number of microorganisms is determined by the number of tubes in each replicate set of tubes in which growth occurred (as evidenced by turbidity) and referring this number to a standard Most Probable Number (MPN) table. This method measures only viable bacteria, and it permits further testing of the cultures for purposes of identification.

Petrifilm Plates

Petrifilm plates are manufactured with a dehydrated nutrient medium on a film. This self-contained, sample-ready approach has been developed as an alternative method to the standard aerobic plate count (SPC) and coliform counts, as determined by violet red bile (VRB) pour plates. Bailey and Cox (1987) reported a correlation coefficient of $\log_{10}$ SPC and Petrifilm counts as 0.93. These scientists concluded that Petrifilm is a potential alternative to SPC. Currently, many other Petrifilms have been investigated that are especially useful to conduct surface contact studies, that is, generic *E. coli, E. coli* O157:H7, and *Listeria* (Edmiston and Russell, 1998).

Cell Mass

The quantifying of cell mass or weight has been used to estimate microbial populations in certain research applications but is not often employed for routine analysis because it can be more time-consuming and less practical than other methods. The fluid that is measured is centrifuged to pack the cells, with subsequent decanting and discarding of the supernatant, or it may be filtered through a bored asbestos or cellulose membrane, which is then weighed.

Turbidity

Turbidity is an arbitrary determinant of the number of microorganisms in a liquid. This technique lacks utility and is rarely used because the food particles in suspension contribute to turbidity and may shadow any increase from microbial growth.

Radiometric Method

In this technique, a sample is introduced into a medium containing a ^{14}C-labeled substrate, such as glucose. The amount of $^{14}CO_2$ produced is measured and related to the quantity of microorganisms present. Because some microorganisms will not metabolize glucose, ^{14}C-glutamate and ^{14}C-formate media have been devised. This technique is limited to applications where data acquisition is required within 8 hours and/or technician labor must be reduced.

Impedance Measurement

Impedance measurements determine the microbial load of a sample by monitoring microbial metabolism rather than biomass. Impedance is the total electrical resistance to the flow of an alternating current being passed through a given medium. Microbial growth in media produce changes in impedance that can be measured by the continuous passage of a small electrical current in as soon as 1 hour. This technique offers potential as a rapid method of determining microbial load. Previous research has revealed a correlation of 0.96 between impedance-detecting time and bacterial counts. The current instrumentation cost for this setup is approximately $70,000, with a cost per sample of approximately $0.50. This arrangement can handle approximately 128 samples at one time with five or more food-related testing methods. Impedance may be used to enumerate Aerobic Plate Count (APC) coliforms, *E. coli*, psychrotrophs, and *Salmonella* organisms to predict shelf life and to do sterility testing.

Endotoxin Detection

The Limulus Amoebocyte Lysate (LAL) assay is for the detection of endotoxins produced by gram-negative bacteria (including psychrotrophs and coliforms). Amoebocyte lysate from the blood of the horseshoe crab forms a gel in the presence of minute amounts of endotoxin. Due to heat stability, both viable and nonviable bacteria are detected, making the test useful in tracing the history of the food supply.

The LAL assay involves placing a sample into a prepared tube of lysate reagent, incubating 1 hour at 37°C, and evaluating the degree of gelation. The cost per test is approximately $2.00 without any initial set-up costs, but because a series of dilutions may be necessary to determine the level of bacteria present, the cost per sample could approximate $8.00.

Bioluminescence

This biochemical method, which has been simplified for easy use, measures the presence of adenosine triphosphate (ATP) by its reaction with the luciferin–luciferase complex. It can be incorporated in the estimation of microbial load of a food sample. The bioluminescent reaction requires ATP, luciferin, and firefly luciferase—an enzyme that produces light in the tail of the firefly. During the reaction, luciferin is oxidized and emits light. A luminometer measures the light produced, which is proportional to the amount of ATP present in the sample. The ATP content of the sample can be correlated with the number of microorganisms present because all microbial cells have a specific amount of ATP. An automated luminometer can detect the presence of yeast, mold, or bacterial cells in liquid samples in 3 minutes. A computer-interfaced luminometer, which employs customized software, a printer, and an automatic sampler, can analyze samples with a sensitivity of 1 microorganism per 200 mL (Anon., 1987). The cost of a bioluminescence instrument can range from $1,500 to $15,000, depending on the degree of automation included, with an average cost of $2.50 per sample. The total test time is approximately 3 minutes.

Use of this method has increased because of the need for more rapid results from product testing. It requires approximately 12 days for products to flow from microbial testing to the distribution center and out to retailers. Use of a rapid method, such as bioluminescence, accelerates product release to less than 24 hours. A surface contamination monitoring test that requires 2 or 3 days using agar-based testing methods can be reduced to 30 seconds.

Although ATP bioluminescence testing has been available since the mid 1980s, advancements in this technology have permitted this method to be applied only since the mid 1990s for broad-based use in manufacturing industries. The incorporation of new, highly sensitive biochemical reagents that emit light when in contact with ATP molecules has permitted rapid microbial screening to detect extremely low levels of microorganisms in food products within 24 hours.

The benefits of rapid methods testing and the reduced risk of contamination have been responsible for the recognition of bioluminescence technology as a reliable rapid test for microbial contamination. Although agar-plate-based technology is perceived as being less expensive than bioluminescence, a cost-analysis study has revealed that the rapid microbial testing process has a savings of approximately 40% over traditional testing methods (Le Coque, 1996/1997).

A limitation to this test is that cleaning compound residues can quench the light reaction to prevent proper response from the assay system. Many commercial bioluminescence detection kits contain neutralizers to combat the effect of detergents/sanitizers. ATP bioluminescence is ineffective in powder plants when milk powder or flour residues exist. Furthermore, organisms that are naturally luminescent exist in seafood plants. This increases the incidence of false-positive results on the surfaces tested. Furthermore, yeasts have up to 20 times as much ATP as bacteria, which complicates enumeration. A major advantage of this test is that ATP from tissue exudate can be detected, whereas other tests do not offer this feature. Furthermore, this test identifies dirty equipment.

Research is being conducted on increasing the sensitivity of bioluminescence reactions through identification of the adenylate kinase enzyme that produces ATP. This approach permits the counting of lower numbers of microorganisms present.

Catalase

This enzyme, which is found in muscle foods, vegetables, milk, and many aerobic bacteria, breaks hydrogen peroxide down into oxygen and water. Because catalase activity increases with the bacterial population, its measurement can estimate the bacterial concentration in certain products. A Catalasemeter utilizes the disc flotation principle to quantitatively measure catalase activity in foods and can detect 10,000 bacteria/mL within minutes. This unit, which incorporates the biochemical method of detection and enumeration, may be used as an on-line monitoring device to detect contamination problems in raw materials and finished products, to control vegetable blanching and milk quality, and to detect subclinical mastitis in cows. The catalase test is quite applicable to fluid products. The kit cost, including instrumentation, is approximately $2,500 and the cost per test is about $0.30.

Direct Epifluorescence Filter Technique (DEFT)

This biophysical technique is a rapid, direct method for counting microorganisms in a sample. This method was developed in England to monitor milk samples and has been applied to other foods, even though it is not used routinely in the food industry. Both membrane filtration and epifluorescence microscopy are used with this technique. Microorganisms are captured from a sample on a polycarbonate membrane. The cells are stained with acridine orange, which causes the viable bacteria to fluoresce orange and the dead bacteria to fluoresce green under the blue portion of the ultraviolet spectrum. The fluorescing bacteria are counted, using an epifluorescence microscope, which illuminates the sample with incident light. This technique has been used to evaluate dairy and muscle foods, beverages, water, and wastewater. The keeping quality of pasteurized milk stored at 5°C and 11°C can be predicted within 24 hours by preincubating samples and counting bacteria by DEFT. Approximate time required per test is 25 minutes, and the cost per sample is $1.00.

The antibody-direct epifluorescent filter technique (Ab-DEFT) has been incorporated in the enumeration of *L. monocytogenes* in ready-to-eat packaged salads and other fresh vegetables, and in the detection of *E. coli* O157:H7 in ground beef, apple juice, and milk. In addition to membrane filtration of food to collect and concentrate microbial cells on the membrane surface, fluorescent antibody staining of the filter surface and epifluorescence microscopy are involved. The fluorescent antibody is added to the filter, placed on a slide, and examined under a microscope. By this method, *L. monocytogenes* can be quantitated in agreement between the three methods (Anon., 1997) and has demonstrated the potential of Ab-DEFT as a rapid alternative for the quantitation of *Listeria* in food. However, nonspecific reactivity of the fluorescent antibodies to indigenous microbial populations has resulted in false-positive reactions using Ab-DEFT.

Remote Inspection Biological Sensor

At this writing, the first of a two-phase test has been completed, and this equipment is expected to be marketed in the near future for the detection of generic *E. coli* and *Salmonella*. The remote inspection biological sensor (RIBS) uses a laser spectrographic technique. A laser beam is directed onto the surface of a carcass. Based on the characteristics of the reflected light, this equipment can make a specific identification of pathogenic bacteria and give a general indication of the number of organisms present (Anon., 1998). The instrument has a detection sensitivity of 5 colony-forming units (CFUs) per square centimeter and is able to effectively discriminate target organisms from background.

RIBS can be integrated effectively into a beef-processing operation. Although it is planned initially for beef carcasses, it should be able to detect foodborne pathogens in poultry, seafood, fruits, and vegetables.

Microcalorimetry

Heat produced as a result of a biological reaction, such as the catabolic processes occurring in growing microorganisms cultured from contaminated samples, can be measured by a sensitive calorimeter called a *microcalorimeter*. This biophysical technique has been applied to enumerate microorganisms in food. The procedure correlates a thermogram (a heat-generation pattern during microbial growth) with the number of microbial cells. After a reference thermogram has been established, others obtained from contaminated samples can be compared to the reference. Various researchers (Anon., 1987) have quantitated microorganisms in meat, milk, molasses, and canned foods.

Radiometry and Infrared Spectrophotometry

The time required for the detection of certain levels of radioactivity by this biophysical technique is inversely related to the number of microorganisms in the sample. This method can be employed for sterility testing of aseptically packaged products. Results are available in 4 to 5 days, compared with 10 days with conventional methods. The enumeration of microorganisms in food samples can be accomplished in less than 24 hours.

Hydrophobic Grid Membrane Filter System

This culturing method is used to detect and enumerate *E. coli* in foods. An ISO-GRID hydrophobic grid membrane filter (HGMF) system is available to detect and enumerate *E. coli*. The sample is filtered through the membrane without use of an enrichment step, and a complex medium (SD-39) is used to detect the target organism. The test can be completed in 48 hours, including biochemical and serological confirmation of presumptive colonies.

DIAGNOSTIC TESTS

Enzyme-Linked Immunoassay Tests

The most common assay for detecting specific pathogens and/or their toxins is immunological (antigen-antibody) reactions. Antigens are the specific constituents of a cell or toxin that induce an immune response and interact with a specific antibody; whereas, antibodies are immunoglobulins that bind specifically to antigens. Either monoclonal or polyclonal antibodies are used in immuno-based assays. Monoclonals are a single type of antibody with a high affinity for a specific target antigen epitope. A polyclonal antibody is a set of different antibodies specific for an antigen but able to recognize different epitopes of the antigen. The advantages of these assays are rapid results, increased sensitivity and specificity, and decreased costs (Phebus and Fung, 1994).

Enzyme-linked immunoassays (ELISAs) have been effective in detecting pathogens and are easy to conduct. These systems are formatted to consist of antibodies attached to a solid support, such as the walls of a microtiter plate or a plastic dipstick. An enrichment culture is added to the solid support, and any target antigens present in the sample will be bound by the antibodies. A sandwich format is used frequently, in which a second enzyme-labeled antibody is added to the sample, followed by a reactive substrate, to produce a positive color reaction. If the target antigens are not present, the labeled antibody will not attach and no color reaction occurs.

An efficient and sensitive method of analyzing samples for pathogens is immunoblotting. The common procedure involves an enrichment culture that is spotted onto a solid support (i.e., nitrocellulose paper), with the remaining protein binding areas of the paper blocked by dipping in a protein solution such as bovine serum albumin or reconstituted dry milk. An enzyme-labeled antibody solution specific for the target pathogen is applied, and a substrate for the enzyme is added after washing to remove the unbound antibody. If the labeled antibody is present, due to attachment to the target antigen, a color reaction will indicate a positive sample. This procedure can be modified for use in conjunction with other methods, such as the HGMF system.

A more recent technique for pathogen detection is the use of superparamagnetic microspheres coated with an antibody specific to a target antigen. The sample is selectively enriched, and a small aliquot (approximately 10 mL) of the enrichment culture is transferred to a test tube. The antibody-coated beads are added and shaken gently for a short period. Then a magnetic particle concentrator is used to separate beads from the sample homogenate. After reconstitution in a buffer, the beads are spread-plated onto a selective agar to observe growth of the target pathogen. If present in the original sample, presumptive colonies must be confirmed (Phebus and Fung, 1994). These beads have been used to detect *E. coli* O157:H7 in foods.

A latex agglutination test can provide quick results with an acceptable degree of specificity for *E. coli* O157 but not for H7 confirmation. An available assay uses a polyclonal O157 antibody coated onto polystyrene latex particles, and a slide agglutination format is incorporated to transfer a suspect culture to a paper card, followed by the addition of the antibody reagent. The presence of the O157 antigen is indicated by agglutination.

A lateral flow immunoprecipitate assay has been developed recently as a screen test for *E. coli* O157:H7. This assay, approved by the Association of Official Analytical Chemists (AOAC), requires an enrichment broth and incubation for 20 hours at 36°C. A 0.1-mL sample of the enrichment broth is then deposited in a test window in a self-contained, single-use test device that contains proprietary reagents. As lateral flow occurs across the reagent zone, the target antigen, if present, reacts with the reagents to form an antigen-antibody–chromgen complex. After approximately 10 minutes of incubation at room temperature, a line will form in the test window, indicating the possible presence of *E. coli* O157:H7. If no line appears, a confirmed negative test results. As flow continues through the test verification zone, all samples will react with reagents, and a line will appear, indicating proper completion of the test. A positive test does not ensure that an *E. coli* O157:H7 strain exists. The suspect sample must be further tested to confirm the presence of the pathogen. This test, which is easy to conduct, incorporates an assay system into a single test unit (Anon., 1998).

Salmonella 1-2 Test

This is a rapid screening test for *Salmonella* that is approved by the AOAC (AOAC, 1990). It is conducted in a single-use, plastic device that contains a nonselective motility medium and a selective enrichment broth. A positive test is indicated by the presence of an immobilization precipitation band that forms in the motility medium from the reaction of motile *Salmonella* with flagellar antibodies.

This test uses a clear plastic device with two chambers. The smaller chamber contains a peptone-based, nonselective motility medium. The sample is added to the tetrathionate-brilliant green-serine broth contained in the inoculation chamber of the 1-2 test unit. After approximately 4 hours of incubation, motile *Salmonellae* move from the selective motility medium. As these organisms progress through the motility medium, they encounter flagellar antibodies that have been diffused into this medium. The reaction of the motile *Salmonellae* with the flagellar antibodies results in the development of the immobilization precipitation band 8 to 14 hours after inoculation of the 1-2 test unit.

IDEXX Bind

The IDEXX Bind for *Salmonella* is based on the use of genetically engineered bacteriophages. The modified bacteriophages attach to *Salmonella* receptors and insert DNA into the bacterial cells. During incubation, the modified DNA causes *Salmonella* to produce ice nucleation proteins. At a specified temperature, the ice nucleation proteins promote the formation of ice crystals. Positive samples will freeze and turn orange at this temperature; whereas negative samples will not freeze.

Random Amplified Polymorphic DNA

The random amplified polymorphic DNA (RAPD) method has achieved promising results, especially to trace *L. monocytogenes* infections in humans. Advantages of this method are the low cost of the multiple DNA primers, discriminating nature of the test, and, probably most importantly, the ability to trace small amounts of *L. monocytogenes*. One problem with this assay is the time required to conduct this test. Thus, it appears to have more utility as a research tool than as a diagnostic test for industry use.

Immunomagnetic Separation and Flow Cytometry

This technique can be used to detect less than 10 *E. coli* O157:H7 cells/g of ground beef after enrichment for 6 hours. The immunomagnetic beads concentrate cells, making it easier to detect, using flow cytometry. Detection limit is not significantly influenced by the presence of other microorganisms. At this point, this method has been used more as a research tool than as a diagnostic tool in the food industry.

Diagnostic Identification Kits

These kits were developed for human clinical medicine but can aid in the identification of various microorganisms. Most of these tests are designed for use with isolated colonies, which may take 1 to 3 days to obtain.

CAMP Test

In this test, a bacterial isolate that is suspected to be *L. monocytogenes* is streaked adjacent to or across a streak of a second, known bacterium on a blood agar plate. At the juncture of the two streaks, the metabolic by-products of the two bacteria diffuse and result in an augmented hemolytic reaction. Hemolysis of blood cells is an important characteristic of pathogenic bacteria such as *L. monocytogenes* because it appears to be closely associated with virulence. Using this method, the virulence of *L. monocytogenes* may be determined.

Fraser Enrichment Broth/Modified Oxford Agar

Kansas State University (Anon. 1991) developed a simple and rapid method for *Listeria* detection using Fraser enrichment broth combined with modified Oxford agar for motility enrichment. The *Listeria* organisms are enriched in Fraser broth and held at 30°C for 24 hours, then 1 mL of the enrichment broth is placed in the Fraser broth in the left arm of a U-shaped tube.

The Fraser broth selectively isolates and promotes *Listeria* growth and precludes the growth of nonmotile microorganisms. The microbes migrate through the modified Oxford agar and arrive as a pure culture in the second branch of the Fraser broth. This becomes the second enrichment necessary for the identification of *Listeria*. An easier indication that *Listeria* organisms are present is the formation of a black precipitate as the bacteria move through the modified Oxford agar. When turbidity develops, a sample can be taken for DNR probe analysis to confirm the presence of *Listeria*. The second enrichment step requires 12 to 24 hours.

Crystal Violet Test

The retention of crystal violet by *Y. enterocolitica* correlates with virulence. Most *Y. enterocolitica* strains isolated from meat and poultry are avirulent. Thus, this rapid test allows samples with virulent strains to be identified and discarded quickly.

Methyl Umbelliferyl Glucuronide Test

Methyl umbelliferyl glucuronide (MUG) is split by the enzyme glucuronidase produced by most *E. coli* and a few other microbes, such as *Salmonella*. When split, MUG becomes fluorescent under ultraviolet illumination of a specific wave length to permit rapid identification of the microorganisms in tubed media or on spread plates for enumeration.

Assay for *E. coli*

At the time of this writing, several techniques are being evaluated for the rapid identification of microorganisms. Many techniques have not been available long enough to establish track records for their efficacy or to achieve AOAC approval. Although several methods are available, most require 24 to 48 hours for incubation of the microorganisms and may need additional testing to confirm the presence of *E. coli*. Many commercial assays for the detection of *E. coli* incorporate membrane filtration technology, and others employ a reagent/sample mixture that is incubated for 24 to 48 hours to obtain a presence/absence result of total *E. coli* contamination.

A new assay for a rapid, inexpensive determination of *E. coli* concentrations in aqueous environments has been developed which is called the *ime. Test-EC Kount Assayer*. This assayer uses a reagent mixture containing an indicator compound that provides a colorimetric (bright blue) indication of *E. coli* concentration in a water-based sample, predicated on cleavage by the beta galactosidase enzyme specific to *E. coli*. This assay provides a simple method for quantifying the concentration of viable *E. coli* in an aqueous sample in 2 to 10 hours.

The procedure involves filling a snapping cup with a sample and introducing it to a vacuum-sealed test ampoule by snapping off the ampoule's sealed tip in one of the holes in the bottom of the cup. The ampoule automatically fills with the aqueous sample. Then the sample is incubated at 35°C and monitored for the production of a blue fluorescence resulting from enzymatic cleavage of the indicator molecule, MUG. The time required for the production of a bright blue color, visualized under long-wave ultraviolet light optically or via instrument, is proportional to the total *E. coli*/mL in the sample. Based on time to positive, a comparison chart provides the corresponding *E. coli* count for the sample. Concentration and detection times are:

E. coli Concentration	Detection Time
9.9×10^6 CFU/mL	2 hours
100 CFU/mL	10 hours

Further incubation of samples that are negative at 12 hours provides a presence/absence determination after 24 hours. Although the track record is not known at this writing, this technique will permit sampling at a remote site and return to a laboratory for analysis. The major limitation of this technique appears to be that not all of the *E. coli* react in the presence of MUG.

Micro ID and Minitek

Micro ID is a self-contained identification unit containing reagent-impregnated paper discs for biochemical testing for the differentiation of Enterobacteriaceae in approximately 4 hours. This technique has provided reliable results. The Minitek system is another miniaturized test kit for the identification of Enterobacteriaceae. This kit also utilizes reagent-impregnated paper discs requiring 24 hours of incubation. It is considered to be accurate and versatile. The

Analytab Products, Inc. (API) strip is the most commonly used identification unit.

DNA Hybridization and Colorimetric Detection

This assay methodology combines DNA hybridization technology with nonradioactive labeling and colorimetric detection. With the appropriate specific DNA probes and enrichment and sample preparation procedures for a particular organism, this basic assay can be applied to the analysis of a wide variety of microbes. The assay can be completed in approximately 2.5 to 3 hours after 2 days of broth culture enrichment of the sample.

An application of this principle is a colorimetric assay, which employs synthetic oligonucleotide DNA probes against ribosomal RNA (rRNA) of the target organism. This approach offers increased sensitivity because rRNA, as an integral part of the bacterial ribosome, is present in multiple copies (1,000 to 10,000) per cell. The number of ribosomes present per cell is dependent on the growth state of the bacterial culture.

Polymerase Chain Reaction

This is a viable technique for the detection of low levels of pathogens found in food products. Polymerase chain reaction (PCR) amplifies very low DNA levels (as low as one molecule) or detectable levels (approximately 10^6) through a series of DNA hybridization reactions and thermocycling. PCR products are detected by various methods, such as gel electrophoreses, calorimetric, or chemiluminescent assays. Its utility is limited currently in day-to-day food analyses, due to the complexity of the procedure and certain equipment requirements (Phebus and Fung, 1994).

Rapid Method Selection

Individual laboratories have different needs for equipment and microbial assays. Recommendations provided by Phebus and Fung (1994) should be considered. These scientists have suggested that laboratories must evaluate their needs and determine their current level of knowledge and instrumentation, and what is being analyzed. If a large number of samples will be evaluated consistently, the speed and costs of supplies and labor may justify an investment in automated instrumentation.

Phebus and Fung (1994) recognized that an extensive amount of effort and money has been devoted to the development of instantaneous or real-time pathogen detection techniques. These tests are not available currently for use in the food industry. However, it is possible to reveal plant sanitation levels quickly and to incorporate these measurements to set high standards for the involved plant. Until more sophisticated technology is available, a pathogen-free status cannot be attained. Even though improved technology may not provide a pathogen-free environment, complementary strategies will contribute to reduced risks.

SUMMARY

The role of microorganisms in food spoilage and foodborne illness must be understood if effective sanitation is to be practiced. Microorganisms cause food spoilage through degradation of appearance and flavor, and foodborne illness occurs through the ingestion of food containing these microorganisms or toxins of public health concern. Microbiology is the science of microscopic forms of life. Control of microbial load from equipment, plants, and foods is part of a sanitation program.

Microorganisms have a growth pattern similar to a bell curve and tend to proliferate and die at a logarithmic rate. Extrinsic factors that have the most effect on microbial growth kinetics are temperature, oxygen availability, and relative humidity. Intrinsic factors that affect growth rate most are A_w and pH levels, oxidation-reduction potential, nutrient requirements, and presence of inhibitory substances.

Chemical changes from microbial degradation occur primarily through enzymes that microorganisms produce, which degrade proteins, lipids, carbohydrates, and other complex molecules into simpler compounds. Foodborne illness may result from microorganisms such as *S. aureus, Salmonella* and *Campylobacter* species, *C. perfringens, C. botulinum, L. monocytogenes, Y. enterocolitica,* and mycotoxins.

The most common methods of microbial destruction are heat, chemicals, and irradiation; whereas, the most common methods for inhibiting microbial growth are refrigeration, dehydration, and fermentation. Microbial load and taxonomy are frequently assessed as measurements of the effectiveness of a sanitation program by the various tests and diagnoses that were discussed in this chapter.

STUDY QUESTIONS

1. What is a microorganism?
2. What is a virus?
3. How does temperature affect the lag phase of the microbial growth curve?
4. What is a psychrotroph?
5. What is A_w?
6. What is a biofilm?
7. What is generation interval?
8. What is a foodborne illness?
9. What is psychosomatic food illness?
10. What microorganism is most likely to cause influenza-like symptoms?
11. What is a mycotoxin?
12. What is cross-contamination?
13. What is thermal death time?
14. What is a D value?

REFERENCES

Anon. 1987. Rapid methods for microbiological analysis of Food. *Food Technol* 41, no. 7: 56.

Anon. 1988. AMI data show frankfurters most likely *Listeria* carrier in meat. *Food Chem News* 30, no. 16: 37.

Anon. 1991. Tube speeds *Listeria* detection. *FSIS Food Safety Rev 1991*: 15.

Anon. 1997. Fresh and fast. *Food Quality* 3, no. 24: 41–43.

Anon. 1998. News and notes. *Food Quality* 5, no. 1: 3.

AOAC. 1990. *Official methods of analysis*. 15th ed. Washington, DC: Author.

Archer, D.L. 1988. The true impact of foodborne infections. *Food Technol* 42, no. 7: 53.

Bailey, J.S., and N.A. Cox. 1987. Evaluation of the petrifilm SM and VRG dry media culture plates for determining microbial quality of poultry. *J Food Prot* 8: 643.

Buchanan, R.L., and L.K. Dagi. 1994. Expansion of resource surface models for the growth of *Escherichia coli* O157:H7 to include sodium nitrate as a variable. *Inst J Food Microbiol* 23: 317.

Buchanan, R.L., and M.P. Doyle. 1997. Foodborne disease significance of *Escherichia coli* O157:H7 and other enterohemorrhagic *E. coli*. *Food Technol* 51, no. 10: 69.

Bullerman, L.B. 1979. Significance of mycotoxins to food safety and human health. *J Food Prot* 42: 65.

Busch, L.A. 1971. Human listeriosis in the United States, 1967–1969. *J Infect Dis* 123: 328.

Celum, C.L., et al. 1987. Incidence of salmonellosis in patients with AIDS. *J Infect Dis* 156: 998.

Cox, L.J. 1989. A perspective on listeriosis. *Food Technol* 43, no. 12: 52.

Doyle, M.P., et al. 1997. *Escherichia coli* O157:H7. In *Food microbiology: Fundamentals and frontiers*, ed. M.P. Doyle, L.R. Beuchat, and T.J. Montville, 171–191. Washington, DC: ASM Press.

Edmiston, A.L., and S.M. Russell. 1998. A rapid microbiological method for enumerating *Escherichia coli* from broiler chicken carcasses. *J Food Prot* (in press).

FDA. 1996a. *FDA Enforcement Reports*. 10 July.

FDA. 1996b. *FDA Enforcement Reports*. 21 August.

Felix, C.W. 1991. Sanitizers fail to kill bacteria in biofilms. *Food Prot Rep* 7, no. 5: 6.

Felix, C.W. 1992. CDC sidesteps *Listeria* hysteria from JAMA articles. *Food Prot Rep* 8 no. 5: 1.

Fratamico, P.M., et al. 1993. Virulence of an *Escherichia coli* O157:H7 sorbitol-positive mutant. *Appl Environ Microbiol* 59: 453.

Gillespie, R.W. 1981. Current status of foodborne disease problems. *Dairy Food Environ Sanit.* 1: 508.

Gravani, R.B. 1987. Bacterial foodborne diseases. *Dairy Food Environ Sanit* 7: 137.

Kramer, D.N. 1992. Myths, cleaning and disinfection. *Dairy Food Environ Sanit* 12: 507.

Kudva, I.T., et al. 1995. Effect of diet in the shedding of *Escherichia coli* O157:H7 in a sheep model. *Appl Environ Microbiol* 61: 1363.

Le Coque, J. 1996/1997. ATP bioluminescence: Increasing quality, lowering costs. *Food Testing Analysis 2,* no. 6: 32.

Mascola, L., et al. 1988. Listeriosis: An uncommon opportunistic infection in patients with Acquired Immunodeficiency Syndrome. *Am J Med* 84: 162.

Phebus, R.K., and D.Y.C. Fung. 1994. Rapid microbial methods for safety and quality assurance of meats. *Proc Meat Ind Res Conf.* Washington, DC: American Meat Institute, 63.

Prestwich, K.R., et al. 1994. The use of electron beams for pasteurization of meats. *Proc Meat Ind Res Conf,* Washington, DC: American Meat Institute, 81.

Pruett, W.P., and J. Dunn. 1994. Pulsed light reduction of pathogenic bacteria on beef carcass surfaces. *Proc Meat Ind Res Conf,* Washington, DC: American Meat Institute, 93.

Russell, S.M. 1997. *Listeria monocytogenes*: One tough microbe. *Broiler Industry* 60, no. 6: 27.

Shapton, D.A., and N.F. Shapton, eds. 1991. Microorganisms—an outline of their structure. In *Principles and practices for the safe processing of foods,* 209. Oxford: Butterworth-Heinemann.

Siegler, R.L, et al. 1993. Recurrent hemolytic uremic syndrome secondary to *Escherichia coli* O157:H7 infection. *Pediatrics* 91: 666.

Smith, G.C. 1994. Fecal material removal and bacterial count reduction by trimming and/or spray-washing beef external fat surfaces. *Proc Meat Ind Res Conf,* 31. Washington, DC: American Meat Institute.

Snyder, O.P. 1992. HACCP—An industry food safety self-control program—Part IV. *Dairy Food Environ Sanit* 12: 230.

Wesley, I. 1997. *Campylobacter* and related microorganisms in cattle. *Proc Beef Safety Symposium.* Greenwood Village, Colorodo: National Cattlemen's Beef Assoc.

Zhao, T., et al. 1995a. Prevalence of enterohemorrhagic *Escherichia coli* O157:H7 in apple cider without preservatives. *Appl Environ Microbiol* 59: 2526.

Zhao, T., et al. 1995b. Prevalence of enterohemorrhagic *Escherichia coli* O157:H7 in a survey of dairy herds. *Appl Environ Microbiol* 61: 1290.

Zottola, E.A. 1972. *Introduction to meat microbiology.* Chicago: American Meat Institute.

Zottola, E.A. 1987. *Introduction to meat microbiology.* 2nd ed. Washington, DC: American Meat Institute.

SELECTED READINGS

Franco, D.A. 1991. Campylobacteriosis—control and prevention. *FSIS Food Safety Rev* 7, Summer.

National Restaurant Association Education Foundation. 1992. *Applied foodservice sanitation.* 4th ed., New York: John Wiley & Sons. In cooperation with the Education Foundation of the National Restaurant Association, Chicago.

Niven, C.F., Jr. 1987. Microbiology and parasitology of meat. In *The science of meat and meat products,* 3rd ed. Westport, CT: Food and Nutrition Press, Inc.

CHAPTER 3

Food Contamination Sources

Food products provide an ideal nutrition source for microorganisms and generally have a pH value in the range needed to contribute to proliferation. They are contaminated with soil, air, and waterborne microorganisms during harvesting, processing, distribution, and preparation. Extremely high numbers of microorganisms are found in meat animals' intestinal tracts, and some of these find their way to the carcass surfaces during slaughter. Some apparently healthy animals may harbor various microorganisms in the liver, kidneys, lymph nodes, and spleen. These microorganisms and those from contamination through slaughtering can migrate to the skeletal muscles via the circulatory system. When carcasses and cuts are subsequently handled through the food distribution channels, where they are reduced to retail cuts, they are subjected to an increasing number of microorganisms from the cut surfaces. The fate of these microorganisms and those from other foods depend on several important environmental factors, such as the ability of the organisms to utilize fresh food as a substrate at low temperatures. In addition, oxygenated conditions and high moisture will segregate those microorganisms most capable of rapid growth under these conditions.

Refrigeration, one of the most important methods for reducing the effects of contamination, is widely applied to food products in commercial food processing and distribution. Its use has prevented outbreaks of foodborne illness by controlling the microbes responsible for this condition. However, correct techniques for cold storage frequently are not followed, and food contamination may result. The growth rate of microorganisms may increase tremendously in an environment slightly above the minimal temperature for growth. Generally, foods cool slowly in air, and the rate of cooling decreases with the increasing size of the container. Therefore, it is difficult to cool large volumes of food properly. Many of the *Clostridium perfringens* foodborne illness outbreaks have been caused by storage of large pieces of meat or broth in slowly cooling containers.

Food products may transmit certain microorganisms, causing foodborne illness that can be classified as either infections or intoxications. Foodborne infections can result in two ways:

1. The infecting microorganism is ingested and then multiplies, as is true for *Salmonella, Shigella,* and some enteropathogenic *Escherichia coli.*
2. Toxins are released as the microorganisms multiply, sporulate, or lyse. Examples of such infections are *C.*

53

perfringens and some strains of entero-pathogenic *E. coli*.

TRANSFER OF CONTAMINATION

Before a foodborne illness can occur, foodborne disease transmission requires that several conditions be met. The presence of only a few pathogens in a food will generally not cause an illness, although regulatory agencies still consider this a potentially hazardous situation. Bryan (1979) cited several models that have been used to support this hypothesis and to illustrate the relationship among factors that cause foodborne illness. Two of the models that will be discussed briefly are the chain of infection and the web of causation.

Chain of Infection

A chain of infection is a series of related events or factors that must exist or materialize and be linked together before an infection will occur. These links can be identified as *agent, source, mode of transmission,* and *host.* The essential links in the infectious process must be contained in such a chain. The causative factors (Figure 3–1) that are necessary for the transmission of a bacterial foodborne disease are:

1. Transmission of the causative agent from the environment in which the food is produced, processed, or prepared to the food itself.

2. A source and a reservoir of transmission for each agent.
3. Transmission of the agent from the source to a food.
4. Growth support of the microorganism through the food or host that has been contaminated.

Conditions such as required nutrients, moisture, pH, oxidation-reduction potential, lack of competitive microorganisms, and lack of inhibitors must also exist for contaminants to survive and grow. Contaminated food must remain in a suitable temperature range for a sufficient time to permit growth to a level capable of causing infection or intoxication.

The infection chain emphasizes the multiple causation of foodborne diseases. The presence of the disease agent is indispensable, but all of the steps are essential in the designated sequence before foodborne disease can result.

Web of Causation

The web of causation as modified by Bryan (1979) is a complex flow chart that indicates the factors that affect the transmission of foodborne disease. This presentation of disease causation attempts to incorporate all of the factors and their complex interrelationships. These webs, generally oversimplified schematic representations of disease transmission processes, will not be illustrated because a very large and comprehensive figure would be required to include all pathogenic microorganisms affecting all foods.

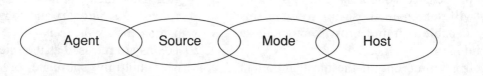

Agent   Source   Mode   Host

Figure 3–1 The chain of infection.

CONTAMINATION OF FOODS

One of the most viable contamination sources is the food product itself. Waste products that are not handled in a sanitary way become contaminated and support microbial growth. Figure 3–2 illustrates potential contamination by humans.

Dairy Products

Equipment with extensively designed sanitary features to improve the hygiene of milk production and to eliminate disease problems in dairy cows has contributed to more wholesome dairy products, although contamination can occur from the udders of cows and milking equipment. The subsequent pasteurization in processing plants has further reduced milk-borne disease microorganisms. Nevertheless, dairy products are vulnerable to cross-contamination from items that have not been pasteurized. Because not all dairy products are pasteurized, the presence of pathogens (especially *Listeria monocytogenes*) in this industry has increased. (Additional discussion related to contamination of dairy products is presented in Chapter 14.)

Red Meat Products

The muscle tissues of healthy living animals are nearly free of microorganisms. Contamination of meat occurs from the external surface, such as hair, skin, and the gastrointestinal and respiratory tracts. The animal's white blood cells and the antibodies developed throughout their lives effectively control infectious agents in the living body. These internal defense mechanisms are destroyed when blood is removed during slaughter.

Initial microbial inoculation of meat results from the introduction of microorgan-

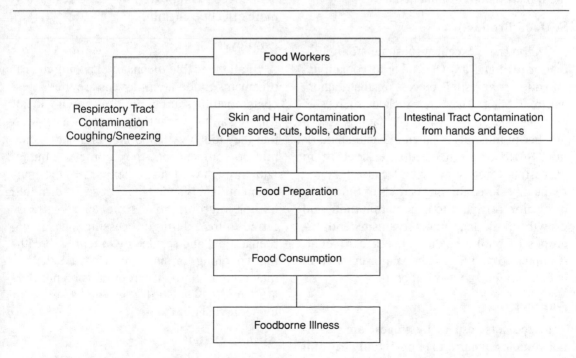

Figure 3–2 Potential contamination of food by humans.

isms into the vascular system when contaminated knives are used for exsanguination. The vascular system rapidly disseminates these microorganisms throughout the body. Contamination subsequently occurs by the introduction of microorganisms on the meat surfaces in operations performed during slaughtering, cutting, processing, storage, and distribution of meat. Other contamination can occur by contact of the carcass with the hide, feet, manure, dirt, and visceral contents from punctured digestive organisms.

Poultry Products

Poultry is vulnerable to contamination by *Salmonella* and *Campylobacter* organisms during processing. The processing of poultry, especially defeathering and evisceration, allows opportunity for the distribution of microorganisms among carcasses. Contaminated hands and gloves and other tools of processing plant workers also contribute to the transmission of salmonellae.

Seafood Products

Seafoods are excellent substrates for microbial growth and are vulnerable to contamination during harvesting, processing, distribution, and marketing. They are excellent sources of proteins and amino acids, B vitamins, and a number of minerals required in bacterial nutrition. Seafoods are handled extensively from harvesting to consumption. Because they are frequently stored for long periods of time without prior refrigeration, contamination and growth of spoilage microorganisms and microbes of public health concern can occur. (Chapter 16 provides additional discussion related to seafood contamination.)

Adjuncts

Ingredients (especially spices) are potential vehicles of harmful or potentially harmful microorganisms and toxins. The amounts and types of these agents vary with place and method of harvesting, type of food ingredient, processing technique, and handling. The food plant management team should be aware of the hazards connected with individual incoming ingredients. Only supplies and materials gathered in accordance with recognized good practices should be used. This requirement also applies to control of testing of critical materials, either by the manufacturing firm, receiving establishment, or both.

OTHER CONTAMINATION SOURCES

Equipment

Contamination of equipment occurs during production, as well as when the equipment is idle. Even with hygienic design features, equipment can collect microorganisms and other debris from the air, as well as from employees and materials during production. Product contamination from equipment can be reduced by improved hygienic design and more effective cleaning.

Employees

Of all the viable means of exposing microorganisms to food, employees are the largest contamination source. Employees who do not follow sanitary practices contaminate food that they touch with spoilage and pathogenic microorganisms that they come in contact with through work and other parts of the environment. The hands, hair, nose, and mouth harbor microorganisms that can be transferred to food during processing, packaging, preparation, and service by touching, breathing, coughing, or sneezing. Because the human body is warm, microorganisms proliferate rapidly, especially in the absence of hygienic practices.

Air and Water

Water serves as a cleaning medium during the sanitation operation and is an ingredient added in the formulation of various processed foods. It can also serve as a source of contamination. If excessive contamination exists, another water source should be obtained, or water from the existing source should be treated by chemicals (such as ultraviolet units) or other methods.

Contamination can result from airborne microorganisms in food processing, packaging, storage, and preparation areas. This contamination can result from unclean air surrounding the food plant or from contamination through improper sanitary practices. The most effective methods of reducing air contamination are through sanitary practices, filtering of air entering food processing and preparation areas, and protection from air by appropriate packaging techniques and materials.

Sewage

Raw, untreated sewage can contain pathogens that have been eliminated from the human body, as well as other materials of the environment. Examples are microorganisms causing typhoid and paratyphoid fevers, dysentery, and infectious hepatitis. Sewage may contaminate food and equipment through faulty plumbing.

If raw sewage drains or flows into potable water lines, wells, rivers, lakes, and ocean bays, the water and living organisms such as seafood are contaminated. To prevent this kind of contamination, privies and septic tanks should be sufficiently separated from wells, streams, and other bodies of water. Raw sewage should not be applied to fields where fruits and vegetables are grown. (Additional discussion related to sewage treatment is presented in Chapter 10.)

Insects and Rodents

Flies and cockroaches are associated with living quarters, eating establishments, and food processing facilities, as well as with toilets, garbage, and other filth. These pests transfer filth from contaminated areas to food through their waste products; mouth, feet, and other body parts; and during regurgitation of filth onto clean food during consumption. To stop contamination from these pests, eradication is necessary, and food processing, preparation, and serving areas should be protected against their entry.

Rats and mice transmit filth and disease through their feet, fur, and intestinal tract. Like flies and cockroaches, they transfer filth from garbage dumps and sewers to food or food processing and food service areas. (Discussion related to control of rodents, insects, and other pests is provided in Chapter 11.)

PROTECTION AGAINST CONTAMINATION

The Environment

Foods should not be touched by human hands when consumed uncooked or after cooking, if such contact can be avoided. If contact is necessary, workers should thoroughly wash their hands prior to and periodically during the time that contact is necessary. Contact with hands can be reduced by the use of disposable plastic gloves during food processing, preparation, and service. A processed or prepared food, either in storage or ready for serving or holding, should be covered by a close-fitting clean cover that will not collect loose dust, lint, or other debris. If the nature of the food does not permit this kind of protection, it should be placed in an enclosed, dust-free cabinet at the appropriate temperature. Foods in small modular wrappers or containers, such as milk and juice, should be disposed directly from those wrappers or containers. If foods are served from a buffet, they should be presented on a steam table or ice tray, depending on temperature requirements, and should be protected during display by a transparent shield over and in front of the food. The shield will protect the food against contamination from the serving area (including ambient air), from handling by those being served, and from sneezes, coughs, or other employee- and customer-originated contamination. Any food that has touched any unclean surface should be cleaned thoroughly or discarded. Equipment and utensils for food processing, packaging, preparation, and service should be cleaned and sanitized between use. Food service employees should be instructed to handle dishes and eating utensils in such a way that their hands do not touch any surface that will be in contact with food or the consumer's mouth.

Storage

Storage facilities should provide adequate space with appropriate control and protection against dust, insects, rodents, and other extraneous matter. Organized storage layouts with appropriate stock rotation can frequently reduce contamination and facilitate cleaning, and can contribute to a tidier operation. In addition, storage area floors can be swept or scrubbed and shelves and/or racks cleaned with appropriate cleaning compounds and subsequent sanitizing. (Chapters 7 and 8 discuss appropriate cleaning compounds and sanitizers.) Trash and garbage should not be permitted to accumulate in a food storage area.

Litter and Garbage

The food industry generates a large volume of wastes: used packaging materials, containers, and waste products. To reduce contamination, refuse should be placed in appropriate containers for removal from the food area. The preferred disposal method (required by some regulatory agencies) is to use containers for garbage that are separated from

those for disposal of litter and rubbish. Clean, disinfected receptacles should be located in work areas to accommodate waste food particles and packaging materials. These receptacles should be seamless, with close-fitting lids that should be kept closed except when the receptacles are being filled and emptied. Plastic liners are inexpensive and provide added protection. All receptacles should be washed and disinfected regularly and frequently, usually daily. Containers in food processing and food preparation areas should not be used for garbage or litter, other than that produced in those areas.

Toxic Substances

Poisons and toxic chemicals should not be stored near food products. In fact, only chemicals required for cleaning should be stored on the same premises. Clean-up chemicals should be clearly labeled. Only cleaning compounds, supplies, utensils, and equipment approved by regulatory or other agencies should be used in food handling, processing, and preparation.

SUMMARY

Food products are rich in nutrients required by microorganisms and may become contaminated. Major contamination sources are water, air, dust, equipment, sewage, insects, rodents, and employees.

Contamination of raw materials can also occur from the soil, sewage, live animals, external surface, and the internal organs of meat animals. Additional contamination of animal foods can come from diseased animals, although advances in health care have nearly eliminated this source. Contamination from chemical sources can occur through accidental mixing of chemical supplies with foods. Ingredients can contribute to additional microbial or chemical contamination. Contamination can be reduced through effective housekeeping and sanitation, protection of food during storage, proper disposal of garbage and litter, and protection against contact with toxic substances.

STUDY QUESTIONS

1. What is the chain of infection?
2. What is the major contamination source of food?
3. Which microorganism is most likely to cause foodborne illness if large pieces of meat or broth have been stored in slowly cooling containers?
4. Which pathogenic microorganism may be found in unpasteurized dairy products that have become cross-contaminated?
5. What is the best way to reduce contamination from food equipment?
6. How can sewage-contaminated water, if consumed, affect humans?

REFERENCE

Bryan, F.L. 1979. Epidemiology of foodborne diseases. In *Food-borne infections and intoxications,* 2nd ed., eds. H. Riemann and F.L. Bryan, 4. New York: Academic Press.

SUGGESTED READINGS

Guthrie, R.K. 1988. *Food Sanitation.* 3rd ed. New York: Van Nostrand Reinhold, New York.

Todd, E.C.D. 1980. Poultry-associated foodborne disease–Its occurrence, cost, sources, and prevention. *J Food Prot,* 43: 129.

CHAPTER 4

Personal Hygiene and Sanitary Food Handling

Food handlers can transmit bacteria causing illness. In fact, humans are the major source of food contamination. Their hands, breath, hair, and perspiration contaminate food, as can their unguarded coughs and sneezes, which can transmit microorganisms capable of causing illness. Transfer of human and animal excreta by workers is a potential source of pathogenic microorganisms that can invade the food supply.

By necessity, the food industry is focusing more on employee education and training and emphasizing that supervisors and workers be familiar with the principles of food protection. Harrington (1992) reported that direct costs of a foodborne illness outbreak can approximate $75,000 per food service establishment, including investigation cleanup, restaffing and restocking, product loss, settlements, and increased regulatory sanctions. In multiunit chain operations, the negative effects of public opinion often spiral outward to uninvolved units. The indirect loss of business was $7 million in an outbreak during the past in a metropolitan area.

PERSONAL HYGIENE

The word *hygiene* is used to describe an application of sanitary principles for the preservation of health. Personal hygiene refers to the cleanliness of a person's body. The health of workers plays an important part in food sanitation. People are potential sources of microorganisms that cause illness in others through the transmission of viruses or through food poisoning.

Employee Hygiene

Ill employees should not come in contact with food or equipment and utensils used in the processing, preparation, and serving of food. Human illnesses that may be transmitted through food are diseases of the respiratory tract, such as the common cold, sore throat, pneumonia, scarlet fever, tuberculosis, and trench mouth; intestinal disorders; dysentery; typhoid fever; and infectious hepatitis. In many illnesses, the disease-causing microorganisms may remain with the person after recovery. A person with this condition is known as a *carrier*.

When employees become ill, their potential as a source of contamination increases dramatically. Staphylococci are normally found in and around boils, acne, carbuncles, infected cuts, and eyes and ears. A sinus infection, sore throat, nagging cough, and other symptoms of the common cold are further signs that microorganisms are increasing in number. The same principle applies to gastrointestinal ailments, such as diarrhea or an upset stomach. Even when evidence of illness passes, some of the causative microorganisms may remain as a source of recontamination. For example, Salmonellae may persist for several months after the employee has recovered. The virus responsible for hepatitis has been found in the intestinal tract over 5 years after the disease symptoms have disappeared. To explain the importance of employee hygienic practices, it is beneficial to look at different parts of the human in terms of potential sources of bacterial contamination.

Skin

This living organ provides four major functions: protection, sensation, heat regulation, and elimination. Protection is an important function in terms of personal hygiene. The epidermis (outer layer of skin) and the dermis (inner layer of skin) are tough, pliable, elastic layers that provide resistance to damage from the environment. The epidermis is less subject to damage than other parts of the body because it does not contain nervous tissue or blood vessels. The outermost layer of the epidermis is called the *corneum*. Cells of the corneum consist of 25 to 30 rows. They tend to be flatter and softer than most other cells and function through the formation of a layer that is impermeable to microorganisms. This layer is important to the distribution of transient and resident microbial flora. These tissues are replaced by newly created cells from the underlying layers every 4 to 5 days as they wear away. These dead cells are 30×0.6 μm in diameter and are easily dislodged in clothing or disseminated into the air. The dermis, an underlying layer of skin, is composed of connective tissue, elastic fibers, blood and lymph vessels, nervous tissue, muscle tissue, glands, and ducts. The glands of the dermis secrete perspiration and

oil. The skin functions as a working organ through constant deposition of perspiration, oil, and dead cells on the outer surface. When these materials mix with environmental substances such as dust, dirt, and grease, they form an ideal environment for bacterial growth. Thus, the skin becomes a potential source of bacterial contamination. As the secretions build up and the bacteria continue to grow, the skin may become irritated. Food handlers may rub and scratch the area, thereby transferring bacteria to food. Improper hand washing and infrequent bathing increases the amount of microorganisms dispersed with the dead cell fragments. Contamination results in shortening the product's shelf life or in foodborne illness.

Foodborne illness may occur if a food handler is a carrier of *Staphylococcus aureus* or *Staphylococcus epidermis*, two of the predominant bacterial species normally present on the skin. These organisms are present in the hair follicles and in the ducts of sweat glands. They are capable of causing abscesses, boils, and wound infections following surgical operations. As secretions occur, perspiration from the eccrine gland, as well as sebum (a fatty material seated into hair follicles) carry bacteria from the gland and deposit them on the skin surface, with subsequent reinfection.

Certain genera of bacteria do not grow on the skin because the skin acts as a physical barrier and also secretes chemicals that can destroy some of the microorganisms that are foreign to it. This self-disinfectant characteristic is most effective when the skin is clean.

The epidermis contains cracks, crevices, and hollows that can provide a favorable environment for microorganisms. Bacteria also grow in hair follicles and in the sweat sebaceous glands. Because hands are very tactile, the opportunity for cuts, calluses, and contact with a wide variety of microorganisms is evident. Hands are in association with so much of the environment that contact with contaminating bacteria is unavoidable.

Resident bacteria of the skin, which are not easily removed, live in microcolonies that are usually buried deep in the pores of the skin and protected by fatty secretions of the setaceous glands. The microorganisms in the resident group are more frequently *Micrococcus luteus* and *S. epidermis*, whereas the bacteria most associated with the transient group are *S. aureus*.

Poor care of the skin and skin disorders, aside from detrimental appearance, may cause bacterial infections, such as boils and impetigo. *Boils* are severe local infections that result when microorganisms penetrate the hair follicles and skin glands after the epidermis has been broken. This damage can occur from excess irritation of clothing. Swelling and soreness result as microorganisms such as staphylococci multiply and produce an exotoxin that kills the surrounding cells. The body reacts to this exotoxin by accumulating lymph, blood, and tissue cells in the infected area to counteract the invaders. A restraining barrier is formed that isolates the infection. A boil should never be squeezed. If it is squeezed, the infection may spread to adjoining areas and cause additional boils. Such a cluster is called a *carbuncle*. If staphylococci gain entrance to the bloodstream, they may be carried to other parts of the body, causing meningitis, bone infection, or other undesirable conditions. Employees having boils should exercise caution if they must handle food because the boil is a prime source of pathogenic staphylococci. An employee who touches a boil or a pimple should use a hand dip for disinfection. Cleanliness of the skin and wearing apparel is important in the prevention of boils.

Impetigo is an infectious disease of the skin that is caused by members of the staphylococci group. This condition appears more readily in young people who fail to keep their skin clean. The infection spreads easily to other parts of the body and may be transmitted by contact. Keeping the skin clean helps to prevent impetigo.

Fingers

Bacteria may be picked up by the hands touching dirty equipment, contaminated food, clothing, or other areas of the body. When this occurs, the employees should use a hand-dip sanitizer to reduce transfer of contamination. Plastic gloves may be a solution (although their use has been considered controversial by sanitation experts who maintain that their use may allow massive contamination). They help prevent the transfer of pathogenic bacteria from the fingers and hands to food and have a favorable psychological effect on those observing the food being handled in this way.

The use of gloves offers both benefits and liabilities. A clean contact surface may be attained initially and bacteria that are sequestered on and in the skin are not permitted to enter foods as long as the gloves are not torn or breached in some way. However, the skin beneath the gloves is occluded, and heavily contaminated perspiration builds up rapidly between the internal surface of the glove and skin. Furthermore, gloves tend to promote complacency that is not conducive to good hygiene.

Fingernails

One of the easiest ways to spread bacteria is through dirt under the fingernails. Employees with dirty fingernails should never handle any food. Washing the hands with soap and water removes transient bacteria, and the use of an antiseptic or sanitizer in hand soap controls resident bacteria. Hospitals have demonstrated that an alcohol containing a humectant can be very beneficial in controlling and removing both transient and resident bacteria without hand irritation (Restaino and Wind, 1990).

Jewelry

Jewelry should not be worn in food processing or food service areas to reduce safety hazards in an environment containing machinery. Also, it may be contaminated and fall into food.

Hair

Microorganisms (especially staphylococci) are found on hair. Employees who scratch their heads should use the hand dip before handling food and should wear a head cover. The necessity of wearing hair coverings in food processing areas should be considered a condition of employment for all new employees and should be made known at the time that they are hired. Disposable hair covers should be worn beneath hard hats. The use of "overseas" type paper hats is not a good sanitation practice as not all of the hair is restrained.

Eyes

The eye itself is normally free of bacteria but mild bacterial infections may develop. Bacteria can then be found on the eyelashes and at the indentation between the nose and eye. By rubbing the eyes, the hands are contaminated.

Mouth

Many bacteria are found in the mouth and on the lips. This fact can be dramatically demonstrated by requesting an employee to press his or her lips on the surface of sterile agar medium in a Petri dish. The subsequent growth of bacterial colonies at the site of this impression (or those made by other parts of the body) provides dramatic evidence of the presence of bacteria. During a sneeze, some of the bacteria are transferred to the air and may land on food being handled. Furthermore, smoking should be prohibited while working. Various disease-causing bacteria, as well as viruses, are also found in the mouth, especially if an employee is ill. These microorganisms can be transmitted to other individuals, as well as to food products, when one sneezes.

Spitting is usually prompted after smoking, due to an irritating taste in the mouth or when one has a head cold. This practice should never be permitted in any food processing establishment. Spitting is unsightly and is a mode of disease transmission and product contamination. Brushing the teeth prevents the buildup of bacterial plaque on the teeth and reduces the degree of contamination that might be transmitted to a food product if an employee gets saliva on the hands or sneezes.

Nose, Nasopharynx, and Respiratory Tract

The nose and throat have a more limited microbial population than does the mouth. This is because of the body's effective filtering system. Particles larger than 7 µm in diameter that are inhaled are retained in the upper respiratory tract. This is accomplished through the highly viscid mucus that constitutes a continuous membrane overlying the surfaces within the nose, sinuses, pharynx, and esophagus. Approximately half of the particles that are 3 µm or larger in diameter are removed in the remaining tract, and the rest penetrate the lungs. Those particles that do penetrate and lodge themselves in the bronchi and bronchioles are destroyed by the body's defenses. Viruses are controlled by virus-inactivating agents found in the normal serous fluid of the nose.

Occasionally, microorganisms do penetrate the mucous membranes and establish themselves in the throat and respiratory tract. Staphylococci, streptococci, and diphtheroids are frequently found in these areas. Other microorganisms occasionally inhabit the tonsils. The *common cold* is one of the most prevalent of all infectious diseases. It is generally accepted that the common cold is caused by rhinoviruses. The initial viral attack is generally followed by the onset of a secondary infection because the initial disease lowers the resistance of the mucous membranes in the upper respiratory tract. The secondary infection may be caused by a variety of agents, including bacteria. Bacteria, especially from employees with a cold, can be transmitted from the nose to hands to food with just a slight scratching of the nose.

Employees who have colds should use a hand-dip sanitizer after blowing their noses. Otherwise, these bacteria can be transferred to the food being handled. The discharge from a sneeze or cough should be blocked by the elbow or shoulder.

Sinus infection results from the infection of the membrane of the nasal sinuses. The mucous membranes become swollen and inflamed, and secretions accumulate in the blocked cavities. Pain, dizziness, and a running nose result from the pressure buildup in the cavities. Precautions should be taken if employees with nasal discharges must handle food products. An infectious agent is present in the mucous discharge, and other organisms, such as *S. aureus,* could be present. For this reason, employees should wash and disinfect their hands after blowing their noses, and all sneezes should be completely blocked.

A *sore throat* is usually caused by a species of streptococci. The primary source of pathogenic streptococci is the human being, who carries this microbe in the upper respiratory tract. "Strep throat," *laryngitis,* and *bronchitis* are spread by the mucous discharge of carriers. Streptococci are also responsible for scarlet fever, rheumatic fever, and tonsillitis. These conditions may be spread by employees with poor hygienic practices.

Influenza, commonly referred to as *flu*, is an acute infectious respiratory disease that occurs in small to widespread epidemic outbreaks. It gains entrance to the body through the respiratory tract. Death may result from secondary bacterial infections by staphylococci, streptococci, or pneumococci.

Most of these ailments are highly contagious. Therefore, employees infected with any of them should not be permitted to work. They endanger the products they handle and fellow employees as well. All coughs and sneezes produce atomized droplets of mucous containing the infectious agents and should be blocked. Hands should be kept as clean as possible by making use of hand dips to prevent contamination of the infectious microorganism.

Excretory Organs

Intestinal discharges are a prime source of bacterial contamination. Approximately 30% to 35% of the dry weight of the intestinal contents of humans is composed of bacterial cells. *Streptococcus fecalis* and staphylococci are generally the only bacteria found in the upper part of the small intestine; however, the species and individual organisms become more numerous in the lower intestine. Particles of feces collect on the hairs in the anal region and are spread to the clothing. When employees go the washroom, they may pick up some of the intestinal bacteria. If the hands are not washed properly, these organisms will be spread to handled products. The bacteria commonly found in this area are frequently found in food products. A lack of personal hygiene is responsible for this type of contamination. For this reason, employees should wash their hands with soap before leaving the washroom and should use a hand-dip sanitizer before handling food.

Both viruses and bacterial disease organisms can be found in food products. Intestinal viruses may be spread by food products. In these cases, the product acts as a carrier for the viruses. Unlike bacterial contaminants, they cannot multiply in the food.

The intestinal tracts of humans and animals carry the most common forms of bacteria, which, when multiplied sufficiently, are toxic or poisonous to the body. The infections or poisons range from slight to severe and may result in death. *Salmonella, Shigella,* and enterococci bacteria causing different types of intestinal disorders are the most common.

Personal Contamination of Food Products

The intrinsic factors that affect microbial contamination by people are as follows.

1. *Body location*. The composition of the normal microbial flora varies depending on the body area. The face, neck, hands, and hair contain a higher proportion of transient microorganisms and a higher bacterial density. The exposed areas of the body are more vulnerable to contamination from environmental sources. When environmental conditions change, the microbial flora adapt to the new environment.
2. *Age*. The microbial population changes as a person matures. This trend is especially true for adolescents entering puberty. They produce large quantities of lipids known as *sebum,* which promotes the formation of acne caused by *Propionibacterium acnes*.
3. *Hair*. Because of the density and oil production, the hair on the scalp enhances the growth of microbes such as *S. aureus* and *Pityrosporum*.
4. *pH*. The pH of the skin is influenced by the secretion of lactic acid from the sweat glands, bacterial production of fatty acids, and diffusion of carbon dioxide through the skin (Nobel and Pitcher, 1978). The approximate pH value for the skin (5.5) is more selective against transient microorganisms than it is against the resident flora. Factors that change the pH of the skin (soap, creams, etc.) alter the normal microbial flora.
5. *Nutrients*. Perspiration contains water-soluble nutrients (i.e., inorganic ions and some acids), whereas sebum contains lipid (oil) -soluble materials such as triglycerides, esters, and cholesterol (Nobel and Pitcher, 1978). The role of perspiration and sebum in the growth of microorganisms is not fully understood (Restaino and Wind, 1990).

Humans are the most common contamination source of food. People transmit diseases as carriers. A carrier is a person who harbors and discharges pathogens but does not exhibit the symptoms of the disease. Carriers are divided into three groups:

1. *Convalescent carriers*. People who, after recovering from an infectious disease, continue to harbor the causative organism for a variable length of time, usually less than 10 weeks.
2. *Chronic carriers*. People who continue to harbor the infectious organism indefinitely, although they do not show symptoms of the disease.

3. *Contact carriers.* People who acquire and harbor a pathogen through close contact with an infected person but do not acquire the disease.

People harbor a number of organisms, including:

- *Streptococci.* These organisms, commonly harbored in the human throat and intestines, are responsible for a wider variety of diseases than other bacteria. They are also frequently responsible for the development of secondary infections.
- *Staphylococci.* The most important single reservoir of staphylococci infection of humans is the nasal cavity. Equally important to the food industries are those who possess the pathogenic varieties of the organism as part of their natural skin flora. These people are a constant threat to consumer safety if they are allowed to handle food products.
- *Intestinal microorganisms.* This group of organisms includes *Salmonella, Shigella, Escherichia coli, Cholera,* infectious hepatitis, and infectious intestinal amoebas. These microorganisms are of public health concern because they can contribute to serious illness.

Hand Washing

Approximately 25% of food contamination is attributable to improper hand washing. Hand washing is conducted to break the transmission route of the microorganisms from the hands to another source and to reduce resident bacteria. *Pseudomonas aeruginosa, Klebsiella pneumoniae, Serratia marcescens, E. coli,* and *S. aureus* can survive for up to 90 minutes when artificially inoculated on fingertips (Filho et al., 1985).

Hand washing with soap and water, which act as emulsifying agents to solubilize grease and oils on the hands, will remove transient bacteria. Increased friction through rubbing the hands together or by using a scrub brush with soap can reduce the number of transient and resident bacteria than can quick hand washing.

Because proper hand washing is essential to attain a sanitary operation, mechanized hand washers are being used (Figure 4–1). A typical unit is located in the processing area. When workers enter the area, they must use the washing unit. This equipment is responsible for increasing hand washing frequency by 300%. The user inserts the hands into two cylinders, passing a photo-optic sensor, which activates

Figure 4–1 Mechanized hand washer. Courtesy of Meritech Handwashing Systems, Englewood, Colorado.

the cleansing action. Jet sprays within each cylinder spray a mixture of antimicrobial cleansing solution and water on the hands, followed by a potable water rinse. The 10-second, massage-like cycle has been clinically proven to be 60% more effective at removing pathogenic bacteria from the hands than the average manual hand washing (Anon., 1997b). Also, this process can remove contamination from gloves and can accomplish hand or glove washing with approximately 2 L of water or only one-third of the amount used in most manual hand washing methods.

Antimicrobial agents exert a continuous antagonistic action on emerging microbes and enhance the effectiveness of ordinary hand soap at the time of application. The overall efficacy of an antimicrobial hand soap depends on continuous use throughout the day. A contact of less than 5 seconds during hand washing has little effect on reducing the microbial load. A cleaning compound will remove more transient bacteria, with subsequent destruction by the sanitizer. Contamination from workers is illustrated in Figure 4–2. Figure 4–3 illustrates the suggested procedures for use of the recommended double hand washing method.

A potential barrier to cross-contamination by the hands is a commercial antibacterial lotion marketed as Bio-Safe. This viscous lotion forms an invisible and undetectable polymer coating that bonds electrochemically to the outermost layer of skin to provide protection from dermal exposure in the workplace (Anon., 1997a). Figure 4–4 illustrates a wall-mounted hand sanitizer to reduce microbial contamination of workers.

Methods of Disease Transmission

The following examples suggest how poor hand washing can cause major outbreaks of foodborne illness.

On a 4-day Caribbean cruise, 72 passengers and 12 crew members had diarrhea, and 13 people had to be hospitalized. Stool samples of 19 of the passengers and 2 of the crew members contained *Shigella flexneri* bacteria. The illness was traced to German potato salad prepared by a crew member who carried these bacteria. The disease spread easily because the toilet facilities for the galley crew were limited.

Over 3,000 women who attended a 5-day outdoor music festival in Michigan became ill with gastroenteritis caused by *Shigella sonnei*. The illness began 2 days after the festival ended, and patients were spread all over the United States before the outbreak was recognized. An uncooked tofu salad served on the last day caused the outbreak. Over 2,000 volunteer food handlers prepared the communal meals served during the festival. Before the festival, the staff had a smaller outbreak of shigellosis. Sanitation at the festival was mostly acceptable, but access to soap and running water for hand washing was limited. Good hand washing facilities could have prevented this explosive outbreak of foodborne illness.

Shigella sonnei caused an outbreak of foodborne illness in 240 airline passengers on 219 flights to 24 states, the District of Columbia, and 4 countries. The outbreak was identified only because it involved 21 of 65 football team players and coaches. Football players and coaches, airline passengers, and flight attendants with the illness all had the same strain of *S. sonnei*. The illness was caused by cold food items served on the flights that had been prepared by hand at the airline flight kitchen. Flight kitchens should minimize hand contact when preparing cold foods or should remove these foods from in-flight menus.

There are viable ways of drying the hands and other skin surfaces. Paper roll and sheet

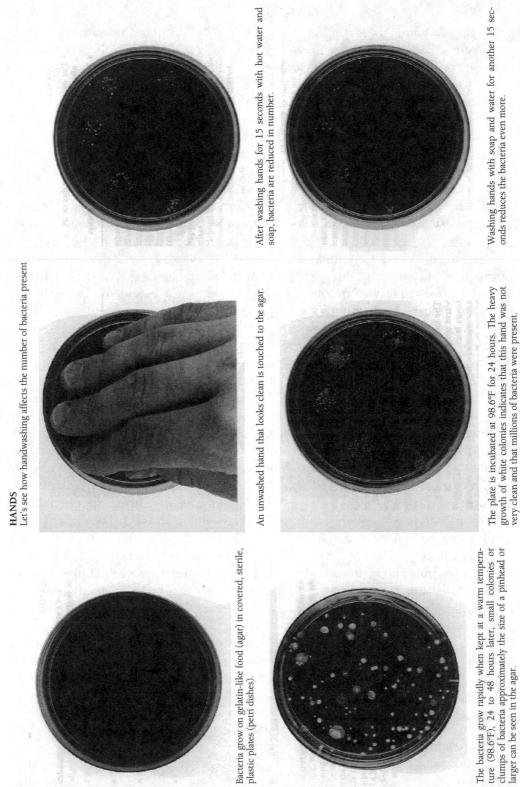

After washing hands for 15 seconds with hot water and soap, bacteria are reduced in number.

Washing hands with soap and water for another 15 seconds reduces the bacteria even more.

HANDS
Let's see how handwashing affects the number of bacteria present

An unwashed hand that looks clean is touched to the agar.

The plate is incubated at 98.6°F for 24 hours. The heavy growth of white colonies indicates that this hand was not very clean and that millions of bacteria were present.

Bacteria grow on gelatin-like food (agar) in covered, sterile, plastic plates (petri dishes).

The bacteria grow rapidly when kept at a warm temperature (98.6°F), 24 to 48 hours later, small colonies or clumps of bacteria approximately the size of a pinhead or larger can be seen in the agar.

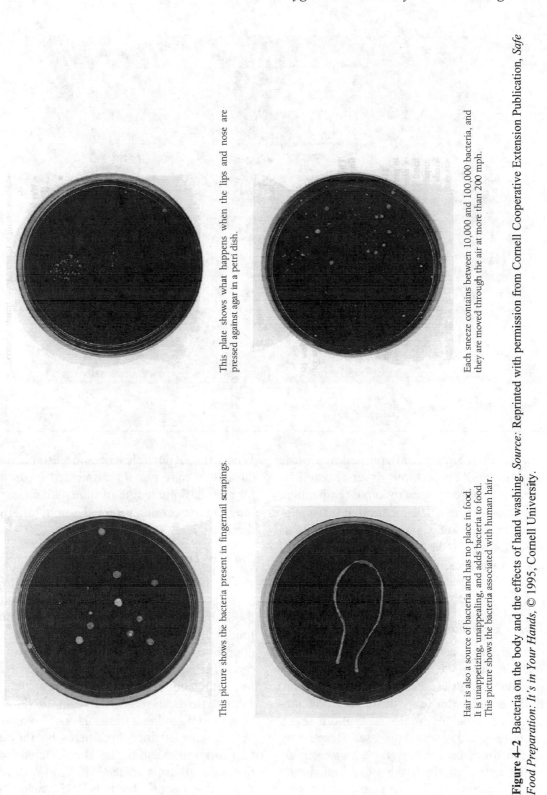

This plate shows what happens when the lips and nose are pressed against agar in a petri dish.

Each sneeze contains between 10,000 and 100,000 bacteria, and they are moved through the air at more than 200 mph.

This picture shows the bacteria present in fingernail scrapings.

Hair is also a source of bacteria and has no place in food. It is unappetizing, unappealing, and adds bacteria to food. This picture shows the bacteria associated with human hair.

Figure 4–2 Bacteria on the body and the effects of hand washing. *Source:* Reprinted with permission from Cornell Cooperative Extension Publication, *Safe Food Preparation: It's in Your Hands,* © 1995, Cornell University.

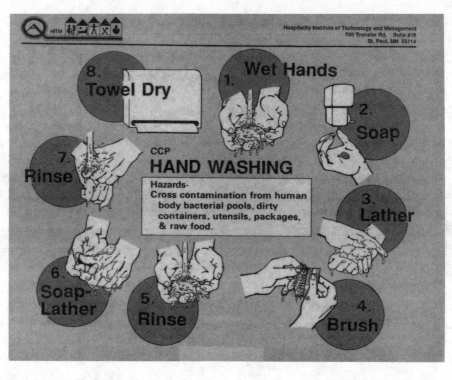

Figure 4–3 Recommended hand washing procedure.

towels are acceptable if deposited in a waste container. Electric blow dryers should be used only in restrooms to avoid temperature rise in other areas. The location of this equipment in processing areas is unacceptable as they can blow dust off of the floor onto food contact surfaces.

Methods of Disease Transmission

Direct Transmission

Many diseases may be transmitted by direct transfer of the microorganisms to another person through close contact. Examples are diphtheria, scarlet fever, influenza, pneumonia, smallpox, tuberculosis, typhoid fever, dysentery, and venereal diseases. Respiratory diseases may be transferred via atomized particles extruded from the nose and mouth when a person talks, sneezes, or coughs.

When these particles become attached to dust, they may remain suspended in the air for an indefinite length of time. Other people may then become infected upon inhaling these particles.

Indirect Transmission

The host of an infectious disease may transfer organisms to vehicles such as water, food, and soil. Lifeless objects, other than food, capable of transmitting infections are doorknobs, telephones, pencils, books, washroom fittings, clothing, money, and knives. Intestinal and respiratory diseases such as salmonellosis, dysentery, and diphtheria may be spread by indirect transmission. To reduce the transfer of microorganisms by indirect transmission, wash basins should have foot-operated controls instead of hand-operated faucets, and doors should be self-closing.

Figure 4–4 Wall mount no-touch hand sanitizer with 8L reservoir. Courtesy of Ecolab, Inc., Mendota Heights, Minnesota.

Requirements for Hygienic Practices

Management must establish a protocol to ensure hygienic practices by employees. Supervisors and managers should set an example for employees by their own high levels of hygiene and good health while conveying the importance of these practices to the employees. They should provide proper laundry facilities for maintenance of cleanliness through clean dressing rooms, services, and welfare facilities.

Management should require employees to have a pre-employment physical examination to verify that they are in good physical, mental, and emotional health. This is an excellent opportunity to impress the importance of good hygienic habits on a new employee and to emphasize how employees "shed" *Salmonella* and *Shigella* (Troller, 1993). Furthermore, those with skin infections may be identified before they handle food. All employees who work with food should be checked regularly for signs of illness, infection, and other unhealthy conditions.

Several countries have a legal requirement for pre-employment health examinations and require that they be repeated at regular intervals. However, this regulation has been challenged because of the expense of routine medical examinations, the difficulty of administering these plans, and because a clear relationship between food handlers and foodborne disease has not been established.

The following practices should be conducted by employees to ensure personal hygiene:

1. Physical health should be maintained and protected through practice of proper nutrition and physical cleanliness.
2. Illness should be reported to the employer before working with food so that work adjustments can be made to protect food from the employee's illness or disease.
3. Hygienic work habits should be developed to eliminate potential food contamination.
4. During the work shift, hands should be washed after using the toilet; handling garbage or other soiled materials; handling uncooked muscle foods, egg products, or dairy products; handling money; smoking; coughing; or sneezing.
5. Personal cleanliness should be maintained by daily bathing and use of deodorants, washing hair at least twice a week, cleaning fingernails daily, using a hat or hair net while handling food, and wearing clean underclothing and uniforms.

6. Employee hands should not touch food service equipment and utensils. Disposable gloves should be used when contact is necessary.
7. Rules such as "no smoking" should be followed, and other precautions related to potential contamination should be taken.

Employers should emphasize hygienic practices of employees as follows:

1. Employees should be provided training in food handling and personal hygiene.
2. A regular inspection of employees and their work habits should be conducted. Violations of practices should be handled as disciplinary violations.
3. Incentives for superior hygiene and sanitary practices should be provided.

Food handlers should be responsible for their own health and personal cleanliness. Employers should be responsible for making certain that the public is protected from unsanitary practices that could cause public illness. Personal hygiene is a basic step that should be taken to ensure the production of wholesome food.

SANITARY FOOD HANDLING

A protective sanitation barrier between food and the sources of contamination should be provided during food handling. Barriers include hair nets, disposable gloves, mouth guards, sneeze guards, and food packaging and containers.

Role of Employees

Food processing and food service firms should protect their employees and consumers from workers with diseases or other microorganisms of public health concern that can affect the wholesomeness or sanitary quality of food. This precaution is important to maintain a good image and sound operating practices consistent with regulatory organizations. In most communities, local health codes prohibit employees having communicable diseases or those who are carriers of such diseases from handling foods or participating in activities that may result in contamination of food or food contact surfaces. Responsible employers should exercise caution in selecting employees by screening unhealthy individuals. Although some areas no longer require health cards because of the expense involved, many local health departments require all employees who handle food to be examined by a physician who will issue a health card only to healthy individuals. Selection of employees should be predicated upon these facts:

1. Absence of communicable diseases should be verified by a county health card or a physician's report.
2. Applicants should not exhibit evidence of a sanitary hazard, such as open sores or presence of excessive skin infections or acne.
3. Applicants who display evidence of respiratory problems should not be hired to handle food or to work in food processing or food preparation areas.
4. Applicants should be clean and neatly groomed and should wear clothing free of unpleasant odor.
5. Applicants should successfully complete a sanitation course and examination such as that provided by the National Restaurant Association.

Required Personal Hygiene

Food organizations should establish personal hygiene rules that are clearly defined and uniformly and rigidly enforced. These regulations should be documented, posted, and/or clearly spelled out in booklets. Policy should address personal cleanliness, working attire, acceptable food-handling practices, and the use of tobacco and other prohibited practices.

Facilities

Hygienic food handling requires appropriate equipment and supplies. Food-handling and food-processing equipment should be constructed according to regulations of the appropriate regulatory agency. Welfare facilities should be clean, neat, well lighted, and conveniently located away from production areas. Restrooms should have self-closing doors. It is also preferred that hand-washing stations have foot- or knee-operated faucets that supply water at 43°C to 50°C. Remotely operated liquid soap dispensers are recommended because bars of soap can increase the transfer of microorganisms. Disposable sanitary towels are best for drying hands. The consumption of snacks, beverages, and other foods, as well as smoking, should be confined to a specific area, which should be clean and free of insects and spills.

Employee Supervision

Employees who handle food should be subjected to the same health standards used in screening prospective employees. Supervisors should observe employees daily for infected cuts, boils, respiratory complications, and other evidence of infection. Many local health authorities require food service and food processing firms to report an employee who is suspected to have a contagious disease or to be a carrier.

Employee Responsibilities

Although the employer is responsible for the conduct and practices of employees, responsibilities should be assigned to employees at the time employment begins.

- Employees should maintain a healthy condition to reduce respiratory or gastrointestinal disorders and other physical ailments.
- Injuries, including cuts, burns, boils, and skin eruptions, should be reported to the employer.
- Abnormal conditions, such as respiratory system complications (e.g., head cold, sinus infection, and bronchial and lung disorder), and intestinal disorders, such as diarrhea, should be reported to the employer.
- Personal cleanliness that should be practiced includes daily bathing, hair washing at least twice a week, daily changing of undergarments, and maintenance of clean fingernails.
- Employees should tell a supervisor when items such as soap or towels in washrooms need to be replenished.
- Habits such as scratching the head or other body parts should be stopped.
- The mouth and nose should be covered during coughing or sneezing.
- The hands should be washed after visiting the toilet, using a handkerchief, smoking, handling soiled articles, or handling money.
- Hands should be kept out of food. Food should not be tasted from the hand, nor should it be consumed in food production areas.
- Food should be handled by utensils that are not touched with the mouth.
- Rules related to use of tobacco should be enforced.

SUMMARY

Food handlers are potential sources of microorganisms that cause illness and food spoilage. *Hygiene* is a word used to describe sanitary principles for the preservation of health. *Personal hygiene* refers to the cleanliness of a person's body. Parts of the body that contribute to the contamination of food include the skin, hands, hair, eyes, mouth, nose, nasopharynx, respiratory tract, and excretory organs. These parts are contamination sources as carriers, through direct or indirect transmission, of detrimental microorganisms.

Management must select clean and healthy employees and ensure that they conduct hygienic practices. Employees must be held responsible for personal hygiene so that the food that they handle remains wholesome.

STUDY QUESTIONS

1. What is hygiene?
2. What is a chronic carrier?
3. What is the difference between direct and indirect transmission of diseases?
4. What is a contact carrier?
5. What are resident bacteria?
6. Which microorganisms cause the common cold?

REFERENCES

Anon. 1995. *Safe food preparation: It's in your hands*. Ithaca, NY: Cornell Cooperative Extension Publication.

Anon. 1997a. Did you wash your hands? *Food Qual III* 19: 52.

Anon. 1997b. Hands-on hygiene. *Food Qual III* 19: 56.

Filho, G.P.P., et al. 1985. Survival of gram negative and gram positive bacteria artificially applied on the hands. *J Clin Microbiol* 21: 652.

Harrington, R.E. 1992. The role of employees in the spread of foodborne disease–Food industry views of the problem and coping strategies. *Dairy Food Environ Sanit* 12: 62.

Nobel, W.C., and D.G. Pitcher. 1978. Microbial ecology of the human skin. *Adv Microbiol Ecol 2*: 245.

Restaino, L., and C.E. Wind. 1990. Antimicrobial effectiveness of hand washing for food establishments. *Dairy Food Environ Sanit* 10: 136.

Troller, J.A. 1993. *Sanitation in food processing*. 2nd ed. New York: Academic Press.

SUGGESTED READINGS

Longree, K., and G. Armbruster. 1996. *Quality food sanitation*. 5th ed. New York: John Wiley & Sons.

National Restaurant Association Education Foundation. 1992. *Applied foodservice sanitation*. 4th ed. New York: John Wiley & Sons, in cooperation with the Education Foundation of the National Restaurant Association, Chicago.

CHAPTER 5

The Role of HACCP in Sanitation

The Hazard Analysis Critical Control Point (HACCP) program is a state-of-the-art preventive approach to consistently safe food production. This program is based on two important concepts of safe food production—prevention and documentation. The major thrusts of HACCP are to determine how and where food safety hazards may exist and how to prevent their occurrence. The important documentation concept is essential to verify that potential hazards have been controlled. HACCP has been recommended and/or required for use throughout the food industry and is the basis for federal food inspection in the United States.

This proactive, prevention-oriented program is based on sound science. HACCP focuses on the prevention or control of food safety hazards that fall in the three main categories of biological, chemical, and physical hazards. The program focuses on safety and not quality and should be considered separate from or a supplement to quality assurance. The objective of HACCP is to ensure that effective sanitation and hygiene and other operational considerations be conducted to produce safe products and to provide proof that safety practices have been followed.

WHAT IS HACCP?

The HACCP concept was developed in the 1950s by the National Aeronautics and Space Administration (NASA) and Natick Laboratories for use in aerospace manufacturing under the name "Failure Mode Effect Analysis." This rational approach to process control for food products was developed jointly by the Pillsbury Company, NASA, and the U.S. Army Natick Laboratories in 1971 as an attempt to apply a zero-defects program to the food processing industry. HACCP was incorporated to guarantee that food used in the U.S. space program would be 100% free of bacterial pathogens. Clark (1991) described HACCP as a simple but very specific method for identifying hazards and for implementing the appropriate control to prevent potential hazards. Because it is designed to prevent rather than detect food hazards, HACCP has been identified by the U.S. Department of Agriculture (USDA) Food Safety and Inspection Service (FSIS) as a tool to prevent food safety hazards during meat and poultry production. The HACCP concept has been embraced and recommended for use by several scientific groups. These include the National

Academy of Sciences (NAS) Committee on the Scientific Basis of the Nation's Meat and Poultry Inspection Program and the NAS Subcommittee on Microbiological Criteria of the Committee on Food Protection. These two committees recognized HACCP as a rational and improved approach to food production control that can determine those areas where control is most critical to the manufacturing of safe and wholesome food.

This technique, which assesses the flow of food through the process, provides a mechanism to monitor these operations frequently and to determine the points that are critical for the control of foodborne disease hazards. *A hazard is the potential to cause harm to the consumer. A critical control point (CCP) is an operation or step by which preventive or control measures can be exercised that will eliminate, prevent, or minimize a hazard (hazards) that has (have) occurred prior to this point.* The HACCP concept has become a valuable program for process control of microbial hazards. This approach is a harbinger of the trend toward more sophistication in food sanitation and inspection. It has been legitimized by governmental regulators and is being adopted by progressive food companies.

The HACCP concept is divided into two parts: (1) hazard analysis and (2) determination of critical control points. Hazard analysis requires a thorough knowledge of food microbiology and a knowledge of which microorganisms may be present and the factors that affect their growth and survival. Food safety and acceptability are most affected by: (1) contaminated raw food or adjuncts, (2) improper temperature control during processing and storage (time-temperature abuse), (3) improper cooling through failure to cool to refrigerated temperature within 2 to 4 hours, (4) improper handling after processing, cross-contamination (between products or between raw and processed foods), (5) ineffective or improper

cleaning of equipment, (6) failure to separate raw and cooked products, and (7) poor employee hygiene and sanitation practices.

The HACCP evaluation process describes the product and its intended use and identifies any potentially hazardous food items subject to microbial contamination and proliferation during food processing or preparation. Then the entire process is observed. Hazard analysis is a procedure for conducting risk analysis for products and ingredients by diagramming the process to reflect the manufacturing and distribution sequence, microbial contamination, survival, and proliferation capable of causing foodborne illness. Critical control points are identified from a flow chart. Any deficiencies that are identified are prioritized and corrected. Monitoring steps are established to evaluate effectiveness. The HACCP program, implemented by the food industry and monitored by regulatory agencies, provides the industry with tools and monitoring points and is used to protect the consuming public effectively and efficiently.

The HACCP concept provides a more rational approach to control microbial hazards in foods than does traditional inspection (Marriott et al., 1991). Although HACCP was developed approximately 25 years ago, this concept did not catch on with other products until 1985, when the NAS recommended HACCP for food processing operations. In later NAS studies, HACCP was recommended for the inspection of meat and poultry products. Seafood inspection has been developed according to HACCP principles. Although HACCP is the current trend in the food industry, this concept may evolve to a portion of a more complete program for total quality management in the future.

HACCP should be incorporated as a quality assurance (QA) function and as a systematic approach to hazard identification, risk assessment, and hazard control in a food processing and/or food service facility and distribution channel to ensure a hygienic operation. Potential product abuse should be considered, and each stage of the process should be examined as an entity and in relationship to other stages. The analysis should include the production environment as it contributes to microbial and foreign material contamination.

HACCP offers benefits to the regulator, processor, and consumer. The regulator and processor are provided with a history of the operations and can concentrate on components related to controlling hazards. Through monitoring of CCPs, both can evaluate the effectiveness of the control methods. Furthermore, the processor can control the operation on a continuous basis and prevent hazards, instead of reacting to what has already happened. Ultimately, the consumer benefits through access to a product manufactured under conditions where hazards have been identified and controlled.

At the CCPs of a sanitation process, specifications must be established and monitored to verify that the procedure meets established criteria to be posted for review by the supervisors and employees and can be used to stress the importance of monitoring and conforming with guidelines.

Monitoring must encompass systematic observation, measurement, and recording of the significant factors for the prevention or control of hazards. It must be followed up to correct any out-of-control processes or to bring the product back into acceptable limits before startup or during the operation. The procedures should define the acceptable performance of a process and should describe how process deviations should be handled. Bauman (1997) suggested that because specifications for producing a product will contain points critical to safety and some critical to quality, it is important that these not be blended together so that plant people will not confuse them.

Although HACCP was implemented by the industry, this program has been monitored by regulatory agencies. The Food and Drug Administration (FDA) has adopted the HACCP philosophy because this systems approach allows it to utilize its resources more efficiently. This program provides management with tools to protect the consumer's health.

A major target of HACCP is *Listeria monocytogenes*. HACCP can help prevent the growth of *L. monocytogenes* because it requires steps to confirm the effectiveness of this concept. Samples should be taken from the food facility environment and product lots to confirm that the control measures are effective. *L. monocytogenes* has been considered to provide the greatest hazard through environmental contamination. Therefore, most samples are tested from environmental sources. Environmental samples should be taken from ceilings, floors, floor drains, water hoses, equipment surfaces, and other areas on a random basis. Floor drains, which can carry organisms from a large area, should be tested routinely, using a rapid microbial method, such as immunoassay technology.

HACCP Development

The essential steps to the development of a HACCP plan are:

1. Assembly of an HACCP team, including the person responsible for the plan. Selections should include employees with expertise in sanitation, quality assurance, and plant operations. Also, it is desirable to have expertise in marketing, personnel management, and communications. HACCP should be organized as a part of the firm's quality assurance program.

2. Description of the food and its distribution. The name and other descriptors including storage and distribution requirements should be provided. All raw materials and adjuncts should be listed.
3. Identification of the intended use and consumers of the food. It is especially important to identify intended consumers if infants and other immuno-compromised people are the targeted customers.
4. Development of a flow diagram (to be discussed later under this topic).
5. Verification of the flow diagram. The HACCP team should inspect the operation to verify the accuracy and completeness of the flow diagram. Modifications should be made as necessary.
6. Conduction of a hazard analysis.
   a. Identify steps in the process where the hazards of potential significance occur.
   b. List all identified hazards associated with each step.
   c. List preventive measures to control hazards.
7. Identification and documentation of the CCPs in the process.
8. Establishment of critical limits for preventive measures associated with each identified CCP.
9. Establishment of CCP-monitoring requirements, including monitoring frequency and person(s) responsible for the specific monitoring activities.
10. Establishment of corrective action to be taken when monitoring reveals that a deviation from an established critical limit exists. The action should include the safe disposition of affected food and the correction of procedures or conditions that caused the out-of-control situation.

11. Establishment of procedures for verification that the HACCP system is working correctly. Responsible company personnel should conduct verification of compliance with the HACCP plan on a scheduled basis.
12. Establishment of effective record-keeping procedures that document the HACCP system and update the HACCP plan when a change of products, manufacturing conditions, and evidence of new hazards occurs.

Steps 6 through 12 are known as the *seven HACCP principles* that will be discussed later.

A food and its raw materials may be classified into categories as follows:

- *Step 1.* Risk assessment as accomplished through examination of the food for possible hazards.
- *Step 2.* Assignment of hazard categories through identification of general food hazard characteristics.

Determination of CCPs is also part of the development process. Not all steps in a process should be considered critical, and it is important to separate critical from noncritical points. A practical approach to determining CCPs consists of utilizing a HACCP worksheet with the following headings:

1. Description of the food product and its intended use.
2. Flow diagram with the following components:
   - raw material handling
   - in-process preparation, processing, and fabrication steps
   - finished product packaging and handling steps
   - storage and distribution
   - point-of-sale handling

As the flow diagram is made, it is easy to identify CCPs. A CCP can be a location, practice, procedure, or process, and, if controlled, it can prevent or *minimize contamination*. Critical control points must be monitored to ensure that the steps are under control. Monitoring may include observation, physical measurements (temperature, pH, A_w), or microbial analysis and most often includes visual and physiochemical measurements because microbial testing is often too time-consuming. Possible exceptions are microbial analysis of the raw materials. Microbial testing may be the only acceptable monitoring procedure when the microbial status of the raw material is a CCP. Microbial methods can be incorporated to determine directly the presence of hazards during processing and in the finished product. They can be used indirectly to monitor effectiveness of control points for cleaning and employee hygiene. Yet, this use of microbiology is a check and does not have to be an ongoing process. Critical limits must be established for each monitoring procedure.

Monitoring must be verified by laboratory analysis to ensure that the process is working. The HACCP concept has been effective because:

1. Cooperation existed between the government and industry to develop monitoring procedures for CCPs.
2. Education of processors is required.
3. Use of HACCP is fostered by government agencies.

Marriott et al. (1991) suggested that the following must be accomplished for HACCP to function effectively in the industry:

1. Food processors and regulators must be educated about HACCP.
2. Technical sophistication must be applied to HACCP by plant personnel.

3. Overuse of HACCP items that are not hazardous should be discontinued.

HACCP Program Implementation

A sequence of events must occur in the implementation of HACCP. The steps for implementation will be discussed briefly.

HACCP Team Assembly

Initial program development involves the designation of an HACCP team, consisting of members with specific knowledge and expertise appropriate to the product and process. Selection criteria should emphasize production and quality assurance knowledge; however, marketing and communication expertise may be appropriate if these employees have an appreciation and understanding of the product and process. The team should include employees who are involved in the daily processing activities as they are more familiar with the variability and limitations of the operation. Furthermore, those involved with the process should be involved to foster a sense of ownership among those who must implement the plan.

Involvement from experts outside of the organization may be beneficial to provide additional expertise, but they must have the support of production employees. Experts who are knowledgeable about the product and process may serve more effectively in verification of the completeness of the hazard analysis and the HACCP plan. According to Stevenson and Bernard (1995), these individuals should have the knowledge and experience to correctly (1) identify potential hazards, (2) assign levels of severity and risk, (3) provide direction for monitoring verification and corrective actions when deviations occur, and (4) assess the success of the HACCP plan.

Food Description and Distribution Method

A separate HACCP plan should be developed for each food product that is manufactured in the plant. Product description should include the name, formulation, method of distribution, and storage requirements.

Intended Use and Anticipated Consumers

The intended use of the food should be based on the normal use by consumers. If the food is targeted for a specific segment of the population, such as infants, immunocompromised people or those in other categories, the intended group should be identified.

Flow Diagram Development Describing the Process

Each step of the operation should be identified through a simple description of the operation that occurs. This diagram is essential for hazard analysis and assessment of CCPs. Furthermore, the diagram serves as a record of the operation and a future guide for employees, regulators, and customers who must understand the process for verification. The flow diagram should include steps that take place before and after the processing that occurs in the plant and should contain words rather than engineering drawings.

Flow Diagram Verification

The HACCP team should check the operation to verify the accuracy and comprehensiveness of the flow diagram. Modifications should be made if and when necessary.

CGMPs—The Building Blocks for HACCP

The Current Good Manufacturing Practices (CGMPs) regulations, as revised in 1986, were promulgated by the FDA to provide criteria for complying with provisions of the Federal Food, Drug, and Cosmetic Act, which mandates that all human foods be free from adulteration. Emphasis is placed on the prevention of product contamination from direct and indirect sources.

Sanitation regulations promulgated by the U.S. Department of Agriculture (USDA) contain identical or similar requirements. Included is a summary of responsibilities imposed on plant management regarding plant personnel. Criteria for disease control, cleanliness (personal hygiene and dress requirements), education, and training are provided. These requirements are designed to prevent the spread of disease among workers in the food processing area and from workers to the food itself. A competent supervisor must be assigned the responsibility for assuring compliance by all personnel.

Good manufacturing practices should be selected and adopted before HACCP is implemented. Without the application of CGMP principles, an effective HACCP program cannot be conducted. Furthermore, CGMPs must be applied in the development of sanitation standard operating procedures (SSOPs). Compliance with specific CGMPs should be included as part of an HACCP program for meat and poultry plants, as CGMP regulations and the USDA sanitation regulations address some biological, chemical, and physical hazards associated with food production. A CGMP compliance program should contain documented plans and procedures.

Good manufacturing practices and SSOPs are interrelated and an important part of process control. CGMPs are the minimum sanitary and processing requirements necessary to ensure the production of wholesome food. The areas that should be addressed through CGMPs are personnel hygiene and other practices, buildings and facilities, equipment and utensils, and production and process controls. CGMPs should be broad in nature.

Sanitation Standard Operating Procedures—The Cornerstones of HACCP

Although SSOPs are interrelated with CGMPs, they detail a specific sequence of events necessary to perform a task to ensure sanitary conditions. Standard operating procedures (SOPs) are either SSOPs or manufacturing SOPs. CGMPs should guide the development of SSOPs. SSOPs contain a description of the procedures that an establishment will follow to address the elements of preoperational and operational sanitation relating to the prevention of direct product contamination.

Federally and state-inspected meat and poultry plants are required to develop, maintain, and adhere to written SSOPs. This requirement was established because the USDA FSIS concluded that SSOPs were necessary in the definition of each establishment's responsibility to consistently follow effective sanitation procedures and to minimize the risk of direct product contamination or adulteration.

In meat and poultry plants, SSOPS cover daily preoperational and operational sanitation procedures that establishments implement to prevent direct product contamination or adulteration. Establishments must identify the officials who monitor daily sanitation activities, evaluate whether the SSOPs are effective, and take appropriate corrective action when needed. Also, daily records that reflect completion of the procedures in the SSOPs are required. Deviations and corrective actions taken must be documented and maintained for a minimum of 6 months and must be made available for verification and monitoring. Corrective actions (1) include procedures to ensure appropriate disposition of contaminated products, (2) restore sanitary conditions, and (3) prevent the recurrence of direct contamination or product adulteration, including the appropriate reevaluation and modification of the SSOPs and the procedures specified therein.

Written SSOPs contain a description of all cleaning procedures necessary to prevent direct contamination or adulteration of products. The frequency with which each procedure in the SSOPs is included along with a designation of the employee(s) responsibility for the implementation and maintenance through actual performance of such activities or that of the designated person responsible for ensuring that the sanitation procedures are executed.

In meat and poultry plants, SSOPs must be signed and dated by the individual with overall authority on site, or by a higher-level official of the establishment, to signify that the establishment will implement the SSOPs. Furthermore, SSOPs must be signed upon initiation or any modification. The establishment must evaluate and modify SSOPs, as necessary, to reflect changes in the establishment facilities, personnel, or operations to ensure that they remain effective in the prevention of direct product contamination and adulteration.

INTERFACE WITH GMPs AND SSOPs

Sanitation SOPs are a prelude to HACCP. The intent of an HACCP plan is to ensure safety at specific CCPs within specific processes. Sanitation SOPs transcend specific processes. Sanitation SOPs are the cornerstones for an HACCP plan and can serve as a preventive approach to direct product contamination and/or adulteration.

HACCP PRINCIPLES

HACCP is a systematic approach to be incorporated in food production as a means to ensure food safety. The basic principles that underlie the HACCP concept include an assessment of the inherent risks that may be present from harvest through ultimate consumption. It is necessary to establish critical limits that must be met at each CCP, appropriate monitoring procedures, corrective action to be taken if a deviation is encountered, record keeping, and verification activities. The following discussion indicates the seven basic principles of HACCP and gives a brief description of each.

1. Conduct a hazard analysis through the identification of hazards and assessment of their severity and risks by listing the steps in the process where significant hazards occur and describing preventive measures. This step provides for a systematic evaluation of a specific food and its ingredients or components to determine the risk from hazardous microorganisms or their toxins. Hazard analysis can guide the safe design of a food product and identify the CCPs that eliminate or control hazardous microorganisms or their toxins at any point during production. Hazard assessment is a two-part process, consisting of characterization of a food according to six hazards followed by the assignment of a risk category based upon the characterization.

Ranking according to hazard characteristics is based on assessing a food in terms of whether (1) microbially sensitive ingredients are contained in the product, (2) the process contains a controlled processing step that effectively destroys harmful microorganisms, (3) a significant risk of postprocessing contamination with harmful microorganisms or their toxins exists, (4) a substantial potential exists for abusive handling in distribution, in consumer handling, or in preparation that could render the product harmful when consumed, or (5) a terminal heat process after packaging or cooking in the home exists.

Ranking according to these characteristics results in the assignment of risk categories. According to the National Advisory Committee on Microbiological Criteria for Foods (1997), the risk categories are utilized for recognizing the hazard risk for ingredients and how they must be treated or processed to reduce the risk of the entire food production and distribution sequence.

The hazard assessment procedure should be conducted after the development of a working description of the product, establishment of the types of raw materials and ingredients required for preparation of the product, and preparation of a diagram for the food production sequence. The two-part assessment of hazard analysis and assignment of risk categories is conducted as will be described.

Hazard Analysis and Assignment of Risk Categories

Food should be ranked according to hazard characteristics A through F, using a plus symbol (+) to indicate a potential hazard. The number of pluses determines the risk category. If a product falls under hazard class A, it should automatically be considered as risk category VI. A description of the six hazards follows. Hazards can also be stated for chemical or physical hazards, particularly if a food is subjected to them.

- *Hazard A:* This hazard applies to a special class of nonsterile products designated and intended for consumption by at-risk populations, e.g., infants or older, infirm, or immunocompromised individuals.
- *Hazard B:* Products that fit this hazard contain "sensitive ingredients" in terms of microbial hazards.
- *Hazard C:* Foods in this hazard group are manufactured by a process that does

not contain a controlled processing step to effectively destroy harmful microorganisms.

- *Hazard D:* Foods that fit this hazard are subject to recontamination after processing and before packaging.
- *Hazard E:* With this hazard, there is substantial potential for abusive handling in distribution or in consumer handling that could render the product harmful when consumed.
- *Hazard F:* Foods in this group have not been subjected to a terminal heat process after packaging or when cooked in the home.

The following risk categories are based on ranking by hazard characteristics.

- *Category O*—No hazard.
- *Category I*—Food products subject to one of the general hazard characteristics.
- *Category II*—Food products subject to two of the general hazard characteristics.
- *Category III*—Food products subject to three of the general hazard characteristics.
- *Category IV*—Food products subject to four of the general hazard characteristics.
- *Category V*—Food products subject to all five of the general hazard characteristics: Hazard classes B, C, D, E, and F.
- *Category VI*—A special category that applies to nonsterile products designated and intended for consumption by at-risk populations, e.g., infants, aged, infirm, or immunocompromised individuals. All hazard characteristics must be considered.

2. Determine which CCPs are required to control the identified hazards. A *CCP* is defined as "a point, step or procedure at which control can be applied and a food safety hazard can be prevented, eliminated, or reduced to an acceptable level." A CCP must be established where control can be exercised. Hazards identified must be controlled at some point(s) in the food production sequence, from growing and harvesting raw materials to the ultimate consumption of the prepared food.

Critical control points are located at any point in a food production sequence where hazardous microorganisms should be destroyed or controlled. An example of a CCP is a specified heat process at a given time and temperature implemented to destroy a specified microbial pathogen. Another temperature-related CCP is refrigeration required to prevent hazardous organisms from growing or the adjustment of the pH of a food to prevent toxin formation. CCPs should not be confused with those points that do not control safety. A *control point* differs from a CCP, in that it is defined as "any point, step, or procedure in a specific food production operation at which biological, physical, or chemical factors can be controlled." Figure 5–1 presents a CCP Decision Tree recommended by the National Advisory Committee on Microbiological Criteria for Foods (1997) to assist in the identification of CCPs.

Information developed during the hazard analysis serves as a guideline to identify which steps in the process are CCPs. CCPs are located at any point where hazards require prevention, elimination, or reduction to acceptable levels. Examples of CCPs may include but are not limited to specific sanitation procedures, cooking, chilling, product formulation, and cross-contamination prevention.

The number of established CCPs should be kept to a minimum to simplify monitoring and documentation and to avoid dilution of the HACCP program effectiveness. CCPs must be carefully developed and documented and must address only product safety.

HACCP: Principles and Application

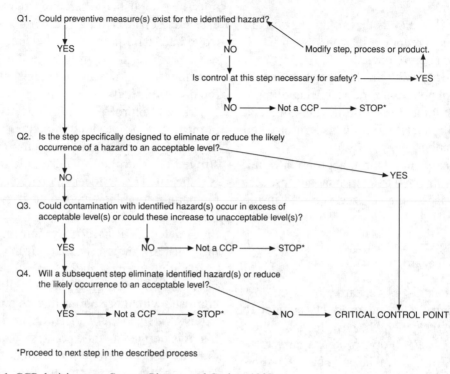

*Proceed to next step in the described process

Figure 5–1 CCP decision tree. *Source:* Pierson and Corlett, 1992.

Food operations may differ in the risk of hazards and the points, steps, or procedures that are CCPs. Differences such as the process, layout, equipment, product manufactured, and ingredients incorporated determine whether a CCP exists. Although general HACCP plans may serve as a guide, each operation should be evaluated before the assignment of CCPs and development of an HACCP plan. In addition to CCPs, nonfood safety concerns may be addressed at control points. CCPs should be kept to a minimum, as an SOP can sometimes be incorporated instead of a CCP. However, if a CCP exists, an SOP may not be an acceptable substitute.

3. Establish critical limits for preventive measures associated with each identified

CCP. A *critical limit* is one or more prescribed tolerance that must be met to ensure that a CCP effectively controls a microbial health hazard to an acceptable level. Various information on critical limits is essential for safe control of a CCP. Each preventive measure has associated with it critical limits that serve as boundaries of safety for each CCP. The critical limits for preventive measures may be specified for time, temperature, physical dimensions, pH, A_w, etc. Development of these critical limits may require determination of probable maximum numbers of microorganisms in the product, as well as sources such as regulatory standards and guidelines. The food industry is responsible for engaging competent authorities to validate that the critical limits will control the identified hazard.

4. Establish procedures to monitor CCPs. Scheduled testing or observation of a CCP and its limits is accomplished through monitoring. Results obtained from monitoring must be documented. From a monitoring standpoint, failure to control a CCP is a critical defect. A critical defect may result in hazardous or unsafe conditions for those who use or depend on the product. Monitoring procedures must be very effective because of the potentially serious consequences of a critical defect.

Monitoring is a planned sequence of observations or measurements to assess whether a CCP is under control and to produce an accurate record for future use in verification. Monitoring is essential to food safety management because it tracks the system's operation. Stevenson and Bernard (1995) suggested that if monitoring reveals that there is a trend toward loss of control, i.e., exceeding a target level, action can be taken to bring the process back into control before a deviation occurs. Monitoring identifies a loss of control or a deviation at a CCP, such as exceeding the critical limit and the need for corrective action. Furthermore, monitoring provides written documentation for use in verification of the HACCP plan.

If feasible, monitoring should be continuous. It is possible to attain continuous monitoring of pH, temperature, and humidity through the use of recorders. If insufficient control is maintained, as recorded on the chart, a process deviation may be identified. When it is impractical to monitor a critical limit continuously, a monitoring interval must be established that will be reliable enough to indicate that the hazard is under control. This can be accomplished through a statistically designed data-collection program or sampling system. Statistical procedures

are useful for measuring and reducing the variation in manufacturing equipment and measurement devices.

Monitoring procedures for CCPs must be designed for rapid results because insufficient time exists for time-consuming analytical testing. Microbial testing is also normally unsatisfactory for monitoring CCPs because of the amount of time involved. Physical and chemical measurements are more viable because they may be done rapidly and can indicate microbial control of the process. Physical and chemical measurements that may be incorporated in monitoring include measurements of pH, time, temperature, moisture level, preventive measures for cross-contamination, and specific food handling procedures.

Random checks may be conducted to supplement the monitoring of certain CCPs. They may be used to check incoming precertified supplies and ingredients, to assess equipment and environmental sanitation, airborne contamination, cleaning and sanitization of gloves, and any area where follow-up is needed. Random checks are normally conducted through physical and chemical testing and microbial tests, as appropriate.

For certain foods, microbially sensitive ingredients, or imports, there may be no alternative to microbial testing. However, a sampling frequency that is adequate for reliable detection of low levels of pathogens is seldom possible because of the large number of samples needed. Microbial testing has limitations in HACCP but is valuable as a means of establishing and randomly verifying the effectiveness of control of the CCPs. All records and documents related to CCP monitoring should be signed by the person executing the monitoring and also by a responsible official of the organization involved.

5. Establish corrective measures to be taken when there is a deviation from an established critical limit. Specific corrective actions must demonstrate that the CCPs have been brought under control. Deviation procedures should be written in the HACCP plan and agreed to by the appropriate regulatory agency prior to approval of the plan. If a deviation occurs, the production facility should place the product on hold until appropriate corrective actions and analyses are completed.

6. Establish procedures for verification that the HACCP plan is working correctly. Verification is accomplished through methods, procedures, and tests to determine that the HACCP system is in compliance with the HACCP plan. Verification confirms that all hazards were identified in the HACCP plan when it was developed. It may be accomplished through chemical and sensory methods and testing for conformance with microbial criteria when established. This activity may include, but is not limited to:

1. scientific or technical process to ensure that critical limits at CCPs are satisfactory
2. establishment of appropriate verification inspection schedules, sample collection, and analysis
3. documented periodic revalidation independent of audits or other verification procedures that must be performed to ensure accuracy of the HACCP plan; revalidation includes a documented on-site review and verification of all flow diagrams and CCPs in the HACCP plan
4. governmental regulatory responsibility and actions to ensure that the HACCP plan is functioning satisfactorily

7. Establish effective record-keeping procedures that document the HACCP plan. The HACCP plan must be on file at the food establishment to provide documentation relating to CCPs and to any action on critical deviations and production disposition. It should clearly designate records that are available for government inspection.

The HACCP plan should contain the following documentation:

1. Listing of the HACCP team and assigned responsibilities.
2. Product description and its intended use.
3. Flow diagrams of the entire manufacturing process with the CCPs identified.
4. Descriptions of hazards and preventive measures for each hazard.
5. Details of critical limits.
6. Descriptions of monitoring to be conducted.
7. Descriptions of corrective action plans for deviations from critical limits.
8. Description of procedures for verification of the HACCP plan.
9. Listing of record-keeping procedures.

ORGANIZATION, IMPLEMENTATION, AND MAINTENANCE

An HACCP plan should be formulated for each specific process or product. The plan should include the objective of the analysis, whether it be safety, spoilage, or foreign control. Documentation should include the objective(s); job title of each employee involved; flow charts of the operations involved, with the CCPs highlighted; hazards and details, with control options, cross-references to equipment maintenance and cleaning schedules, and procedures or CGMPs that apply to the process; and summary and conclusions, including action to be taken as a result of the analysis. Shapton and Shapton (1991) suggested that the HACCP report forms the record of the plan and should be presented in a way that is readily available to anyone who needs to use the report. It is an important resource when any changes are proposed for the process or specification concerned. A matrix has been suggested (Shapton and Shapton, 1991) to allow this feature, with these suggested column headings:

1. CCP number
2. Process/storage state of this CCP
3. Description of the state
4. Hazards associated with the state
5. Hazards controlled
6. Control limits
7. Deviations and how they have been or may be corrected
8. Planned improvements

To ensure success, employees should be educated, trained, and retrained in the use of HACCP. Employee turnover rate necessitates that continuous education be provided so that plant personnel will understand HACCP and the need for various controls that have been established. This approach can be responsible for the reduction of foodborne illness outbreaks and the replacement of costly crisis management with cost-effective control.

Effective implementation of the HACCP concept encompasses education of employees, especially workers in the production areas where problems can occur. An effective approach contains the following steps:

1. *Management education.* Quality assurance personnel and higher management need to understand the HACCP concept so that an effective program can be instituted that relies on total commitment of all personnel. Training courses for the management team are important to create the awareness that is the basis of the entire program. Furthermore, plant managers and supervisors should set a positive example.
2. *Operational steps.* The plant design and operating procedures may require

change to avoid interference with a hygienic operation. Experienced, properly trained personnel should be assigned to critical operations.

3. *Employee motivation.* Improvement of working conditions can be a motivating force in the implementation of HACCP. Task redesign may be a helpful tool in attaining success. All employees must feel a sense of personal responsibility for the quality and safety of food products.
4. *Employee involvement.* To ensure commitment of the workers, they must be involved in problem solving. Consultation groups (quality circles) should be instituted and their recommendations considered. Management should guide, not administer, the HACCP program. Furthermore, this program requires total commitment to a long-term undertaking by all levels of management and production employees.

Because HACCP represents a structured approach to control the safety of food products, it must be organized and managed in a way to ensure that the plan is operating correctly and will be sustained and maintained in the future (Stevenson and Bernard, 1995). The most viable source for leadership to organize and implement HACCP is the quality assurance group within a firm or a similar group with past and/or present responsibilities for the food safety functions in the organization. The challenges that must be accepted to ensure proper implementation include execution of the 12 developmental steps discussed previously.

Stevenson and Bernard (1995) identified the two deficiencies in a HACCP plan most frequently detected as: (1) documentation of the HACCP plan (insufficient "background" documentation on decision making and inadequate documentation of actual processes) and (2) management of the HACCP program. Ineffective management is most likely to be failure to ensure that a comprehensive plan is in place to yield safe products and inadequate review mechanisms to prove that the HACCP plan is being applied correctly.

A clear commitment to food safety and HACCP concepts by the management team is essential. Success depends on management commitment, detailed planning, appropriate resources, and employee empowerment. A corporate statement of support for HACCP is an effective tool for communicating the importance of HACCP to all employees. Furthermore, management should establish specific objectives and implementation schedules for additional support.

Management and Maintenance of HACCP

Management support is essential to the maintenance of an acceptable HACCP plan. One person within an organization should be responsible for the maintenance of HACCP. This responsibility includes coordination of input from others, monitoring of activities, review, validation, verification, and documentation. Furthermore, the coordinator should ensure that the HACCP team has access to the variety of information required to conduct the various assignments. Each individual assigned to HACCP-related tasks should be provided appropriate written instructions and descriptions of responsibilities and tasks. Reporting structures and the relationships of those involved should be determined, and the appropriate forms must be developed and provided for employees.

An HACCP plan should be evaluated frequently and revised as needed. Evaluation should involve the review and interpretation of results and verification and validation of the plan. Proposed changes to the plan should be evaluated. A mandatory evaluation process guarantees that a systematic evaluation will be made of any changes in the process, thus assuring that any revisions affecting product safety will be evaluated before implementation (Stevenson and Bernard, 1995). Verification assures that the HACCP plan will be evaluated and revised as needed.

HACCP Auditing and Validation

After an HACCP plan has been developed and implemented, it should be audited within the first year to determine its effectiveness. Verification should have been accomplished to review those activities, other than monitoring, that determine the adequacy of and compliance with the plan. Auditing may be conducted by the HACCP team, by management, or by a consultant and/or food scientist. Auditing should include a comprehensive review of the entire plan with evaluation and documented observations, conclusions, and recommendations. Auditing serves as a report card for the plan and provides future direction. Furthermore, auditing contributes to validation of the plan. Validation, as defined by the National Advisory Committee on Microbiological Criteria for Foods (1997), is that element of verification that focuses on the collection and evaluation of information to determine whether the HACCP plan, when implemented properly, will effectively control the significant hazards.

SUMMARY

Hazard Analysis Critical Control Points is a state-of-the-art preventive approach to safe food production. This concept is based on the application of prevention and documentation. HACCP is a proactive prevention program based on sound science. The essential steps for HACCP plan development are: assembly of an HACCP team; description of the food and its intended use; identification of the consumers of the food; development and verification of a process flow diagram; conduction of a hazard analysis; identification of critical control points; and establishment of critical limits, monitoring requirements, corrective actions for deviations, procedures for verification, and record-keeping procedures.

Good manufacturing practices are considered the building blocks of HACCP, and sanitation operating procedures are the cornerstones for an HACCP plan. Documentation needed for an effective plan includes descriptions of HACCP team-assigned responsibilities, product description and intended use, flow diagram with identified CCPs, details of significant hazards with information concerning preventive measures, critical limits, monitoring to be conducted, corrective action plans in place for deviations from critical limits, procedures for verification of the plan, and record-keeping procedures. Periodic auditing is necessary for validation and to provide a report card for the program.

STUDY QUESTIONS

1. What is HACCP?
2. What is a hazard?
3. What is a critical control point?
4. What are CGMPs?
5. What are sanitation SOPs?
6. What are the seven HACCP principles?
7. What are the five steps necessary to develop a HACCP plan prior to conducting a hazard analysis?
8. What is monitoring?
9. What is a control point?
10. What is critical limit?
11. What is HACCP verification?
12. What is HACCP plan validation?

REFERENCES

Bauman, H.E. 1987. The hazard analysis critical control point concept. In *Food protection technology,* ed. C.W. Felix, 175. Chelsea, MI: Lewis Publishers.

Clark, D. 1991. FSIS studies detection of food safety hazards. *FSIS Food Safety Rev* 4: Summer.

Marriott, N.G., et al. 1991. *Quality assurance manual for the food industry.* Virginia Cooperative Extension, Virginia Polytechnic Institute and State University, Publication No. 458-013.

National Advisory Committee on Microbiological Criteria for Foods. 1997. *Hazard analysis and critical control point principles and application guidelines.*

Pierson, M.D., and Corlett, D.A., Jr. 1992. *HACCP principles and applications.* New York: Van Nostrand Reinhold.

Shapton, D.A. and N.F. Shapton, eds. 1991. Establishment and implementation of HACCP. *Principles and practices for the safe processing of foods,* 21. Oxford: Butterworth-Heinemann.

Stevenson, K.E. and D.T. Bernard. 1995. *HACCP: Establishing hazard analysis critical control point programs–A workshop manual.* Washington, DC: National Food Processors Institute.

CHAPTER 6

Quality Assurance

Since the late 1970s, the food industry has emphasized an organized sanitation program that monitors the microbiology of raw ingredients in production plants and the wholesomeness and safety of the finished products, in an effort to maintain or upgrade the acceptability of its food products. As consumers become better informed and more sophisticated, it is even more vital for the food industry to develop an effective quality assurance (QA) and sanitation program. The efforts of regulatory agencies in the field of sanitation and food microbiology have been responsible for the food industry's implementation of voluntary QA programs. Food scientists have also had a positive impact on QA programs because many of these professionals have joined various companies in the food industry. Their efforts have been instrumental in the adoption and/or upgrading of QA programs for the organizations that they represent.

In its initial stages, QA was primarily a quality control (QC) function, acting as an arm of manufacturing. It has now evolved to a formidable force within the executive structure of large food firms and has emerged into a broad spectrum of activities. A QA program provides the avenue to establish checks and balances in the areas of food safety, public health, technical expertise, and legal matters affecting food manufacturing firms. Activities related to food sanitation include sanitation inspections, product releases and holds, packaging sanitation, and product recalls and withdrawals.

A QA program that emphasizes sanitation is vital to the growth of a food establishment. If foods are to compete effectively in the marketplace, established hygienic standards must be strictly maintained. However, it is sometimes impractical for production personnel to measure and monitor sanitation while maintaining a high level of productivity and efficiency. Thus, an effective QA program should be available to monitor, within established priorities, each phase of the operation. All personnel should incorporate the team concept to attain established sanitary standards, ensuring that food products in the marketplace are safe.

It is important to recognize that QA is an investment. It may cost not to have QA. A company with a QA program can offset the cost with improved product image, reduced likelihood of product liability suits, consumer satisfaction with a uniform and wholesome product, and improved sales. In practical terms, it makes good sense to have a QA program.

THE ROLE OF TOTAL QUALITY MANAGEMENT

An effective sanitation program is a segment of total quality management (TQM), which must be applied to all aspects of the operation within an organization. Total quality management applies the "right first time" approach. The most critical aspect of TQM is food safety. Thus, sanitation is an important segment of TQM. Additional discussion of TQM will be provided in Chapter 20.

The successful implementation of TQM requires that management and production workers be motivated to improve product acceptability. Furthermore, all involved must understand the TQM concept and possess skills to make the program successful. Computer software is available for training, implementation, and monitoring of TQM programs.

QUALITY ASSURANCE FOR EFFECTIVE SANITATION

Quality is the degree of acceptability. Component characteristics of quality are both measurable and controllable.

A sanitation QA program can achieve the following goals:

- Identify raw material suppliers that provide a consistent and wholesome product
- Make possible stricter sanitary procedures in processing to achieve a safer product, within given tolerances
- Segregate raw materials on the basis of microbial quality to allow the greatest value at the lowest price

By tradition, the food industry has applied QA principles to ensure effective sanitation practices, among them, inspection of the production area and equipment for cleanliness. If evidence of poor cleanup is reported, necessary action is taken to correct the problem. More sophisticated operations frequently incorporate use of a daily sanitation survey with appropriate checks and forms. Visual inspection should include more than a superficial examination, because a film buildup that can harbor spoilage and food-poisoning microorganisms can occur on equipment.

Major Components of Quality Assurance

The following tasks should be included as components of a sanitation QA program:

1. Clear delineation of objectives and policies
2. Establishment of sanitation requirements for processes and products
3. Implementation of an inspection system that includes procedures
4. Development of microbial, physical, and chemical product specifications
5. Establishment of procedures and requirements for microbial, physical, and chemical testing
6. Development of a personnel structure, including an organizational chart for a QA program
7. Development, presentation, and approval of a QA budget for required expenditures
8. Development of a job description for all positions
9. Setup of an appropriate salary structure to attract and retain qualified QA personnel
10. Constant supervision of the QA program with written results in the form of periodic reports

The Major Function of Quality Assurance

The major thrust of a QA organization is one of education and surveillance to ensure that regulations and specifications defined by the organization are implemented. Those involved with the QA program should be responsible for checking the wholesomeness and uniformity of raw materials assigned to manufacturing and for informing production personnel of these results. Further monitoring involves checks for good manufacturing practices and the finished products to ensure that they comply with specifications established under the QA program. These specifications should have been previously agreed upon by those in production or sales. If compliance is not attained, QA personnel should advise those who can suggest corrections.

Quality assurance is generally a function of corporate management, which sets the policies, programs, systems, and procedures to be executed by those assigned to quality control. The major internal responsibility is working with the various functional departments of the company.

Quality control, as currently structured by many firms, is closely related to manufacturing activities at the plant level. Those assigned to QC normally report to QA. Sometimes QC employees report to manufacturing, but they should never be totally independent of QA. Regardless of the organization structure, QA should have the ultimate responsibility for implementing and maintaining an effective sanitation program. The QA organization should also be responsible for improving the sanitation program to keep current with trends, new regulations, and technical expertise. Not all problems can be ascribed to poor sanitation. All QC procedures should be formulated and followed precisely. Quality control differs from TQM in that it is only a segment of the latter and is not a comprehensive management approach.

ORGANIZATION FOR QUALITY ASSURANCE

Large-volume plants should place enough emphasis on process control to form a QA department. Those involved with QA have the obligation to respond to technical requests, interpret results in practical and meaningful terms, and assist with corrective actions. A QA department should be structured as a corporate function so that it is directly responsible for the establishment, organization, execution, and supervision of an effective QA program that is integrated into corporate strategy.

Major Responsibilities of a Sanitation Quality Assurance Program

Before a QA program is effected, these requirements must be established:

1. Criteria for measuring acceptability (e.g., microbial levels) should be determined.
2. Appropriate control checks should be selected.
3. Sampling procedures (e.g., sampling times, numbers to be sampled, and measurements to be made) should be determined.
4. Analysis methods should be selected.

The major responsibilities of sanitation QA are:

- Perform facility and equipment sanitation inspections at least daily.
- Prepare sanitation specifications and standards.
- Develop and implement sampling and testing procedures.

- Implement a microbial testing and reporting program for raw products and manufactured products.
- Evaluate and monitor personnel hygiene practices.
- Evaluate compliance of the QA program with regulatory requirements, company guidelines and standards, and cleaning equipment.
- Inspect production areas for hygienic practices.
- Evaluate performance of cleaning compounds, equipment, and sanitizers.
- Implement a waste product handling system.
- Report and interpret data for the appropriate area of management so that corrective action, if necessary, can be taken.
- Incorporate microbial analyses of ingredients and the finished product.
- Educate and train plant personnel in hygienic practices, sanitation, and quality assurance.
- Collaborate with regulatory officials on technical matters when necessary.

The Role of Management in Quality Assurance

The success or failure of a sanitation program is attributable to the extent that it is supported by management. Management can be the major impetus or deterrent to a QA program. Managers are often uninterested in QA because it is considered a long-term program. Quality assurance programs have not been consistently supported by management because they reflect a cost where dividends cannot always be accurately measured in terms of increased sales and profits. Frequently, lower and middle management have difficulty selling QA when top management does not fully comprehend the concept.

Some of the more progressive management teams have been enthusiastic about QA. They have recognized that a QA program can be used in promotional efforts and can improve sales and product stability. Other managers have been able to improve sales and product stability, and some have been able to improve the image of their organization through sanitary practices and QA laboratories.

One of the limitations of viewing quality as conformance to specifications is its effect on management. When all specifications are met, the perception is that all is well and that management is not compelled to take immediate corrective action through the issue of orders down the hierarchy until results are obtained. This management style leads to a "fire-fighting" approach to problem solving and consumes valuable resources, is very costly, and frustrates people because problems, at best, go away only temporarily.

Quality Assurance and Job Enrichment

Because many employees, including managers and supervisors, fail to recognize the importance of QA, all employees must be made aware of the importance of their responsibilities. Through effective management, QA can be glamorized and made exciting. Although it is beyond the scope of this text to provide specific guidelines for the implementation of a job enrichment program for QA, it is suggested that this concept be considered. An effective job enrichment program can make work for employees more interesting and rewarding. This program also includes employees more as a part of the operation and can actually be more demanding of personnel through assignment of more responsibilities. If more information regarding this concept is desired, the reader is referred to a management text or technical journals related to management.

Quality Assurance Program Structure

Before organizing a QA program, it is important to determine who is responsible for QA and how the chain of command will operate. In the most successful efforts, the QA program is part of top management, not under the jurisdiction of production. Under this arrangement, the QA people report directly to top management and are not responsible to production management. However, a close working relationship must be maintained between QA and the production departments. The QA organization is responsible for ensuring that deviations in sanitation practices are corrected, in addition to checking the final product and determining the stability or keeping quality. Figure 6–1 illustrates areas of responsibility of the administrator of the QA program.

Responsibility for the daily functions of a QA program related to sanitation should be delegated to a designated sanitarian, who should be provided with the time and means to keep abreast of methods and materials necessary to maintain sanitary conditions. The role and position of the sanitarian within the processing firm should be made clear to all personnel. Management should clearly define parameters of responsibility by a written job description and an organizational chart. The sanitarian should report to the level of management with authority over general policy. This position should be equal to that of man-

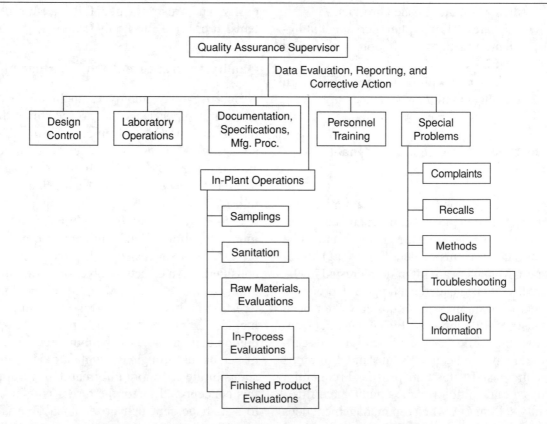

Figure 6–1 Organizational structure for specific QA tasks. *Source:* Adapted from Webb, 1981.

agers of production, engineering, purchasing, and comparable departments to command respect and maintain adequate status to administer an effective sanitation program. Although smaller-volume operations may necessitate a combination of responsibilities, they should be clearly defined. The sanitarian should have a clear understanding of the appropriate responsibilities and how the position fits in the company structure so that assignments can be performed properly. Figures 6–2 and 6–3 show examples of how the plant sanitarian should fit in the QA program of large and small processing organizations.

A high-caliber QA program requires one or more technically trained employees to administer it. The QA director or manager should have experience in food processing and/or preparation. Some of the QA staff can come from the ranks, provided that they show interest, leadership, and initiative. Work-

shops, short courses, and seminars often are available to help train new workers.

ESTABLISHMENT OF A QUALITY ASSURANCE PROGRAM

Preparation of personnel for a more unified system of control requires a change in attitude, which must be handled diplomatically. To reduce resistance, all personnel should be told why the changes are being made. Company philosophy should be developed as part of the program to help establish the new attitude and new responsibilities that personnel need to attain the desired goals.

Elements of a Total Quality Assurance System

For each production area, one person should be responsible for the controls or in-

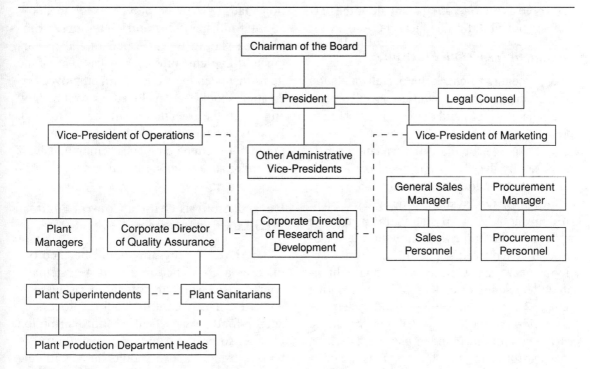

Figure 6–2 Chart reflecting status of a plant sanitarian in a large organization.

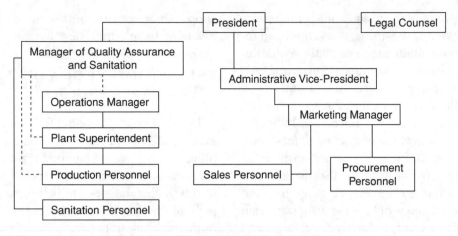

Figure 6–3 Chart reflecting status of a plant sanitarian in a small organization.

spection—either a plant employee or an outside contractor. The frequency of the control or inspection check must be noted, as must the records to be kept. An outline can be converted into a written format, as though it were a set of instructions for plant employees. It can serve as the operating manual for the persons responsible for conducting QA.

Sanitation Inspection Procedure

A procedure to check the overall sanitation of plant facilities and operations, including outside adjacent areas and storage areas on plant property, should be included in a total QA system. A sample for the daily sanitation report proposed by the U.S. Department of Agriculture (USDA) is illustrated in Exhibit 6–1.

In a total QA system, a designated plant official should make the sanitation inspection and record the findings. If sanitation deficiencies are discovered, a plan for corrective action is necessary. Corrective action might include recleaning or closing off an area until a repair is completed. Frequent and systematic sanitation inspection procedures should be used when product contamination is possible, such as from container failure, moisture dripping, or grease escaping from machinery onto

the product or surfaces that come into contact with the food.

New Employee Training

Instruction should include basic information that any new employee needs to know about food handling and cleanliness. Employees should be informed of the importance of hygienic practices. A list of all of the items that need to be covered in employee orientation should be developed, as well as how and when the orientation will be performed. Employee training should include an ongoing program to remind employees continuously of the importance of good sanitation.

Hazard Analysis Critical Control Point (HACCP) Approach

An HACCP program should be incorporated as a QA function and as a systematic approach to hazard identification, risk assessment, and hazard control in a food processing and/or food service facility and distribution channel to ensure a hygienic operation. Potential product abuse should be considered, and each stage of the process should be exam-

Exhibit 6–1 Sample Daily Sanitation Report

U.S. Department of Agriculture Animal and Plant Health Inspection Service Meat and Poultry Inspection Program DAILY SANITATION REPORT	Establishment Name	Est. No	Date

INSTRUCTIONS: Prepare original and one copy signed by inspector and plant management. Inspector files
original and gives copy to plant management.

Under the abbreviations "Pre-Op." (Observations made prior to the start of operations) and "Oper." (Observations made after operations have begun), record as appropriate the following codes. "N.O." [Not Observed],
"A.C." [Acceptable], "Def." [Deficiency(s)].

GENERAL AREA	PRE-OP.	OPER.	REMARKS (Enter "General Area:" No., specific description of deficient areas, equpment, etc.)	ACTIONS TAKEN AND DOWNTIME (Enter "General Area" No.)
1. Ante-Mortem Areas				
2. Outside Premises				
3. Floors				
4. Walls				
5. Windows, Screens, etc.				
6. Ceilings and Overhead Structures				
7. Doors				
8. Rails and Shackles				

9. Equipment:
  a. Product Zone
  b. Nonproduct Zone				

10. Freezers and Coolers				
11. Ice Facilities				
12. Dry Storage Areas				
13. Lights				
14. Welfare Facilities				

15. Employee:
  a. Dress
  b. Hygiene
  c. Work Habits				

16. Handwashing and Sanitizing				
17. Rodent and Insect Control				

18. General Housekeeping
  a. Production area
  b. Nonproduction area				

19. Production Practices				
20. Other				

RECEIVED BY ESTABLISHMENT OFFICIAL *(Signature)*	INSPECTOR(s) SIGNATURE	Page No. _____ of _____

MP FORM 455   PREVIOUS EDITIONS OBSOLETE   Feb. 1974

Source: USDA, 1974.

ined as an entity and in relation to other stages. The analysis should include the production environment as it contributes to microbial and foreign material contamination. Additional information on HACCP may be found in Chapter 5.

Program Evaluation

It is essential to evaluate the sanitary phase of a QA program through reliance either on the senses or on microbial techniques. Most inspectors rely on appearance as an evaluation technique for cleanliness. To the average inspector, a production area with walls, floors, ceilings, and equipment that looks clean, feels clean, and smells clean is satisfactory for production. But an effective QA program must use more than the human senses. It should incorporate a concrete method to evaluate hygiene conditions. To more objectively evaluate sanitation effectiveness, microbial testing methods should be incorporated to detect and enumerate microbial contamination. Also, knowledge of the quantity and genera of microorganisms is important in the control of product wholesomeness and spoilage.

Various techniques are available to evaluate the degree of cleanliness of equipment and foodstuffs and the effectiveness of a sanitation program. However, QA specialists do not always accurately determine or interpret results. Selection of the most appropriate technique should be based on the desired accuracy and precision, desired results, and the amount of effort and expenses available. Generally, the less complicated techniques are less accurate and precise. However, many measurements need not be exceptionally accurate and precise, as long as the degree of sanitation can be determined. Sanitation can be evaluated by the use of contact plates.

However, various thermally processed products may require very sensitive techniques for determining the amount and genera of microorganisms present throughout the finished product and on the processing equipment.

Assay Procedures for Evaluation of Sanitation Effectiveness

Laboratory methods are an important part of the entire scenario. Because these methods play an important role, they should be:

- accurate
- reproducible
- clearly described
- safe
- easy to conduct
- rapid (in turnaround time)
- efficient
- available commercially (all components)
- officially recognized (Association of Official Analytical Chemists [AOAC], U.S. Food and Drug Administration [FDA], USDA)

A brief discussion of the most viable assay procedures follows, according to category. Additional information about microbial determination is discussed in Chapter 2.

Direct Contact Contamination Removal

With this method, plates that contain agar are pressed against a surface to determine the amount of contamination. Variation among similar areas is reduced by swabbing several locations instead of smaller areas in localized sites.

Modifications of the contact plate method include the agar slice (syringe extrusion and agar sausage) and the use of selective and differential media. Another assay technique is the impression method. This technique involves a piece of sterile cellophane tape, which functions as a replicator to transfer cells to a growth-support agar, with subsequent incubation and counting. This approach, as in other contact methods, serves as only an approximate representation of the contamination and does not distinguish between particulate contamination containing one cell or more.

Surface Rinse Method

This method uses elution of contamination by rinsing to permit a microbial assay of the resultant suspension. A sterile fluid is manually or mechanically agitated over an entire surface. The rinse fluid is then diluted and subsequently plated. When applicable, it is more precise than the swab method because a larger surface can be tested. The membrane filter is an aid to the surface rinse method if contamination is not excessive. The membrane, bearing microorganisms, can be incubated on a nutrient pad, stained in 4 to 6 hours, and examined under a microscope with 8-100X magnification. Although the surface rinse method is more accurate and precise than the direct contact method and has a higher recovery rate (approximately 70%) and the flexibility of interfacing with the membrane fiber, it is restricted to horizontal surfaces and usually limited to container-type equipment.

Direct Surface Agar Plating (DSAP) Technique

This technique has utility for examining contamination on surfaces in situ. Eating utensils can be tested by pouring a melted medium into a cup and allowing the agar to solidify. The agar is transferred aseptically to sterile culture plates, with subsequent

overlayering and incubation. Also, the agar slab can be protected by a cover, left in place, and counted after 28 to 48 hours.

Vacuum Method

This method has been utilized sparingly in the food industry because of the complex procedure and sophisticated equipment required. However, it has wide use in the space industry.

Interpretation of Data from QA Tests

Microbial tests to evaluate hygienic conditions of equipment and foodstuffs are discussed in Chapter 2. Additional information on the tests that can be performed is discussed in the following paragraphs.

Importance of a Monitoring Program

A monitoring program should be established and implemented to provide an internal method of evaluating the overall wholesomeness of the finished product and the degree of sanitation. The main purpose is to avoid problems related to product safety and acceptability. The development of a program should include determination of objectives, techniques, and evaluation procedures. In testing, the overall effectiveness of sanitation—not just the quantitation of microorganisms on food contact surfaces—should be considered.

Products and surfaces to be tested should be determined by type of products produced, production steps, and the importance of the designated surface to sanitation practices and to the safety and/or overall acceptability of the food product. The monitoring program should be based on desired accuracy, time requirements, and costs. The type of food-contact surface to be tested should also be considered in determination of the monitoring technique.

To reduce the possibility of incorrect interpretation of results, the monitoring program should be designed so that data can be statistically analyzed. Further misunderstanding can be avoided by a thorough understanding of the benefits and limitations of the test procedures—for example, recognizing that bacterial clumps from the contact method of sampling should yield lower counts than the swab method, which breaks up cell clumps. The incorporation of 0.5% Tween 80 and 0.07% soy lecithin into media for RODAC plates is suggested if sampling is to be conducted on surfaces previously treated with a germicide.

In addition to analyzing data, the monitoring program should include a means of evaluating the information generated by the sampling technique. Acceptable and unacceptable guidelines should be determined under practical operating conditions. Repeated monitoring of given surfaces over time under given conditions (such as after cleaning and sanitizing, and during manufacture) can provide a trend. The QA manager can use this information to establish realistic guidelines for the production operation. The guidelines specifying the amount of contamination should be predicted based on the stage of production, amount of food surface exposed, and the length of contact time between the food and surface. Graphs that display daily counts of microorganisms and the established guidelines can be posted for review by the supervisors and employees, and can be used to stress the importance of monitoring and conforming with guidelines.

Microbial monitoring of food contact surfaces with techniques that have been discussed can be an effective tool to measure and evaluate the effectiveness of a QA program. Furthermore, a monitoring program can isolate potential problem areas in the production operation and serve as a training device for the sanitation crew, supervisors, and QA employees.

Recall of Unsatisfactory Products

Product recall is bringing back merchandise from the distribution system because of one or more unsatisfactory characteristics. Every food business is susceptible to potential product recall. A satisfactory public image of businesses can be preserved during a recall if a well-organized plan is implemented.

During a recall, products are recovered from distribution as a result of voluntary action by a business firm or involuntary action due to FDA action. The basic reasons for recall are best described in the FDA recall classifications:

> *CLASS I*: As a result of a situation where there is a reasonable probability that the use of or exposure to a defective product will cause a serious public health hazard including death.

> *CLASS II*: As a result of a situation where the use of or exposure to a defective product may cause a temporary adverse health hazard, or where a serious adverse public health hazard (death) is remote.

> *CLASS III*: As a result of a situation where use of or exposure to a defective product will not cause a public health hazard.

An example of a Class I product recall would be contamination with a toxic substance (chemical or microbial). A Class II product recall involves products contaminated with food infection microorganisms. A Class III example is products that do not meet a standard of identity.

If the monitoring program shows that products are being produced that are unsafe, a recall plan should be considered. A recall plan for unsafe products caused from poor sanitation and lack of hygienic handling should:

1. Collect, analyze, and evaluate all information related to the product.
2. Determine the imminence of the recall.
3. Notify all company officials and regulatory officials.
4. Provide operating orders to company staff needed to execute the recall.
5. Issue an immediate embargo on all further shipments of involved product lots.
6. If determined appropriate, issue news releases for consumers on the specifics of the product.
7. Notify customers.
8. Notify and assist distributors in tracking down the product.
9. Return all products to specific locations and isolate them.
10. Maintain a detailed log of recall events.
11. Investigate the nature, extent, and causes of the problem to prevent recurrence.
12. Provide progress reports to company and regulatory officials.
13. Conduct an effectiveness check to determine the amount of questionable product recalled.
14. Determine the ultimate disposition of the recalled product.

Sampling for a Quality Assurance Program

A sample is part of anything that is submitted for inspection or analysis that is a representative of the whole population. For the sample to be appropriate, it must be statistically *valid*. Validity is achieved by selecting the sample so as to ensure that each unit of material in a lot being sampled has an equal chance of being chosen for examination. This process is called *randomization.*

A sample must be representative of the population to ensure integrity of results. A suggested sample number is the square root of the total number that would be sampled. Representative samples are not only random samples, but must constitute a proportionate amount of each part of the population. For example, the sample must be identical with that of the gross material from which it was selected. A major concern of the QA organization should be the collection, identification, and storage of a sufficient sample for inspection and/or analysis. A statistically valid sample is important because:

- A sample is the basis for establishing the condition of the entire item or lot. A larger sample size increases the integrity that can be placed on findings.
- Submitting the entire item or lot for inspection is expensive and usually impractical.
- Sampling is used for the establishment of data for utilization in the development of standards and product acceptance. It is widely accepted to obtain regulatory control.

The integrity of collected samples can be diminished by inaccurate and incomplete information. Forms should contain all of the information necessary for the particular type of sampling and subsequent type of analysis. Sample cases should be insulated to ensure temperature maintenance during the period of transit to the point of inspection or analyses.

Samples must be kept in the 0°C to 4.5°C range. Commercially sealed refrigerants, which come in several different forms and temperature ranges, are available. Ice may be used in a sealed waterproof container or loose, if there is no chance of contaminating the sample. If maintenance in the zero to sub-zero temperature range is essential, dry ice should be used.

Sampling Procedures

Defined sampling procedures for solid, semisolid, viscous, and liquid samples must be determined at the time of sampling. The following sampling procedures is an example:

1. Identify and collect only representative samples.
2. Record product temperature, where applicable, at the time of sampling.
3. Maintain collected samples at the correct temperature. Nonperishable items and those normally at ambient temperature may be maintained without refrigeration. Perishable and normally refrigerated items should be held at 0°C to 4.5°C; normally frozen and special samples should be maintained at −18° C or below.
4. After collection, protect the sample from contamination or damage. Do not label certain plastic sample containers with a marking pen; ink can penetrate the contents.
5. Seal samples to ensure their integrity.
6. Submit samples to the laboratory in the original unopened container whenever possible.
7. When sampling homogeneous bulk products or products in containers too large to be transported to the laboratory, mix, if possible, and transfer at least 100 g of the sample to a sterile sample container, under aseptic conditions. Frozen products may be sampled with the aid of an electric drill and 2.5-cm auger.

Basic QA Tools

Depending on the food product area, items from the following equipment and supplies should be considered for sampling and product evaluation.

Measurement Apparatus

These include a centigrade thermometer, headspace gauge, vacuum gauge, titration burettes, filtering apparatus, 0.1- to 10.0-mL sterile disposable pipettes, micrometer.

Lab Supplies

Suggested supplies include petri dishes or petrifilm, crucibles, aluminum moisture pans, glass microscope slides, desiccator, can opener, record forms, marking tape, pencils, pens, aluminum foil, sterile cotton swabs, paper towels, microbial media, Bunsen burner, forceps, spoons, knives, and inoculation tubes.

Clerical Supplies

Supply list depends on what tests are being conducted. Necessary basic sanitation QA tools are:

1. Ingredient specifications
2. Approved supplier list
3. Product specifications
4. Manufacturing procedures
5. Monitoring program (analyses, records, reports)
6. Good Manufacturing Practices (GMP) requirements
7. Cleaning and sanitizing program
8. Recall program

Role of Statistical Quality Control

A portion of the discussion about statistical quality control (SQC) is taken from Marriott et al. (1991). Statistical quality control is the application of statistics in controlling a process. Measurements of acceptability attributes are taken at periodic intervals during production and are used to determine whether or not the particular process in question is under control—that is, within certain predetermined limits. A statistical QA program is a most useful tool for management. It enables management to control a product as the work is being done. This program also furnishes an audit of products as they are manufactured.

The samples taken for analysis are destroyed; thus, only SQC is practical for monitoring food safety. The greatest advantage of an SQC program is that it enables management to monitor an operation continuously and to make operating a closely controlled production process.

Sample selection and sampling techniques are the critical factors in any QC system. Because only small amounts (usually less than 10 g) of a product are used in the final analysis, it is imperative that this sample be representative of the lot from which it was selected.

Statistical quality control, also referred to as *operations research, operations analysis,* or *reliability,* is the use of scientific principles of probability and statistics as a foundation for decisions concerning the overall acceptability of a product. Its use provides a formal set of procedures in order to conclude what is important and how to perform appropriate evaluations. Various statistical methods can determine which outcomes are most probable and how much confidence can be placed in decisions.

Central Tendency Measurements

Three measurements are commonly used to describe data collected from a process or lot. These are the arithmetic mean or average, mode or modal average, and median. The mean is the sum of the individual observations divided by the total number of observations. The mode is the value of observations that occurs most frequently in a data set. The median is the middle value present in collected data. By using these values, the manufacturer can represent characteristics of central tendencies of the measurements taken. Table 6–1 illustrates calculated values for the mean, mode, and median from a collection of sample data.

Variability

Even though the measurements shown in Table 6–1 provide a description of where most sample values tend to occur, it alone does not give a complete picture of how consistently the product is being manufactured. There must be a uniformity and minimal variation in microbial load or other characteristics between the products manufactured. Two measures of variation are the range and standard deviation. Measuring variability by means of the range is accomplished by subtracting the lowest observation from the highest.

$$R = X_{max} - X_{min}$$

From Table 6–1 the calculation would be:

$$R = 20 - 11 = 9$$

Because the range is based on just two observations, it does not provide a very accurate picture of variation. It is sometimes used for small sample sizes; however, as the number of samples increases, the range tends to in-

Table 6–1 Central Tendency Values

Data	Mean	Mode	Median
11,12,14,14,16,17,18,19,20	15.67	14	16

crease because there is an increased chance of selecting an extremely high or low sample observation. The standard deviation is a more accurate measurement of how data are dispersed because it takes into account all of the values in the data set. The formula for calculating the standard deviation is:

$$S = \sqrt{\frac{(x - \bar{x})^2 + (x_2 - \bar{x})^2 + ... + (x_n - \bar{x})^2}{n - 1}}$$

Although this formula is more complicated than the range calculation, it can be determined easily by using a personal computer. As the standard deviation increases, it reflects increased variability of the data. In order to maintain uniformity, the standard deviation should be kept to a minimum.

Displaying Data

It is beneficial to represent data in a frequency table. This is especially effective when a relatively large sample of numbers must be analyzed. A frequency table aids in displaying data for better understanding of sample tendencies. A frequency table displays numerical classes that cover the data range of sampling and list the frequency of occurrence of values within each class. Class limitations must be selected in order to make the table easy to read and graph. The frequency table of microbial load from raw materials (Table 6–2) displays how data are divided into each class.

To help visualize how these data are arranged, one can graph it in the form of a histogram. Figure 6–4 takes the information from Table 6–2 and displays it graphically.

The histogram in Figure 6–4 depicts an important curve common to statistical analysis—the normal curve or normal probability density function. Many events that occur in

Table 6–2 Frequency Table for Microbial Load (CFUs/g)

Class in CFUs	Frequency
0-100	5
100-1,000	10
1,000-10,000	22
10,000-100,000	13
100,000-1,000,000	3

nature approximate the normal curve. The normal curve has the easily recognizable bell shape and is symmetrical about the center (see Figure 6–5). The area underneath the curve represents all the events described by the frequency distribution.

From Figure 6–5, the mean is represented by the highest point on the curve. The variation of the curve is represented by the standard deviation. It can be used to determine various portions underneath the curve. This is illustrated in the figure where one standard deviation to the right mean represents roughly 34% of the sample values. Consequently, 68.27% of the values fall within ±1 standard deviation from the mean. Similarly 95.45% fall within ±2 standard deviations. Virtually all of the area (99.75) is represented by ±3 standard deviations. The information thus far can be used to establish control limits in order to determine whether a process is in a state of statistical control.

Control Charts

Control charts offer an excellent method of attaining and maintaining a satisfactory level of acceptability for a process. The control chart is a widely used industry technique for on-line examination of materials produced. In addition to providing a desired safety level, it can be useful in improving sanitation and in providing a sign of impending trouble. The

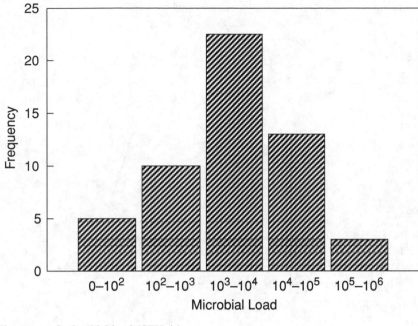

Figure 6–4 Histogram of microbial load (CFUs/g).

primary objective is to determine the best methodology, given the available resources, then to monitor control points. This variation can be classified as either chance-cause variation or assignable-cause variation.

In chance-cause variation, the end products are different because of random occurrences natural to the process. They are relatively small and are unpredictable in occurrence. There is a certain degree of chance-cause variation present in all products manufactured.

Assignable-cause variation is just what the name implies. Cause can be "assigned" to a contributing factor, such as a difference in microbial load of raw materials, process and machine aberration, environmental factors, or operational characteristics of individuals involved along the production line. This kind of variation, once determined, can be controlled through appropriate corrective action. When a process shows only variation due to

chance causes, it is "under control." Quality control charts were developed in order to differentiate between the two types of variation and to provide a method to determine whether a system is under control. Figure 6–6 illustrates a typical control chart for a quality characteristic. The y axis represents the characteristic of interest plotted against the x axis, which can be a sample number or time interval. The center line represents the average or mean value of the quality trait established by the manufactured product when the process is under control. The two horizontal lines above and below the center line are labeled so that as long as the process is in a state of control, all sample points should fall between them. The variation of the points within the control limits can be attributed to chance cause, and no action is required. An exception to this rule would apply if a substantial number of data points fall above or below the center line instead of being randomly scattered. This

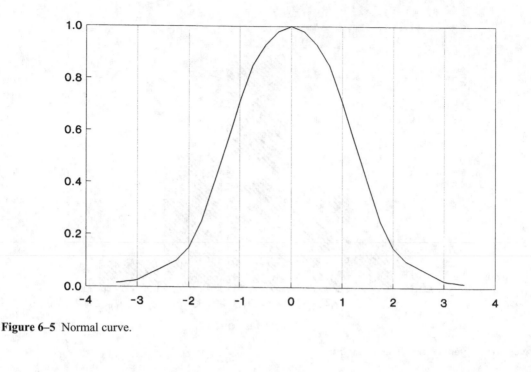

Figure 6–5 Normal curve.

would indicate a condition that is possibly out of control and would warrant further investigation. If a point falls above or below the out-of-bounds lines, one can assume that a factor has been introduced that has placed the process in an out-of-control state, and appropriate action is required.

Control charts can be divided into two types:

1. Control charts for measurement
2. Control charts for attributes

Measurement Control Charts

Measurement of variable control charts can be applied to any characteristic that can be measured. The X chart is the most widely used chart for monitoring central tendencies, whereas the R chart is used for controlling process variation. The following examples show how both of these control charts are used in a manufacturing environment.

A food manufacturer may be interested in monitoring the pH value of the finished product in order to satisfy shelf life requirements. Five samples may be pulled from the production line every hour during an 8-hour shift and analyzed for pH. The results are indicated in the Table 6–3.

First calculate the average (X) and range (R) for each inspection sample. For example, sample calculations for sample 1 are:

$$X = \frac{4.6 + 4.4 + 4.1 + 4.8 + 4.5}{5} = 4.48$$

R is the highest value minus the smallest value of the five samples.

After all of the sample Xs and Rs are calculated, take the average of the Xs and Rs to obtain X and R.

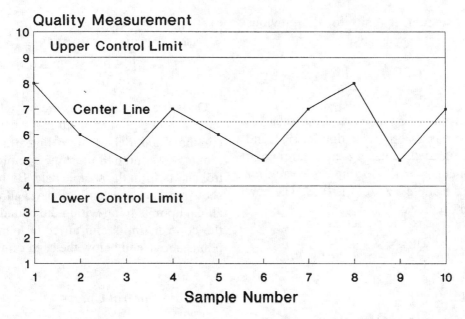

Figure 6–6 Typical control chart.

$$\overline{X} = \frac{\text{Sum of all } \overline{X}s}{\text{number of sample lots}} = \frac{35.58}{8} = 4.4475$$

$$\overline{R} = \frac{\text{Sum of all } Rs}{\text{number of sample lots}} = \frac{4.7}{8} = 0.5875$$

From the calculated, the center line for the X and R chart can be defined to be:

X Chart center line = 4.4475
R Chart center line = 0.5875

In order to calculate the upper control limits (UCL) and lower control limits (LCL), the standard deviation for each sample lot must be determined. Rather than perform the lengthy calculation needed for this value, another method can be used to determine these

Table 6–3 X and R Values for pH Measurements

Sample	pH Measurements					X	R
1	4.6	4.4	4.1	4.8	4.5	4.48	0.7
2	4.1	4.2	4.3	4.6	4.6	4.36	0.5
3	4.6	4.6	4.3	4.2	4.5	4.44	0.4
4	4.7	4.8	4.5	4.5	4.3	4.56	0.5
5	4.1	4.1	4.0	4.6	4.9	4.32	0.8
6	4.2	4.2	4.6	4.6	4.9	4.50	0.7
7	4.6	4.5	4.6	4.7	4.7	4.62	0.2
8	4.0	3.9	4.8	4.4	4.4	4.30	0.9
					Average:	4.4475	0.5875

values. The control limits for the previous charts were represented by:

$$UCL = \overline{X} + 3\,\overline{\delta}$$
$$LCL = \overline{X} + 3\,\overline{\delta}$$

By substituting a factor (A_2) from a statistical table into the above equation for UCL and LCL, the needed values for the control point can be obtained. In this example, the value for (A_2) for a sample size of 5 is 0.58. The new equation becomes:

$$UCL = \overline{X} + A_2\overline{R}$$
$$LCL = \overline{X} - A_2\overline{R}$$

substituting,

$$UCL = 4.4475 + 0.58(0.5875) = 4.7883$$
$$LCL = 4.4475 - 0.58(0.5875) = 4.1067$$

The control limits for the R chart are determined similarly, using factors D_4 and D_3 from the statistical reference table.

$$D_4 = 2.11, \ D_3 = 0$$
$$UCL = D_4\overline{R} = 2.11(0.5875) = 1.2396$$
$$LCL = D_3\overline{R} = 0(0.5875) = 0$$

Once these calculations are complete, the values can be plotted on an X-Y chart to obtain the X and R charts for pH measurements. Figures 6–7 and 6–8 illustrate complete control charts from the sample data. Both graphs show a process currently under control, with all data points lying within the boundaries of the control limits and an equal number of points above and below the center line.

Attribute Control Charts

Attribute control charts differ from measurements charts in that one is interested in an acceptable or unacceptable classification of products. The following charts are commonly used for attribute testing:

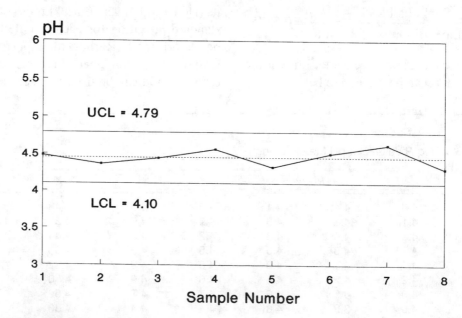

Figure 6–7 X chart for pH measurements.

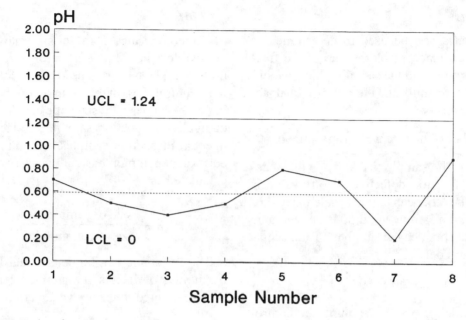

Figure 6–8 *R* chart for pH measurements.

1. p charts
2. np charts
3. c charts
4. u charts

p Charts

The p chart, one of the more useful attribute control charts, is used for determining the unacceptable (p) fraction. It is defined as the number of unacceptable items divided by the total number of items inspected. For example, if a producer examines five samples per hour (for an 8-hour shift) from the production line and finds a total of eight unacceptable units, p would be calculated as follows:

Total number of unacceptable = 8
Total number inspected = 5(8) = 40

$$p = \frac{\text{number of unacceptable}}{\text{total number inspected}} = \frac{8}{40} = 0.20$$

Sometimes this value is represented as percentage unacceptable. In this example, percentage defective would be:

$$0.20 \times 100 = 20\%$$

An attributable control chart can be constructed from a sampling schedule by obtaining an average fraction unacceptable (p) value from a data set and using the formula $p \pm 3\delta$, or the desired control limits. Because attribute testing follows a binomial distribution, the standard deviation would be calculated:

$$\delta = \sqrt{\frac{\bar{p}(1-\bar{p})}{n}},$$

where *n* is the number of items in a sample.
Control limits would be obtained by:

$$UCL = \bar{p} + 3\delta$$
$$LCL = \bar{p} - 3\delta$$

When these data are plotted and no points are outside of the control limits, it can be assumed that the process is in a state of statistical control, and any variation can be attributed to natural occurrences.

np Charts

np charts can be used to determine the number of unacceptable instead of the fraction defective, and the sampling lots are constant. The formula for the number of unacceptable (np) is:

$$\text{number of unacceptable (np)} = n \times p,$$

where n is the sample size and p is the unacceptable fraction defective. If one value is known, the other can be easily calculated. For example, if a sample lot of 50 is known to be 2% unacceptable, the number of unacceptable should be:

$$np = 50 \times 0.02 = 1$$

The calculation for determining the control limits would be the same as for the p chart, except that the standard deviation would be:

$$\delta = \sqrt{n\rho(1-\rho)}$$

c Charts

These charts are used when the concern is the number of defects per unit of product. They are not as heavily used as are the p and np charts but can be effective if applied correctly. Assume that a manufacturer examines 10 lots and discovers 320 defects. The equations for the average (c) and standard deviation required for a c chart would be:

$$\bar{c} = \frac{320}{10} = 32$$

$$\delta = \sqrt{c} = \sqrt{32} = 5.66$$

The control limits would be:

$$UCL = \bar{c} + 3\sqrt{c} = 32 + 3(5.66) = 48.97$$
$$LCL = \bar{c} - 3\sqrt{c} = 32 - 3(5.66) = 15.03$$

u Charts

Sometimes, a constant lot size may not be attainable when examining for defects per unit area. The u chart is used to test for statistical control. By establishing a common unit in terms of a basic lot size, one can determine equivalent inspection sample lot sizes from unequal inspection samples. The number of equivalent common basic lot sizes (k) can be calculated as:

$$k = \frac{\text{size of sample lot}}{\text{size of common lot}}$$

The u statistic can be determined from c, the number of defects of a sample lot, and the k value defined in the above equation.

$$u = \frac{c}{k}$$

From these values, the upper and lower control limits for the u chart can be defined.

$$UCL = \bar{u} + 3\sqrt{\frac{\bar{u}}{k}}$$

$$LCL = \bar{u} - 3\sqrt{\frac{\bar{u}}{k}}$$

In addition to charting, a manufacturer may introduce other statistical analyses, such as modeling, variable correlations, regression, analysis of variance, and forecasting to the production area. These techniques provide additional statistical methods for examining processes in order to ensure maximum production efficiency.

Explanation and Definition of Statistical Quality Control Program Standards

The following terms apply to maintenance of standards:

- *Standard:* The level or amount of a specific attribute desired in the product.
- *Quality attribute:* A specific factor or characteristic of the food product that determines a proportionate part of the acceptability of the product. Attributes are measured by a predetermined method, and the results are compared against an established standard and lower and upper control limits to determine if the product attribute is at the desired level in the food product.
- *Retained product:* A product that is not to be used in production or sold until corrective action has been taken to meet the established standards. Retained products should not be released for production or sales use until the problem is corrected.

Rating Scales

Two rating scales have been devised for evaluation of attributes:

1. *Exact measurement:* For attributes that can be measured in precise units (bacterial load, percentage, parts per million, etc.).
2. *Subjective evaluation:* Used when no exact method of measurement has been developed. The evaluation must be made by an individual making a sensory judgment (taste, feel, sight, smell). This is usually described numerically. Two scales have been developed for evaluating acceptability:

Scale 1	Scale 2
7—Excellent	4—Extreme
6—Very good	3—Moderate
5—Good	2—Slight
4—Average	1—None
3—Fair	
2—Poor	
1—Very poor	

The number of samples to conduct at any point during production to evaluate the sanitation operation also depends on the variations of analysis of the samples. A minimum of three to five samples of approximately 2 kg each should be selected and pooled from each lot of incoming raw material. After a sufficient number of samples has been analyzed, control charts can be constructed for each raw material.

Sampling of the finished product should be conducted at a special step in the production sequence, such as at the time of packaging. Sampling at this stage does not need to be done on individual products for inspection or regulatory purposes because it is directed at monitoring process control, not individual product analysis. However, to be familiar with the wholesomeness and overall acceptability of each product, the preferred procedure is to analyze and maintain control charts on all products.

Sample size usually consists of three to five specimens that serve as a representative of the population sampled. Another guideline for sample size is the square root of the total units, and, for large lots, an acceptable size may be the square root of the total units divided by 2. Daily sampling is necessary to monitor process control effectively. Action limits for finished products should be as outlined under the analysis program and should be used in determining whether the process conforms to the designated specifications. If three consecutive samples exceed the maximum limit for contamination, production should cease, and further cleaning and sanitizing should be conducted.

Cumulative Sum (CUSUM) Control Charts

Data can be plotted where greater sensitivity in detecting small process changes is required by use of the CUSUM chart. This chart is a graphic plot of the running summation of deviation from a control value. These differences are totaled with each subsequent sampling time to provide the CUSUM values. This monitoring technique can be incorporated in sanitation operations that require a higher degree of precision than obtained from a regular statistical QC chart. The CUSUM chart gives a more accurate account of real changes, faster detection and correction of deviation, and a graphical estimation of trends. It enhances an optimum process control for various applications. Webb and Price (1987) suggested that the CUSUM chart was not developed for multiple levels and is not practical for use on production processes that drift over an extended period of time. If used, it is important that the results of the CUSUM system be kept current so that immediate corrective action be taken.

The personal computer or central computer can rapidly perform the statistical computations and identify the points that require corrective action, thus reducing the burden of processing large quantities of data. These data can be available to promptly expedite corrective actions, project future performances, and determine when and where preventive QC procedures are necessary.

SUMMARY

Product wholesomeness and uniformity can be more effectively maintained through a QA program that incorporates available scientific and mechanical tools. Quality is considered to be the degree of acceptability by the user. These characteristics are both measurable and controllable. The major ingredients needed for a successful QA program are education and cooperation. The HACCP approach can be incorporated in a QA program because it applies to a zero-defects concept in food production. Effective surveillance of a QA program can detect unsanitary products and variations in production. Statistical QC techniques make inspection more reliable and eliminate the cost of 100% inspection. The principal tool of a statistical QC system is the control chart. Trends of control charts provide more information than do individual values. Values outside the control limits indicate that the production process should be closely observed and possibly modified.

STUDY QUESTIONS

1. What is quality?
2. What is total quality management?
3. Why should QA personnel not be placed under the supervision of production management?
4. What is SQC?
5. What is CUSUM?
6. What are Class I, II, and III recalls?

REFERENCES

Marriott, N.G., Boling, J.W., Bishop, J.R., and Hackney, C.R. 1991. *Quality assurance manual for the food industry.* Virginia Cooperative Extension, Virginia Polytechnic Institute and State University, Blacksburg. Publication No. 458-013.

USDA. 1974. Sample daily sanitation report. Washington, DC: U.S. Department of Agriculture.

Webb, N.B. 1981. Organization of quality control programs and personnel training. In *Quality assurance short course,* 2. Washington, DC: American Meat Institute.

Webb, N.B., and Price, J.F. 1987. Quality control concepts and systems. In *The science of meat and meat products*, 3rd ed., 607. Westport, CT: Food and Nutrition Press.

SUGGESTED READINGS

Baldock, J.D. 1979. Quality assurance for effective sanitation. In *Sanitation notebook for the seafood industry,* eds. G.J. Flick Jr., et al., III–59. Blacksburg: Department of Food Science and Technology, Virginia Polytechnic Institute and State University.

Bauman, H.E. 1987. The hazard analysis critical control point concept. In *Food protection technology*, ed. C.W. Felix, 175. Chelsea, MI: Lewis Publishers.

Pedraja, R.R. 1979. How to develop an effective quality assurance and sanitation program. In *Sanitation notebook for the seafood industry,* eds. G.J. Flick, Jr., et al., III–45. Blacksburg: Department of Food Science and Technology, Virginia Polytechnic Institute and State University.

Taylor, R.W. 1987. *Statistical quality control.* Evansville, IN: Koch Label Co.

CHAPTER 7

Cleaning Compounds

Cleaners are compounded specifically for performing certain jobs, such as for washing floors and walls, use in a high-pressure washer, cleaning in place (CIP), and other purposes. Good cleaners must be economical, nontoxic, noncorrosive, noncaking, nondusting, easy to measure or meter, stable during storage, and easily and completely dissolved.

Cleaning compound requirements vary according to the area and equipment to be cleaned. The selection of compounds for blending to form a satisfactory cleaner requires specialized and technical knowledge. Major considerations in cleaning compound selection are the nature of the soil to be cleaned, water characteristics, application method, and area and kind of equipment to be cleaned.

SOIL CHARACTERISTICS

Chemical Characteristics

Potential contamination sources from chemicals that can be found in foods are those used in food production and food preparation areas, and they include cleaning compounds, sanitizers, insecticides, rodenticides, and air fresheners. These substances may contaminate equipment, utensils, or surfaces, serving as a vehicle for transfer of the contaminants to food. This statement can be verified by those who have drunk from a glass or cup that imparts a distinct taste of dishwashing soap. Insecticides, rodenticides, air fresheners, and deodorizers may accidentally contaminate foods if applied by a spray or vapor. This can be prevented by use of a paint or solid insecticide or pesticide. Other potential chemical contaminants could be particulate rather than soluble chemicals.

People involved with sanitation can most effectively protect against chemical contamination by establishing rigid housekeeping methods to be used by production and clean-up employees. In addition to ordinary care and attention to detail, personal hygiene practices can prevent contamination by debris from food containers, glass, metal, plastic, paper, cardboard, and foreign materials. Such contamination can be reduced or even eliminated if carelessness and sloppy personal habits of all employees are abolished.

Physical Characteristics

Soil is material in the wrong location. It consists of dirt and dust materials with discrete particles in three dimensions, organic materials with discrete particles in three dimensions, and organic materials that could be encountered in a food service or processing facility. Examples of soil are fat deposits on a cutting board, lubricant deposits on a moving conveyor belt, and other organic deposits on processing equipment.

Soils can be classified according to the method of removal from the object to be cleaned:[1]

Soils soluble in water (or other solvents) containing no cleaner. These soils will dissolve in tap water and in other solvents that do not contain a cleaning compound. They include many inorganic salts, sugars, starches, and minerals. Soils of this type present no technical problem because their removal is merely a dissolving action.

Soils soluble in a cleaning solution that contains a solubilizer or detergent: Acid-soluble soils are soluble in acidic solutions with a pH below 7.0. Deposits include oxidized iron (rust), zinc carbonates, calcium oxalates, metal oxides (iron and zinc), films on stainless steel, water stone (reaction between various alkaline cleaners and chemical constituents of water having noncarbonate hardness), hard-water scale (calcium and magnesium carbonates), and milk stone (a water stone and milk film interaction, precipitated by heat on a metal surface). *Alkali-soluble soils* are basic media with a pH above 7.0. Fatty acids, blood, proteins, and other organic deposits are solubilized by an alkaline solution. Under alkaline conditions, a fat reacts with the alkali to form a soap. This reaction is called *saponification*. The soap formed from the reaction is soluble and will act as a solubilizer and dispersant for the remaining soil.

Soils insoluble in the cleaning solution: These soils are insoluble throughout the range of normal cleaning solutions. However, they must be loosened from the surface on which they are attached and subsequently suspended in the cleaning media.

A soil that falls into one class for one type of cleaning compound may fall into another class if another cleaner is applied. For example, sugar is soluble in water when an aqueous detergent system is used but it is insoluble in the organic solvents used in the dry-cleaning industry and, therefore, falls in another class. It is important to select the appropriate solvent and the correct cleaning compound for removing a specific soil. Table 7–1 summarizes the solubility characteristics of various kinds of soil. Soils are further classified as inorganic soils. An acid cleaning compound is most appropriate for the removal of inorganic deposits. An alkaline cleaner is more effective in removing organic deposits. If these classes are subdivided, it is easier to determine the specific characteristics of each type of soil and the most effective cleaning compound. Table 7–2 gives a breakdown of soil subclasses, with examples of certain deposits.

Soil deposits are characteristically complex in nature and are frequently complicated by organic soils being protected by deposits of inorganic soils, and vice versa. Therefore, it is important to identify correctly the type of deposit and to use the most effective cleaning compound or combination of compounds to effectively remove soil deposits. It is frequently essential to utilize a two-step cleaning procedure that contains more than one cleaning compound to remove a combination of inorganic and organic deposits. Table 7–3 illustrates the types of cleaning compounds applicable to the broad categories of soil previously discussed.

Chemical Characteristics

Surface attachment is influenced by the chemical and physical properties of soil, such

---

[1]Soil classifications to be discussed are those reported in *Plant Sanitation for the Meat Packing Industry* (Anon., 1976).

Table 7–1 Solubility Characteristics of Various Soils

Type of Salt	Solubility Characteristics	Removal Ease	Changes Induced by Heating the Surface
Monovalent Salts	Water-soluble, acid-soluble	Easy to difficult	Interaction with other constituents with removal difficulty
Sugar	Water-soluble	Easy	Carmelization and removal difficulty
Fat	Water-insoluble, alkali-soluble	Difficult	Polymerization and removal difficulty
Protein	Water-insoluble, slightly acid-soluble, alkali-soluble	Very difficult	Denturation and extreme difficulty in removal

as surface tension, wetting power, and chemical reactivity with the surface of attachment; and by physical characteristics, including particle size, shape, and density. Some soils are held to a surface by adhesion forces, or *dispersion forces*. Certain soils are bonded to the surface activity of the adsorbed particles. Adsorption forces must be overcome by a surfactant that reduces surface energy of the soil and subsequently weakens the bond between the soil and surface of attachment.

Physical characteristics of soil can also affect adhesion strength, which is directly related to environmental humidity and time of contact. Adhesion forces are also dependent on geometric shape, particle size, surface irregularities, and plastic properties. Mechanical entrapment in irregular surfaces and crevices contributes to the accumulation of soils on equipment and other surfaces.

EFFECTS OF SURFACE CHARACTERISTICS ON SOIL DEPOSITION

Surface characteristics should be considered when selecting a cleaning compound and cleaning method (Table 7–4). Clearly, the equipment and building material used affects soil deposition and cleaning requirements.

Sanitation specialists should be thoroughly familiar with all finishes used on equipment and areas in the food facility and should know which cleaning chemicals will attack surfaces. If the local management team is unfa-

Table 7–2 Classification of Soil Deposits

Type of Soil	Soil Subclass	Deposit Examples
Inorganic soil	Hard-water deposits	Calcium and magnesium carbonates
	Metallic deposits	Common rust, other oxides
	Alkaline deposits	Films left by improper rinsing after use of an alkaline cleaner
Organic soil	Food deposits	Food residues
	Petroleum deposits	Lubrication oils, grease, and other lubrication products
	Nonpetroleum deposits	Animal fats and vegetable oils

Table 7–3 Types of Cleaning Compounds for Soil Deposits

Type of Soil	Required Cleaning Compound
Inorganic soil	Acid-type cleaner
Organic soil	
(Nonpetroleum)	Alkaline-type cleaner
(Petroleum)	Solvent-type cleaner

miliar with the cleaning compounds and surface finishes, a consultant or reputable supplier of cleaning compounds should be sought to provide technical assistance, including recommending chemicals and sanitation procedures.

SOIL ATTACHMENT CHARACTERISTICS

Soils deposited in cracks, crevices, and other uneven areas are difficult to remove, especially in hard-to-reach areas. Ease of soil removal from a surface depends on surface characteristics such as smoothness, hardness, porosity, and wettability. Soil removal from a surface consists of three subprocesses (Anon., 1976).

First is *separation of the soil from the surface,* material, or equipment to be cleaned. Soil separation can occur through mechanical action of high-pressure water, steam, air, and scrubbing; through alteration of the chemical

Table 7–4 Characteristics of Various Surfaces of Food Processing Plants

Material	Characteristics	Precautions
Wood	Previous to moisture, fats, and oils; difficult to maintain; soften by alkali; destroyed by caustics.	Wood should not be used because of its unsanitary features. Stainless steel, polyethylene, and rubber materials should be used instead of wood.
Black Metals	Rust may be promoted by acidic and chlorinated detergents.	Because these metals are prone to rust, they are often tinned or galvanized. Neutral detergents should be used in cleaning these surfaces.
Tin	May be corroded by strong alkaline and acid cleaners.	Tin surfaces should not come in contact with foods.
Concrete	May be etched by acid foods and cleaning compounds.	Concrete should be dense, acid-resistant, and nondusting. Acid brick may be used in place of concrete.
Glass	Smooth and impervious; may be etched by strong alkaline cleaning compounds.	Glass should be cleaned with moderately alkaline or neutral detergents.
Paint	Surface quality depends on the method of application etched by strong alkaline cleaning compounds.	Certain edible paints are satisfactory for food plants.
Rubber	Should be nonporous, nonspongy; not affected by alkaline detergents; is attacked by organic solvents and strong acids.	Rubber cutting boards can warp, and their surface dulls knife blades.
Stainless Steel	Generally resistant to corrosion; smooth-surfaced and impervious (unless corrosion occurs); resistant to oxidation at high temperatures; easily cleaned; nonmagnetic).	Stainless steel is expensive and may be less plentiful in the future. Certain varieties are attacked by halogens (cholorine, iodine, bromine, and fluorine).

nature of soil (e.g., reaction of an alkali with a fatty acid to form a soap); or without alteration of the chemical nature of the soil (e.g., surfactants that reduce surface tension of the cleaning medium, such as water, to allow more intimate contact with the soil).

The soil and surface must be thoroughly wet for a cleaning compound to aid in separating the soil from the surface. The cleaning compound reduces the energy binding the soil to a surface, permitting the soil to be loosened and separated. The effectiveness of energy reduction and reduced binding may be increased through increased temperature of the cleaning compound and water or high-pressure spray, which can aid in cutting heavy soil deposits from the surface.

The second subprocess is *soil dispersion in the cleaning solution*. Dispersion is the dilution of soil in a cleaning solution. Soil that is soluble in a cleaning solution is dispersed if an adequate dilution of cleaning medium is maintained and if the solubility limits of the soil in the media are not exceeded. The use of fresh cleaning solution or the continuous dilution of the dispersed solution with fresh solution will increase dispersion.

Some soils that have been loosened from the surface being cleaned will not dissolve in the cleaning media. Dispersion of insoluble soils is more complicated. It is important to reduce soil to smaller particles or droplets with transport away from the cleaned surface. In this application, mechanical energy supplied by agitation, high-pressure water, or scrubbing is needed to supplement the action of cleaning compounds in breaking down the soil into small particles. A synergistic action of the energy reduction activity of the cleaning compound and the mechanical energy can break the soil into small particles and separate it from the surface.

The last subprocess is the *prevention of redeposition of dispersed soil*. Redeposition

can be reduced by removal of the dispersed solution from the surface being cleaned. Other reduction methods are continued agitation of the dispersed solution while still in association with the surface to stop settling of the dispersed soil; prevention of any reaction of the cleaning compound with water on the soil (note that soft water containing sequestering agents will reduce the possibility of forming hard-water deposits from soap present in the cleaning compound or formed through fat saponification); elimination of any residual solution and dispersed soil that may have collected on the surface by flushing or rinsing the cleaned surface; and maintenance of soil in a finely dispersed condition to avoid further entrapment on the cleaned surface.

Adsorption of surface-active agents on the surface of soil particles causes similar electrical charges to be imparted to the particles. This condition prevents aggregation of larger particles because like-charged particles repel each other. Surface redeposition is minimized because a similar repulsion exists between surfactant-coated particles and the surfactant-coated clean surface.

A systems approach to cleaning encompasses equipment for mechanical energy, cleaning compounds to reduce the energy holding the soil to the surface, and sanitizing compounds to destroy microbial contamination associated with soil deposits. Successful soil removal depends on cleaning procedures, cleaning compounds, water quality, high-pressure application of the cleaning media, mechanical agitation, and temperature of cleaning compounds and media.

The Role of Cleaning Media

Water is the cleaning medium most frequently used for soil removal. Other cleaning media may include *air* for removal of packaging material, dust, and other debris where water is not an acceptable cleaning medium. Additional media may include *solvents*, which are incorporated in the removal of lubricants and other similar petroleum products.

The major functions of water as a cleaning medium include:

- prerinse for the removal of large soil particles
- wetting (or softening) of soils on the surface where removal is essential
- transport of the cleaning compound to the area to be cleaned
- suspension of soil to be removed
- transport of suspended soil from the surface being cleaned
- rinsing of the cleaning compound from the area being cleaned
- transport of a sanitizer to the cleaned area

Satisfactory water is required to complement the cleaners. The water should be free of microorganisms, clear, colorless, noncorrosive, and free of minerals (known as *soft water*). Hard water, which contains minerals, may interfere with the action of some cleaning compounds, thereby limiting their ability to perform effectively (although some cleaning compounds can counteract the adverse effects of hard water).

CLEANING COMPOUND CHARACTERISTICS

Food particles and other debris provide the nutrients required for microorganisms to proliferate. Microorganisms are protected during a cleaning operation by neutralizing the effects of chlorinated cleaning compounds and sanitizers, thereby preventing penetration to the microbes. Soil must be removed thoroughly through use of mechanical energy and cleaning compounds to provide a microbially clean environment.

How Cleaning Compounds Function

The major functions of a cleaning compound are to *lower the surface tension of water so that soils may be dislodged and loosened and to suspend soil particles for subsequent flushing away*. To complete the cleaning process, a sanitizer is applied to destroy residual microorganisms that are exposed through cleaning.

One of the oldest and best-known cleaning compounds is plain soap. However, it has limited utility in food processing and food service units and is rarely used because it does not clean well and reacts with hard water to form an insoluble curd (such as a bathtub ring). A basic soap contributes to cleaning through the removal of fats, oils, and greases by suspending particles of these water-insoluble materials. After fat or oil has been suspended, removal by flushing is easy, although a residual film will exist. The suspension process of water-insoluble materials through interaction with a soap is called *emulsification*.

In emulsification, the cleaning compound interacts with water and the soil. Figure 7–1 illustrates that the hydrophilic portion of a cleaning compound molecule is soluble in water. The hydrophobic portion is soluble in the soil. When the cleaning compound molecules surround the soil, a suspended soil particle results by micelle formation (Figure 7–2).

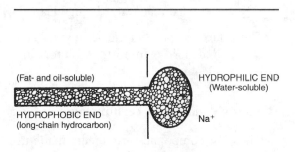

Figure 7–1 Anionic surfacant molecule.

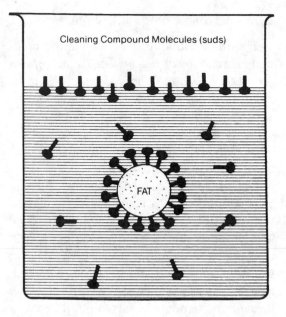

Figure 7–2 Soil particle suspended by micelle formation.

Factors Affecting Cleaning Performance (Anon., 1996)

- Time: contact time on the surface being cleaned
- Action: physical force exerted onto the surface (velocity or flow)
- Concentration: amount of cleaner used
- Temperature: amount of energy (as heat) used in the cleaning solution
- Water: used to prepare cleaning solution
- Individual: worker performing clean-up operation
- Nature: composition of the soil
- Surface: what material is being cleaned

These factors spell out the acronym *TACT WINS* and describe important factors involved in cleaning.

Cleaning Compound Terminology

Cleaning compounds are agents made up of a variety of compounds. Each manufacturing firm has its own brand name and codes. This text provides a basic understanding of the various agents that make up cleaning compounds without recommending or endorsing any branded products. The best "rule of thumb" to consider when selecting a cleaning compound is that "like cleans like." Therefore an acid soil requires an acid cleaner, and an alkaline soil should be removed with an alkaline cleaning compound.

To further understand the properties of cleaning compounds, the following terms are important:

- *Chelating agent (frequently called* sequestering agent *or* sequestrant*):* An additive used in cleaning compounds that prevents hardness constituents and salts of calcium and magnesium from depositing on equipment surfaces by binding these salts to their molecular structure. These agents can also bind other ions.
- *Emulsification:* A complex action consisting of a physical breakdown of fats and oils into smaller particles that are dispersed throughout the medium. The soil is still present but is reduced in physical size.
- *Peptizing:* A process that involves the formation of a colloidal solution from a material that is partially soluble, by the action of alkaline materials on protein soils.
- *Rinsibility:* The ability of a cleaning compound to be removed easily from a surface with a minimal amount of residue.
- *Saponification:* The action of an alkaline material on an insoluble soil (i.e., animal fat or vegetable oil) to produce a soluble, crude soap.
- *Sequestrant (sometimes called* chelating agent*):* An inorganic ingredient that is blended with cleaning compounds to prevent the precipitation of unstable salts that contribute to water hardness.

These unstable salts will break down in the presence of alkaline compounds or at a high temperature. Many alkaline cleaning compounds are more effective with an elevated temperature; however, a high-temperature cleaning solution contributes to precipitation of calcium and magnesium carbonates, commonly known as a *scale*. A sequestrant is a chemical agent that ties up calcium and magnesium ions in a solution to prevent the ions from forming insoluble curds with the cleaning detergent, which result in precipitation deposits.

- *Surfactant:* A complex molecule that, when blended with a cleaning compound, reduces the surface tension of water to permit closer contact between the soil deposit and cleaning medium.
- *Suspension:* A process by which a cleaning compound loosens, lifts, and holds soil particles in solution.
- *Water hardness:* The amount of salts such as calcium chloride, magnesium chloride, sulfates, and bicarbonates present in water. *Permanent hardness* is frequently used when referring to calcium and magnesium chlorides and sulfates in the water. These salts are rather stable and soluble under most conditions, causing minimal problems with cleaning. *Temporary hardness* is caused by the presence of calcium and magnesium bicarbonates, which are relatively soluble but unstable. The unstable condition of calcium and magnesium bicarbonates contributes to white deposits on equipment, heat exchangers, and water utensils. The combined amount of permanent and temporary hardness is referred to as *total hardness*.
- *Water softening:* A condition caused by the removal or inactivation of the calcium and magnesium ions in water. This is accomplished by chelation, precipitating calcium and magnesium as insoluble salts through a precipitating agent, such as trisodium phosphate, and by ion exchange involving replacement of calcium and magnesium, as is accomplished by commercial water softeners.
- *Wetting (penetration):* Caused by the resultant action of a surfactant that, due to its chemical structure, is capable of wetting or penetrating the soil deposit to start the loosening process from the surface.

CLASSIFICATION OF CLEANING COMPOUNDS

Most cleaning compounds that are used in the food industry are classified as blending products. Ingredients are combined to produce a single product with specific characteristics that performs a given function for one or more cleaning applications. The following classes of cleaning compounds are most frequently used in connection with food service facilities and processing plants.

Alkaline Cleaning Compounds

The term *pH* is frequently used in the food sanitation industry to describe the nature of the cleaning solution. The pH is a logarithmic measurement of hydrogen ion concentration. A pH of 7 to 14 is classified as being alkaline; a pH ranging from 0 to 7 is acidic. Acidity decreases from 0 to 7, with 7 being a neutral pH. As pH increases from 7 to 14, alkalinity increases. Alkaline cleaners are divided into subclasses with characteristics as discussed.

Strongly Alkaline Cleaners

These cleaners have strong dissolving powers and are very corrosive. They can burn, ulcerate, and scar skin. Prolonged contact may permanently damage tissue. Inhalation of the fumes or mist may cause respiratory tract damage. Mixing strong alkaline cleaners with water causes an exothermic reaction; the heat generated may cause the solution to boil or vaporize. Such explosive boiling may spray nearby personnel with the caustic compound.

Examples of strongly alkaline compounds are sodium hydroxide (caustic soda) and silicates having high $N_2O:SiO_2$ ratios. The addition of silicates tends to reduce the corrosiveness and to improve the penetrating and rinsing properties of sodium hydroxide. These cleaners are used to remove heavy soils, such as those from commercial ovens and smokehouses, and have little effect on mineral deposits. Caustic soda, which has highly germicidal activity, protein dissolution, and deflocculation/emulsifying properties, is used for removing heavy soils. Because of its potential damage to humans and equipment, caustic soda is not used as a manual cleaner.

Heavy-Duty Alkaline Cleaners

These compounds have moderate dissolving powers and are generally slightly corrosive or noncorrosive. However, prolonged contact with body parts may remove necessary oils from the skin, leaving it vulnerable to infections.

The active ingredients of these cleaners may be sodium metasilicate (a good buffering agent), sodium hexametaphosphate, sodium pyrophosphate, sodium carbonate, and trisodium phosphate, which is known for its good soil-emulsification activity. The addition of sulfites tends to reduce the corrosion attack on tin and tinned metals. These cleaners are frequently used with high-pressure or other mechanized systems. They are excellent for removing fats but have no value for mineral deposit control. Sodium carbonate, which is one of the oldest alkaline cleaners,

functions primarily as a buffering agent. Borax may be added as a buffering agent. Sodium carbonate, which is relatively low in cost, is used as a buffering agent in many formulations and has a wide range of uses in heavy-duty and manual cleaning applications.

Mild Alkaline Cleaners

Mild cleaners frequently exist in solution and are used for hand cleaning of lightly soiled areas. Examples of mild alkaline compounds are sodium bicarbonate, sodium sesquicarbonate, tetrasodium pyrophosphate, phosphate water conditioners (sequesters), and alkyl aryl sulfonates (surfactants). These compounds have good water-softening capabilities but exhibit no value for mineral deposit control.

Table 7–5 summarizes cleaning characteristics of commonly used alkaline cleaners. Comparisons of emulsifying properties, detergency, and corrosiveness are also provided.

Acid Cleaning Compounds

Acid cleaning compounds are used for removing encrusted surface materials and dissolving mineral scale deposits. They are effective, especially in removing mineral deposits formed as a result of using alkaline cleaning compounds or other cleaners. A portion of the minerals found in water may be deposited when heated to 80°C or higher. These deposits adhere to metal surfaces and appear as a rusty or whitish scale. Activity of acid cleaners is expressed through chemical action with minerals found in deposits, making them water soluble for easy removal.

Organic acids, such as citric, tartaric, sulfamic, and gluconic acid, are also excellent water softeners, rinse easily, and are not corrosive or irritating to the skin. Although inorganic acids are excellent for removing and controlling mineral deposits, they can be

Table 7–5 Cleaning Characteristics of Commonly Used Alkaline Cleaning Compounds

Alkaline Detergent	pH of 0.5% Solution	Detergency*	Corrosiveness*	Emulsifying Property*
Sodium hydroxide (caustic soda)	12.7	2.5	3.5	2.0
Sodium orthosilicate	12.6	3.0	4.0	3.0
Sodium sesquisilicate	12.6	2.0	3.2	2.5
Sodium metasilicate	12.0	3.8	0.8	4.0
Trisodium phosphate	11.8	3.5	4.0	3.5
Sodium carbonate (Soda ash)	11.3	1.5	4.0	2.8
Tetrasodium pyrophosphate	10.1	3.5	3.0	0.0
Sodium sesquicarbonate	9.7	1.3	3.2	2.5
Sodium tripolyphosphate	8.8	2.0	2.0	0.0
Sodium tetraphosphate	8.4	3.0	1.0	0.0
Sodium bicarbonate	8.2	1.5	2.3	1.5

*Based on a 4.0 scale, where 0 = no property and 4 = excellent property.
Source: Anon., 1976.

extremely corrosive and irritating to the skin. Acid cleaning compounds are a specialized type of cleaner and are not recognized as effective, all-purpose cleaning compounds. They are not nearly as effective against soil caused by fats, oils, and proteins, which acts as a binder, as are alkaline cleaning compounds. Alkaline cleaning compounds chemically attack the binder of organic soils, which releases the retaining or tenacious forces. Acid cleaning compounds are not capable of this function.

Strongly Acid Cleaners

These compounds are corrosive to concrete, most metals, and fabrics. Some of these cleaners, when heated, produce corrosive, toxic gases, which can ulcerate lungs. Strongly acid cleaners are used in cleaning operations to remove the encrusted surface matter and mineral scale frequently found on steam-producing equipment, boilers, and some processing equipment. When the solution temperature is too high, the mineral scale may redeposit and form a tarnish or whitish film on the equipment being cleaned.

Strongly acid agents used for cleaning operations in food plants are hydrochloric (muriatic), hydrofluoric, sulfamic, sulfuric, and phosphoric acids. Nitric and sulfuric acids are not used in manual cleaners because of their corrosive properties. Corrosion inhibitors, such as potassium chromate for nitric acid solutions or butylamine for hydrochloric acid detergents, may be added.

Phosphoric acid and hydrofluoric acid both clean and brighten certain metals. However, hydrofluoric acid is corrosive to stainless steel and dangerous to handle because of the tendency toward hydrogen evolution during use. Phosphoric acid is widely used in the United States. It is relatively low in corrosive properties, compatible with many surfactants, and is used in manual and heavy-duty formulations.

Mildly Acid Cleaners

These compounds are mildly corrosive and may cause allergenic reactions. Some acid cleaners attack skin and eyes. Examples of mildly acid cleaning compounds are levulinic, hydroxyacetic, acetic, and gluconic acids. Wetting agents and corrosion inhibitors (i.e., 2-naphtoquinoline, acridine, 9-phenylacridine) may be added. The organic acids, which are used as manual cleaning products, are higher in cost than are the other acid cleaning compounds. These mild compounds can also function as water softeners.

Solvent Cleaners

Solvent cleaners are normally used on petroleum-based soils and greases in the maintenance area. Their use should be strictly controlled.

Cleaners with Active Chlorine

Wyman (1996) reported that cleaners containing active chlorine, such as sodium or potassium hypochlorite, are effective in the removal of carbohydrate and/or proteinaceous soils because they aggressively attack such materials and chemically modify them to render them more susceptible to interaction with the balance of the components. Active chlorine-containing products are especially valuable when cleaning a surface in which the soil is derived from a food source comprised of some form of starch or protein. Also, they are effective in removing molds from surfaces.

Because of a form of chemical bonding known as *cross-linking*, many carbohydrates are such that a large number of the "big" molecules are bonded together (Wyman, 1996). In this instance, they cannot dissolve, which makes cleaning them from a surface very difficult. According to Wyman (1996), heat history imparted whenever carbohydrate-containing materials are heated increases the number or cross-links and complicates cleaning. Active chlorine-containing cleaners have the ability to break chemical bonds, leading to the formation of smaller, more soluble molecules and an increase in cleaning speed and efficacy.

Active chlorine, such as hypochlorite, attacks the large, complex carbohydrate molecules and degrades them to smaller, more soluble and readily removed derivatives (Wyman, 1996). Because active chlorine acts quickly, only portions of the molecules need be modified for the change in ease of removability to occur. Small amounts of active chlorine give effective cleaning results.

In the reaction of sodium hypochlorite with carbohydrates, the former can reduce the molecular weight of starch and increase its solubility. As with most cases, the reaction rates increase with elevated temperature. Because hypochlorite is an effective biocide at pH values lower than 8.5, the cleaning reaction rate of this compound is faster at a pH of 8 than at 10. A lower pH accounts for more of the hypochlorite in the form of hypochlorous acid, which diffuses into bacteria and carbohydrate residues faster than the hypochlorite ion, to increase the cleaning reaction rate.

Proteins are crosslinked by chemical bonding and bonds that tie the large molecules together. Hydrogen bonding occurs because certain atoms in the molecule have a stronger attraction for electrons than do others. This reaction generates an electrostatic interaction, which complicates the removal of proteins by conventional means (Wyman, 1996). Furthermore, proteins can interact through hydrogen bonding to decrease their solubility. Active chlorine-containing cleaners react with the insoluble proteins and render them soluble and/or readily dispersible through degradation by rapid oxidation of sulfide crosslinks that are present. Because the degradation need not be complete for solubilization to occur, a small amount of hypochlorite will remove a relatively large quantity of protein.

Hydrogen atoms attached to nitrogen in amides are replaced by chlorine when such molecules are allowed to react with hypochlorite. Wyman (1996) has hypothesized that this reaction occurs with proteins. Thus, the replacement of nitrogen-bonded hydrogens with chlorine will reduce hydrogen bonding and will improve solubility. This further explains why active chlorine degrades proteins to render them soluble and to enhance their removal from soiled surfaces, or at least modifies them enough for accelerated interaction with and removal by the rest of the cleaning components. However, cleaners that contain hypochlorite should be applied soon after they are made-up as they lack stability during storage.

Synthetic Detergents

The major components of synthetic detergents serve essentially the same function as soap—emulsification of fats, oils and greases—except that there is no reaction to cause a curd formation. The hydrophilic end of soap curds in hard water, whereas this end of a synthetic detergent surfactant does not have this characteristic. Synthetic detergents are effective because their addition lowers the surface tension of the solution, promotes wetting of particles, and deflocculates and suspends soil particles (Anon., 1976).

Wetting agents may be divided into three major categories:

1. *Cationic wetting agents* (such as quaternary ammonia) are normally considered sanitizers rather than wetting agents. They produce positively charged active ions in an aqueous solution. Detergents in this category are poor wetting agents, although they are strong bactericides.
2. *Anionic wetting agents* have a negatively charged active ion when in solution. They are the most commonly used wetting agents in cleaning compounds because of their compatibility with alkaline cleaning agents and good wetting qualities. Anionic agents differ from cationic agents by not being associated with any bactericidal properties.
3. *Nonionic wetting agents* have no charge associated with them when in aqueous solution. Therefore, they are effective under both acid and alkaline conditions. Wetting agents are also responsible for suds formation produced by a detergent. Their main problem is that they produce foam, which can cause complications in drainage and sewage systems. A cleaning compound does not have to foam to be an effective cleaner. One advantage of nonionic wetting agents is that they are not affected by water hardness.

Wetting agents serve an important function as components in cleaning compounds. Most have strong emulsifying, dispersion, and wetting capabilities. They are noncorrosive and nonirritating, and, normally, they are rinsed easily from equipment and other surfaces.

The properties of synthetic cleaning compounds are influenced by the water-soluble portion of the molecule (hydrophile) and by the water-insoluble segment. Table 7–6 summarizes the properties of various synthetic compounds.

Alkaline Soaps

Soaps, created by the reaction of an alkali compound with a fatty acid, are considered alkaline salts of carboxylic acids. Most commercial soaps are made for lauric (C_{12}) to stearic (C_{18}) of the fatty acid series, napthenic acids, rosin and the monovalent alkalis (such as sodium, potassium, ammonium), or amine salts. Soaps are not used much in industrial cleaning because they are less effective in hard water and are generally inactivated by acid solutions.

Table 7–6 Properties of Synthetic Detergents

	Surface Activity			Stability		
	Foaming	Emulsification	Deterging	Acid	Alkali	Water
Anionic						
Soaps	Good	Fair	Excellent	None	Good	Poor
Sulfated alcohols	High	Fair	Excellent	Fair	Fair	Fair/Good
Sulfated olefins	Good	Good	Good	Fair	Fair	Fair
Sulfated oils	Low	Good	Poor	Excellent	Fair	Good
Sulfated monoglycerides	Good	Good	Good	Fair	Fair	Good
Sulfated amides	Good	Good	Good	Poor	Good	Good
Alkyl aryl polyether sulfates	High	Good	Excellent	Excellent	Excellent	Good
Alkyl sulfonates						
"Petronates"	Low	Good	Poor	Excellent	Good	Fair
"Nytron"	Good	Fair	Excellent	Excellent	Excellent	Good
Sulfonated amides	High	Good	Excellent	Excellent	Excellent	Excellent
Sulfosuccinate	Good	Fair	Good	Excellent	Excellent	Excellent
Sulfonated ethers	High	Good	Excellent	Good	Good	Good
Alkyl aryl sulfonates	High	Fair	Excellent	Good	Good	Good
Heterocyclic sulfonates	Good	Fair	Good	Excellent	Excellent	Excellent
Cationic						
Tertiary amines	Fair	Fair	Fair	Poor	Good	Poor
Heterocyclic sulfonates	Fair	Good	Fair	Poor	Good	Good
Quaternary	Good	Good	Good	Fair	Fair	Poor
Nonionic						
Amine fatty acid condensate	High	Excellent	Good	Poor	Good	Poor
Ethylene oxide fatty acid	Low	Excellent	Excellent	Fair	Fair	Excellent
Alkyl aryl polyether alcohols	High	Excellent	Excellent	Good	Fair	Excellent
Ethylene oxide fatty alcohol condensate	Low	Good	Good	Excellent	Excellent	Excellent
Miscellaneous						
Thioethics	Low	Good	Excellent	Excellent	Excellent	Excellent
Pluronics	Low	Good	Excellent	Excellent	Excellent	Excellent

Source: Anon., 1976.

Enzyme-Based Cleaners

Increased knowledge of bacterial attachment suggests that enzyme-based cleaners merit consideration because they break soil down into smaller pieces and aid in its removal by destroying its attachment sites. These cleaners are classified as proteases as they break down protein and work best on the alkaline side. These cleaners must be used at 60°C or lower. They offer future utility in that they contain no chlorine or phosphates and are less corrosive than are chlorine sanitizers. They can lower the pH of effluent. The disadvantages of enzyme-based cleaners are that liquid detergents require injection equipment and a two-part system activation, and they are not as effective on all types of other soils as are chlorine sanitizers.

Phosphate Substitutes for Laundry Detergents

The use of phosphates in laundry detergents has been prohibited in certain areas of the United States. Some of the substitutes for phosphates approved for use, such as carbonates and citrates, have provided less acceptable results. Lovingood et al. (1987) reported that unbuilt liquids and phosphate-built powders were more effective in soil removal and whiteness retention than were the carbonate-built powders. Carbonate-built detergents, although less expensive, tend to give less acceptable results because of deposit buildup on washed materials and on parts of the washer, especially when hard water is used.

Solvent Cleaners

Solvent cleaners are ether- or alcohol-type materials capable of dissolving soil deposits. These compounds are most frequently used to clean soils caused by petroleum products, such as lubricating oils and greases. Because most organic soils can be saponified by alkaline cleaners, an alkaline or a neutral cleaning compound is more frequently used. However, solvent cleaners are frequently used if large amounts of petroleum deposits exist. A solvent-type cleaner is frequently required to remove this type of soil deposit from equipment. This type of soil will not usually be found directly on processing equipment surfaces, but rather in the general area.

Solvent cleaners are derived from various volatile materials from the petroleum industry and combined with wetting agents, water softeners, and other additives. Heavy-duty solvent cleaners are immiscible with water and frequently form an emulsion when water is added. Heavy-duty solvent cleaners are manufactured for use without water, whereas some solvent cleaners with low solvent content can be combined with water and still exhibit the grease-cutting action expected from a solvent.

Detergent Auxiliaries

Detergent auxiliaries are additives included in cleaning compounds to protect sensitive surfaces or to improve the cleaning properties of the compound.

Protection Auxiliaries

Acid Compounds

Acids may be used with synthetic cleaning compounds for cleaning alkaline-sensitive surfaces—for example, surfaces coated with alkaline-sensitive paints or varnishes, and light metal cleaning. The following acids are useful in protecting sensitive surfaces:

- Phosphoric acid, used to clean metals before painting, because it removes rusts and metal scales and subsequently passivates the surface.
- Oxalic acid, which effectively removes iron oxide rust without attacking the metal, although precautionary steps are necessary because this acid can react with hard-water constituents to form calcium oxalate, a poisonous precipitate.
- Citric acid, which does not produce toxic compounds but is not as efficient as oxalic acid in rust removal.
- Gluconic acid, which removes alkali and protein films through sequestering power without a toxic effect and may be used as a water conditioner.
- Sodium bisulfate, a low-cost course for heavy-duty powdered acid cleaners.

Protective Colloids and Suspending Agents

Hydrophilic colloids that prevent particle redeposition on the cleaned surface are commonly referred to as *protective colloids, thickeners*, and *suspending agents*. Examples are gelatin, glue, starch, sodium cellulose sulfate, hydroxyethyl cellulose, and carboxymethyl cellulose. Other agents with protective properties are:

- Low-alkali, high-silica compounds, such as glassy or colloidal silicates, metasilicates, and sodium chromates (and gelatin), which inhibit tin and aluminum spangling.
- Sodium chromate or dichromate, borax, and sodium nitrate in neutral detergent systems, which are efficient inhibitors of steel and iron corrosion.
- Metasilicates and colloidal silicates, which protect glass and enamel surfaces from caustic etching.
- Sodium sulfite, sodium fluorosilicate, and metabisulfite, which are reducing agents in the detergent system and protect tin and tin-plated surfaces by removing dissolved oxygen from the wash solution.

CLEANING AUXILIARIES

Various auxiliaries protect sensitive surfaces or improve the cleaning properties of a compound. Some are described below.

Sequestrants

These auxiliaries, which are also called *chelating agents* and *sequestering agents*, chelate by complexing with magnesium and calcium ions to produce compounds. This action effectively reduces the reactivity of water hardness constituents. Sequestrants consist of polyphosphates or organic amine derivatives. Phosphates differ in heat stability, wetting and rinsing properties, water conditioning, hardness, and sequestering power.

Cleaning detergents consist of a surfactant and a builder. *Builders* increase the effectiveness of a cleaner by controlling properties of the cleaning solution that tend to reduce the surfactant's effectiveness. Phosphates are considered excellent builders, especially for heavy-duty cleaning compounds. Phosphates serve as builders in cleaning compounds by providing:

- Enhancement of the wetting effect and resultant cleaning efficiency of cleaning compounds.
- Sufficient alkalinity necessary for effective cleaning without being hazardous.
- Maintenance of the proper alkalinity in the cleaning solution through buffering ability.
- Emulsification of oily, greasy soil by degradation and subsequent release from the surface to be cleaned.
- Loosening and suspension of soil with the ability to prevent redeposition on the clean surface.
- Water softening by keeping minerals dissolved to prevent settling on what is being cleaned.
- Reduction in numbers of bacteria associated with a clean surface.

There are a number of polyphosphates of special significance. *Sodium acid pyrophosphate* has excellent buffering and peptizing properties, with limited capability for sequestering water hardness constituents. *Tetrasodium pyrophosphate*, which does not sequester calcium as do the higher phosphates, is very stable above 60°C and in alkaline solutions.

Sodium tripolyphosphate and *sodium tetraphosphate* have calcium-sequestering power superior to that of tetrasodium pyrophosphate but tend to revert to orthophosphate and pyrophosphate when held above 60°C or in the alkalinity of pH 10 or higher.

Sodium hexametaphosphate (Calgon) is an effective calcium sequestrant with limited magnesium-sequestering power. *Amorphous phosphates* are complex glassy phosphates with excellent calcium-sequestering power.

Organic chelating agents, which are used in formulation in water conditioners, are more efficient than are phosphates in sequestering calcium and magnesium ions and in minimizing scale buildup. Most organic agents are salts of ethylenediaminetetraacetic acid (EDTA). The chelating agents are stable above 60°C and in solution for extended periods of storage. These chelating properties for EDTA salts improve as pH increases. They may be used in conveyor lubricant formulations.

Surfactants

These surface-active agents function to facilitate the transport of cleaning and sanitizing compounds over the surface to be cleaned. Although the major functions of surfactants are wetting and penetrating, detergency characteristics, such as emulsification, deflocculation, and suspension of particles, contribute to their effectiveness.

Surfactants are classified as synthetic detergents because of their numerous properties. As auxiliaries, they are also classified in the same three groups, according to their wetting properties and active components in solution. These auxiliaries are classified as cationic surfactants, which ionize in solution to produce active positively charged ions and serve as excellent bactericidal agents and ineffective detergents; anionic surfactants, which ionize in solution to produce active negatively charged ions and are generally excellent detergents and ineffective bactericides; and nonionic surfactants with no positive and negative ions in solution or bactericidal properties but with excellent wetting and penetrating characteristics. In ad-

dition, the *amphoteric* surfactants have a positive or negative charge, depending on the pH of the solution.

The general structure for anionic surfactants is $Q–X^-M^+$, where Q is the hydrophobic portion of the molecule, X^- is the anionic or hydrophilic portion, and M^+ is the counterion in solution. The hydrophobic portion of the molecule is normally a hydrocarbon chain of the form C_nH_{2n+1}, which is usually designated as R. Q may represent an alkyl-substituted aromatic molecule, an amide, an ether, a fatty acid, an oxyethylated alcohol, a phenol, an amine, or an olefin. The two most familiar anionic surfactants are soaps and linear alkylbenzene sulfonates.

The hydrophobic group forms a part of the cation dissolved in water in the cationic surfactants, whereas the hydrophobic portion of an anionic surfactant forms a part of the anion in aqueous solution. A cationic compound is formed by reacting a tertiary amine with an alkyl halide to form a quaternary ammonium salt, $R_1 R_2 R_3 + R_{4X} \rightleftharpoons R_1 R_2 R_3 R_4N^+ + X^-$. At least one of the R substituents is a hydrophobic group, such as dimethylammonium chloride, a germicidal agent.

The hydrophilic portion of nonionic surfactants often is composed of one or more condensed blocks of ethylene oxide. The hydrophobic portion can be any of several groups, including those named for the anionic types. The bond between the hydrophobe and the hydrophile may be an ether grouping or an amide or ester grouping. Other nonionic surfactants are alkanolamides and amine oxides.

The behavior of amphoteric surfactants is a result of two different functional groups in the molecule. The principal amphoteric surfactants are alkyl betaine derivatives, imidazole derivatives, amine sulfonates, and fatty amine sulfates.

Surfactants exhibit certain characteristics, such as:

- solubility in a least one phase of a liquid system
- amphipathic structure with opposing solubility tendencies; i.e., hydrophilic, lipophilic, or hydrophobic
- orientation of monolayers at phase interfaces formed by ions of surfactant molecules
- equilibrium concentration of a surfactant solute at a phase interface greater than the concentration in the bulk of either of the solutions
- micelle formation when the concentration of the solute in the bulk of the solution exceeds a limiting value that is a fundamental characteristic of each solute-solvent system
- exhibition of one or more functional properties; i.e., detergency, wetting, foaming, emulsifying, solubilizing, dispersion, demulsifying, and defoaming.

SCOURING COMPOUNDS

Scouring compounds, also known as *chemical abrasives*, are normally manufactured from inert or mildly alkaline materials. These abrasives are generally compounded with various soaps and are provided for scouring with brushes or metal sponges. Neutral scouring compounds are frequently compounded with acid cleaners for removal or alkaline deposits and encrusted materials. Abrasive cleaning compounds should be used carefully when cleaning stainless steel to avoid scratching.

Slightly Alkaline Scouring Compounds

Scouring compounds that are made from mildly alkaline materials and used for light deposits of soil are borax and sodium bicarbonate. These compounds have limited detergency and emulsifying capabilities.

Neutral Scouring Compounds

These compounds are made from earth, including volcanic ash, seismotites, pumice, silica flours, and feldspar. They may be found in cleaning powders or pastes used in manual scrubbing and scouring operations.

Water Quality Considerations

The chemical properties of water should be considered as this is a cleaning medium basic to most cleaning compounds. Water with varying amounts of calcium, magnesium and other alkali metals (hard water) interferes with the effectiveness of cleaning compounds (especially bicarbonates), contributing to precipitate formation. Such precipitates serve as sites for accumulation of organic debris and microorganisms, and make effective sanitation more difficult. The United States Geological Survey (USGS) definitions for water hardness are provided in Table 7–7.

If hard water exists, it may be more economical to use a water softener than to include chelators that mitigate the problem. With few exceptions, hot water causes less scale formation than does cold water. However, where hard water is used, maximum scale formation occurs at 82°C.

Table 7–7 U.S. Geological Survey Definitions for Water Hardness

Hardness	Parts per million (mg/L)
Very hard	>180
Hard	120–180
Moderately Hard	60–120
Soft	0–60

Source: Reprinted from U.S. Geological Survey.

CLEANING COMPOUND SELECTION

The type of soil determines which cleaning compound can be used most effectively. As previously emphasized, "like cleans like." In general, organic soils are most effectively removed by alkaline, general-purpose cleaning compounds. Heavy deposits of fats and proteins require a heavy-duty alkaline cleaning compound. Mineral deposits and other soils that are not successfully removed by alkaline cleaners require acidic cleaning compounds. The most frequently used types of cleaner-sanitizers are phosphates complexed with organic chlorine. A discussion of other factors that are also important in determining which cleaning compound is most effective will follow. Table 7–8 illustrates appropriate compound application and the prevention of various soils.

Soil Deposition

The amount of soil to be removed affects the alkalinity or acidity of the cleaning compound used and determines which surfactants and sequestrants may be needed. The extent of soil deposition and the selection of an appropriate cleaning compound affect the degree of cleaning.

The kind of soil deposit also dictates which class of cleaning compounds should be used. Soil characteristics also indicate which protection auxiliaries and cleaning auxiliaries are needed, which ultimately determines the degree of cleaning.

Temperature and Concentration of Cleaning Compound Solution

As the temperature and concentration of the cleaning compound solution increase, the activity of the compound increases. How-ever, an extreme temperature (above 55°C) and concentration exceeding recommendations of the manufacturer or supplier can cause protein denaturation of the soil deposits, which can reduce the effectiveness of soil removal.

Cleaning Time

As the length of time that the cleaning compound is in direct contact with the soil increases, the surface becomes cleaner. The method of cleaning compound application and the characteristics of the cleaner affect this exposure time.

Mechanical Force Used

The amount of mechanical energy in the form of agitation and high-pressure spray will affect the penetration of the cleaning compound and the physical separation of soil from the surface. The amount of agitation also helps in soil removal. Chapter 9 discusses further the role of mechanical energy (cleaning equipment) in soil removal.

HANDLING AND STORAGE PRECAUTIONS

Careless use of cleaning compounds poses a health and safety threat. Sanitation personnel should be trained for the proper use of these chemicals and supplied with appropriate safety clothing (gloves, boots, glasses, etc.). Furthermore, U.S. safety regulations require that Material Safety Data Sheets (MSDS) be available to all employees involved in these operations.

Most cleaners, except the liquid materials, are classified as hygroscopic in nature. They will absorb moisture when left exposed; thus, the product will deteriorate or cake in the

Table 7–8 Common Detergent Ingredients

Ingredients	Emulsi-fication	Saponifi-cation	Wetting	Dispersion	Suspension	Water Softening	Mineral Deposit Control	Rinsability	Suds Formation	Noncor-rosive	Non-irritating
Basic alkalis											
Caustic soda	C	A	C	C	C	C	D	D	C	D	D
Sodium metasilicate	B	B	C	B	C	C	C	B	C	B	D
Soda ash	C	B	C	C	C	C	D	C	C	C	D
Trisodium phosphate	B	B	C	B	B	A	D	B	C	C+	C-
Complex phosphates											
Sodium tetraphosphate	A	C	C	A	A	B	B	A	C	AA	A
Sodium tripolyphosphate	A	C	C	A	A	A	B	A	C	AA	B
Sodium hexametaphosphate	A	C	C	A	A	B	B	A	C	AA	A
Tetrasodium pyrophosphate	B	B	C	B	B	A	B	A	C	AA	B
Organic compounds											
Chelating agents	C	C	C	C	C	AA	A	A	C	AA	A
Wetting agents	AA	C	AA	A	B	C	C	AA	AAA	A	A
Organic agents	C	C	C	C	C	A	AA	B	C	A	A
Mineral acids	C	C	C	C		A	AA	C	C	D	D

Note: A = high value; B = medium value; C = low value; D = negative value.
Source: Anon., 1979.

container. Containers must be resealed properly after use to prevent contamination and to keep the materials free from moisture.

Cleaning compounds should be stored in the area remote from normal plant traffic, with dry floors, moisture-free air, and moderate temperature (to prevent freezing of liquid products). This area should be equipped with pallets, skids, or storage racks to keep the containers off of floors. The storage area should be locked to prevent theft.

Use of an inventory sheet is recommended as an aid for reordering and pointing out irregularities in product consumption. The control and supply of these cleaning materials should be handled by one person appointed by the facility's management to minimize product waste and to ensure that sufficient quantities of each cleaning material are available when required. This worker should be familiar with each cleaning operation so that he or she can instruct other employees in the correct techniques of any specific cleaning operation or use of cleaning equipment.

Selection of the correct cleaning material and its proper application is sometimes complicated. Suppliers of the cleaning compound can provide specific directions for both the compound and its use. Clear instructions will ensure that the product is used effectively without damaging the surface being cleaned. Supplier instructions for cleaning specific equipment with commercial cleaning compounds should be reviewed. Compounds from different suppliers should not be mixed.

Various areas in food plants require different cleaning mixtures. Large plants normally purchase basic cleaning compounds and blend them into concentrated batch lots. Many processing plants may devise 12 to 15 formulations to do specific jobs around the plant. Smaller facilities frequently purchase formulated cleaners in drum lots.

Regardless of how cleaning compounds are procured and blended, these materials should be handled with caution. Strong chemical cleaners can cause burns, poisoning, dermatitis (inflammation of the skin), and other problems to workers handling them. Since the use of stronger compounds has become prominent, there has been an increase in vulnerability to injuries.

Alkali Hazards

Strong alkaline cleaning compounds, in both solid form and in solution, have a corrosive action on all body tissue, especially the eyes. Irritation from exposure to the material is usually evident immediately. Damage frequently includes burns and deep ulceration, with ultimate scarring. Prolonged contact with dilute solutions may have a destructive effect on tissue. Dilute solutions may gradually degrease the skin, leaving vulnerable tissue exposed to allergens or other dermatitis-promoting substances. It is important to be aware that dry powder or particles can get inside a glove or a shoe and cause a severe burn. Inhalation of the dust or concentrated mist of alkaline solutions can cause damage to the upper respiratory tract and lung tissue.

Many alkaline materials react violently when mixed with water. The heat of reaction upon mixing may elevate the temperature above the boiling point, and large amounts of a hazardous mist and vapor may erupt.

Acid Cleaner Hazards

Sulfamic Acid

This compound, one of the safer acid cleaners, is a crystalline substance that can be stored easily with a minimal hazard from decomposition. However, it should be stored in a location protected from fire because it emits toxic oxides of sulfur when heated to decomposition.

Acetic Acid

This acid attacks the skin and is especially hazardous to the eyes. It presents a greater fire hazard than do many other common acids used in cleaners and should be stored in areas designed for flammable materials.

Citric Acid

This compound is one of the safer acids. Although allergenic reactions may be anticipated from prolonged exposure, it presents only a slight fire hazard. However, acid fumes are emitted when it is heated to decomposition.

Hydrochloric Acid (Muriatic Acid)

Misuse of this acid can easily result in injury. The maximum allowable concentration of vapor in air for an 8-hour exposure period has been previously reported as 5 parts per million (ppm). After a short exposure, 35 ppm will cause throat irritation. This acid is frequently used in cleaners intended for descaling metal equipment because it reacts with tin, zinc, and galvanized coatings. It loosens the outer layers of material and carries soil and stain away. Hydrochloric acid will roughen the surface of concrete floors through an etching effect to produce a slip-resistant surface. When heated or contacted by hot water or steam, this acid will produce toxic and corrosive hydrogen chloride gas.

Sodium Acid Sulfate and Sodium Acid Phosphate

These cleaners will cause skin irritation or chemical burns with prolonged exposure. Water solutions of these compounds are strongly acidic and will damage the eyes if flushing is not immediate.

Phosphoric Acid

This acid is used in metal cleaners and metal brighteners. In a concentrated state, it is extremely corrosive to the skin and eyes. Phosphoric acid and sulfuric acid remove water from tissues. When heated, phosphoric acid emits toxic fumes of oxides of phosphorous. When compounded with other chemicals for use as a metal cleaner, only small amounts should be used to minimize the hazard.

Hydrofluoric Acid

Use of hydrofluoric acid in compounds helps to clean and brighten metal. Aluminum can be cleaned effectively with small amounts of this ingredient. In its pure state, hydrofluoric acid is extremely irritating and corrosive to the skin and mucous membranes. Inhalation of the vapor may cause ulcers of the respiratory tract. This material, even in very dilute amounts, should be used with caution. When heated, it emits a highly corrosive fluoride vapor, and it will react with steam to produce a toxic and corrosive mist. Ordinarily, it is used in small amounts because larger quantities can cause hydrogen evolution if in contact with metal containers. It must be stored in a safe environment, such as those used for flammable liquids.

Acid cleaners of this nature do not always attack the skin or eyes as quickly as do alkaline cleaning compounds. A severely exposed person may not realize the extent of injury until serious damage has occurred. This acid can penetrate the oil barrier of the skin to the point at which washing and flushing the area may be of little value. Hydrofluoric acid is especially hazardous because it gives little warning of injury until extensive damage has been done. Inhaled fluoride can cause damage to bones. This acid should not be confused with other acids because its action and indicated medical treatment are specific.

Soaps and Synthetic Detergents

Chemical builders used to increase the cleaning effectiveness of these substances in mixtures are usually alkaline compounds. Alkalis and alkaline substances are sometimes called *caustics* but are more correctly designated by the general term *bases*. They emulsify fats, oils, and other types of soil, which can then be washed away. Soaps and detergents for household cleaning use generally have a pH of 8 to 9.5. Continuous exposure to them can cause harmful degreasing of the skin, but they are safe in ordinary use. Detergents can either remove the natural oils from the skin or set up a reaction with the oils of the skin to increase susceptibility to chemicals that ordinarily do not affect the skin. Some slightly acid cleaners with a pH of 6 (the pH of the skin) are used for removing heavy, adherent grime from the body. These hand soaps usually contain solvents that suspend greasy soil without materially degreasing the skin.

Protective Equipment

Sanitation workers should wear waterproof, knee-high footwear to maintain dry feet. Trouser legs should be worn on the outside of the boots to prevent entry of powdered material, hot water, or strong cleaning solutions. Strap-top boots are recommended where trouser legs may be worn inside boots.

Protective equipment requirements vary with the strength of solution and method of use. Where cleaning materials are dispersed through spray and brush form for overhead cleaning, protective hoods, long gloves with gauntlets turned back to prevent the cleaner from running up the arms, and long aprons should be worn. Proper respiratory protective devices approved for the specific exposure should be worn where mists or gases are encountered during mixing or use. Supervisors should be made aware of the proper size and type of respiratory equipment and must ensure that this equipment is used and maintained properly.

Chemical goggles or safety glasses should be used when handling even mild cleaning compounds. Soaps of the strength of an ordinary hand soap can cause severe eye irritation (even though these materials are considered relatively mild) as their average pH is 9.0. Constant contact with even milder cleaning solutions can cause dermatitis due to chemical reaction, degreasing effects on the skin, or both. A person wearing contact lenses should not work in any area where dangerous chemicals are handled.

Mixing and Using

An apron, goggles, rubber gloves, and dust respirator must be worn when mixing or compounding dry ingredients. Cleaners should be mixed and dispensed only by experienced, well-trained personnel. The sanitation supervisor should have knowledge of chemical fundamentals of cleaning ingredients and should provide workers with the knowledge required to prevent accidents. They should know the hazards of each individual compound and how compounds are likely to react when mixed. Safety information on new compounds put in use should be made available. Workers should be instructed that cleaning compounds are not simply soaps, but strong and potentially dangerous chemicals that require protective measures. Protective equipment must be cleaned after use.

Most cleaning solutions should be compounded with cold water only. A few must be mixed with hot water to go into solution. These materials must be limited to those that do not produce a heat reaction during mixing with water. Cold water should be added during mixing to keep the solution below the boiling point or the point at which obnoxious vapors are emitted.

All cleaning compounds should be used in recommended concentrations. Once a dry cleaner is mixed or compounded, it should be stored in an identified container indicating its commonly used name, ingredients, precautions, and recommended concentration. Proper supervision is essential—sanitation workers are frequently prone to take the attitude that "if a little is good, a lot is better." The result is concentrations that are too strong for safe use. Workers must be impressed with the importance of not mixing cleaning ingredients once they are compounded. They should be warned not to place small amounts of dry chemicals back in a barrel or to blend them with unknown chemicals.

Storage and Transport

Cleaning ingredients and batches of compounded cleaners must be kept in locked storage and dispensed only with supervision. A system of inventory control should be maintained to aid in supervision and to discover deficiencies in dispensing.

Bulk storage of cleaning ingredients should be in areas designated for whatever hazard might be characteristic of that material. Reactive, basic, and acidic materials should be segregated. All bulk materials should be stored in fire-safe areas. Lids should be tightly in place, especially if the containers are stored under automatic sprinklers. Special chemicals should bear their own particular warnings that should be observed.

Containers of alkaline material should be kept tightly sealed because these materials generally take up water from the air. They should be reclosed as soon as possible after opening to protect the material from atmospheric moisture.

First Aid for Chemical Burns

Whenever an employee is splashed with cleaning chemicals, *flush the individual with a large amount of water immediately. Keep flushing for 15 to 20 minutes. Do not use materials of opposing pH to neutralize contaminated skin or clothing.* Such material may merely aggravate the condition through effects of its own properties.

Workers can carry a buffered solution for the eyes, which is sold in sealed containers. If water is unavailable, this liquid can be used to dilute and wash away chemicals from the eye. This emergency measure must be followed at once by washing the eyes for approximately 15 to 20 minutes as soon as the worker can be reached. The eyes should be checked by a physician after this injury. Instead of, or in addition to, the buffered solution, a plastic squeeze bottle of sterile water may be carried. Although these emergency measures are available, workers should not be allowed to regard eye contact accidents lightly. *The use of eye protection devices should be firmly enforced, especially where flushing water is not readily available.*

An injured employee should not be released from first aid or medical treatment until all of the chemical is removed. Speed is the most important factor in first aid for chemical exposures. An employee who is severely burned may act confused and need help. *Prompt flushing of chemicals from the skin, including the removal of contaminated clothing, is the most important factor in the handling of such chemical burns.* Insufficient flushing with water is very little better than none at all. Sources of water such as chemical burn showers or eye wash stations are best. However, any other source of water, regardless of its cleanliness, should be used for speed. An ample supply of water must be available near all locations where workmen may be exposed to corrosive chemicals. An ordinary shower head or garden hose spray nozzle does not supply water at a fast enough rate to flush a chemical. A flood of water is required. A satisfactory type of shower bath is one with a quick-opening valve that operates as soon as a person steps on a platform or works some other type of readily accessible control.

Everyone concerned with the chemical exposure problem should be thoroughly familiar with the following steps:

1. Anyone exposed to a strong chemical must be helped by others.
2. Flush the employee immediately at the nearest source of water. A shower is best, but any source will do. The eyes should be held open, and an extensive amount of water should be thrown into the eyes if necessary.
3. Remove all clothing.
4. After preliminary flushing, if a better source of water is near, get to it quickly and continue flushing all parts of the body thoroughly for at least 15 minutes. Secondary first-aid treatments, after flooding the victim's injury with water, should be kept to a minimum. Laymen should not attempt treatments with which they are not familiar or which they are not authorized to give.
5. If the injured person is confused or in shock, immobilize him or her immediately, apply warm clothing, then cover and transfer the individual to a medical facility by stretcher.
6. All but the most minor chemical burns should be treated by a medical doctor with specific knowledge of such burns. Some chemicals may have an internal toxic action, and the danger of bacterial infection exists when the skin has been eroded by a chemical.

Dermatitis Precautions

The industrial physician has the primary responsibility for determining whether an individual may be predisposed to skin irritations and for recommending suitable placement on the basis of these findings. When dermatitis suddenly develops among individuals on a job, the affected employees should be sent immediately to an experienced physician for examination and tests to determine whether they have acquired a sensitivity to the substance or substances being handled. If sensitivity has developed, the physician may decide that the affected worker should be removed from the exposure. Chemical compounds used in the cleaning operation should be listed and posted with the suggested treatment for exposure in the first-aid and supervisor's offices. Area physicians and medical centers should be listed.

SUMMARY

An effective sanitation program includes knowledge of soil deposits and use of the appropriate, versatile cleaning compound for the specific cleaning application. Soil characteristics determine the most appropriate cleaning compound. Generally, an acidic cleaning compound is most effective for removal or inorganic deposits, an alkaline cleaner for removing nonpetroleum organic soils, and a solvent-type cleaner for removal of petroleum soils.

The major function of cleaning compounds is to lower the surface tension of water so that soils may be loosened and flushed away. Detergent auxiliaries are included in cleaning compounds to protect sensitive surfaces or to improve the cleaning properties. Knowledge of how to handle cleaning compounds is essential to reduce the potential for injury of employees. If a worker is accidentally splashed with a cleaning compound, the affected area must be flushed with a large amount of water immediately.

STUDY QUESTIONS

1. What does soil mean to those involved with cleaning a food facility?
2. How does a cleaning compound function?
3. What is emulsification?
4. What is a chelating agent?
5. What does suspension mean to those cleaning a food facility?
6. What is a surfactant?
7. What is a sequestrant?
8. What is a builder?
9. What are cleaning auxiliaries?
10. Which two acid cleaning compounds are considered to be among the safest to use?
11. What treatment should be given to an employee who is splashed with cleaning chemicals?
12. What three words state a rule of thumb in cleaning compound solutions?

REFERENCES

Anon. 1976. *Plant sanitation for the meat packing industry*. Office of Continuing Education, University of Guelph and Meat Packers Council of Canada.

Anon. 1979. *Common detergent ingredients*. St. Paul, MN: Klenzade, Division Ecolab, Inc.

Anon. 1996. *The role of cleaning compounds*. Wyandotte, MI: Diversey Corp.

Lovingood, R.P., et al. 1987. *Effectiveness of phosphate and nonphosphate laundry detergents*. Extension Division. Virginia Polytechnic Institute and State University, Publication No. 354-062.

Wyman, D.P. 1996. Understanding active chlorine chemistry. *Food Qual* 2, no. 18: 77.

SUGGESTED READINGS

Anon. 1991. *The fundamentals of cleaning or TACT WINS*. Wyandotte, MI: Diversey Corp.

Marriott, N.G. 1990. *Meat sanitation guide II*. Blacksburg: American Association of Meat Processors and Virginia Polytechnic Institute and State University.

Moody, M.W. 1979. How cleaning compounds do the job. In *Sanitation notebook for the seafood industry*, eds. G.J.

Flick et al., II–68. Blacksburg: Department of Food Science and Technology, Virginia Polytechnic Institute and State University.

Zottola, E.A. 1973. How cleaning compounds do the job. In *Proceedings of the conference on sanitation and safety*, 53. Blacksburg: Extension Division, State Technical Services, and Department of Food Science and Technology, Virginia Polytechnic Institute and State University.

CHAPTER 8

Sanitizers

Soil that remains on food processing equipment after use is usually contaminated with microorganisms that are nourished by the nutrients of the soil deposits. Thus, the soil provides a medium for microbial proliferation. A sanitary environment is obtained by thoroughly removing soil deposits and subsequently applying a sanitizer to destroy residual microorganisms. If soil residues are present, they protect microorganisms from contact with chemical sanitizing agents. Soil deposits can reduce the effectiveness of a sanitizer through a dilution effect and reaction of the organic matter in the soil with the sanitizing compound.

SANITIZING METHODS

Thermal Sanitizing

Thermal sanitizing is relatively inefficient because of the energy required. Its efficiency depends on the humidity, temperature required, and length of time a given temperature must be maintained. Microorganisms can be destroyed with the correct temperature if the item is heated long enough and if the dispensing method and application design, as well as equipment and plant design, permit the heat to penetrate to all areas. Temperature should be measured with accurate thermometers located at the outlet pipes to ensure effective sanitizing. The two major sources for thermal sterilization are steam and hot water.

Steam

Sanitizing with steam is expensive because of high energy costs, and it is usually ineffective. Workers frequently mistake water vapor for steam; therefore, the temperature usually is not high enough to sterilize that which is being cleaned. If the surface being treated is highly contaminated, a cake may form on the organic residues and prevent sufficient heat penetration to kill the microbes.

Experience in the industry has shown that steam is not amenable to continuous sanitizing of conveyors. In fact, condensation from this operation and other steam applications has actually complicated cleaning operations.

Hot Water

Immersion of small components (i.e., knives, small parts, eating utensils, and small containers) into water heated to 80°C or higher is another thermal method of sterilization. The microbicidal action is thought to be the denaturation of some of the protein molecules in the cell. Pouring "hot" water into containers is not a reliable sterilizing method because of the difficulty of maintaining a water temperature high enough to ensure adequate sterilization. Hot water can be an effective, nonselective sanitizing method for food-contact surfaces; however, spores may survive more than an hour at boiling temperature. This sterilizing method is frequently used for plate heat exchangers and eating utensils.

The temperature of the water determines the time of exposure needed to ensure sterilization. An example of time-temperature relationships would be combinations adopted for various plants that utilize 15 minutes of exposure time at 85°C or 20 minutes at 80°C. A shorter time requires a higher temperature. The volume of water and its flow rate will also influence the time taken by the components to reach the required temperature. If water hardness exceeds 60 mg/L, water scale is frequently deposited on surfaces being sanitized unless the water is softened. Hot water is readily available and nontoxic. Sanitizing can be accomplished either by pumping the water through assembled equipment or by immersing equipment in the water.

Radiation Sanitizing

Radiation at a wave length of approximately 2,500 Å in the form of ultraviolet light or high-energy cathode or gamma rays will destroy microorganisms. For example, ultraviolet light has been used in the form of low-pressure mercury vapor lamps to destroy microorganisms in hospitals and homes. Ultraviolet light units are now commonly used in Europe to disinfect drinking and food processing waters and are being installed in the United States. However, this method of sanitizing has been restricted to fruits, vegetables, and spices and has not been useful in food plants and food service facilities because of its limited total effectiveness. The effective killing range for microorganisms through use of ultraviolet light is short enough to limit its

utility in food operations. Bacterial resistance determines the lethal exposure time. The light rays must actually strike the microorganisms, and they can be absorbed by dust, thin films of grease, and opaque or turbid solutions. Also, radiation controls the infestation of insects; regardless of the stage of their life cycle.

Chemical Sanitizing

The chemical sanitizers available for use in food processing and food service operations vary in chemical composition and activity, depending on conditions. Generally, the more concentrated a sanitizer, the more rapid and effective its action. The individual characteristics of each chemical sanitizer must be known and understood so that the most appropriate sanitizer for a specific sanitizer application can be selected. Because chemical sanitizers lack penetration ability, microorganisms present in cracks, crevices, pockets, and in mineral soils may not be totally destroyed. For sanitizers to be effective when combined with cleaning compounds, the temperature of the cleaning solution should be 55°C or lower, and the soil should be light. The efficacy of sanitizers (especially chemical sanitizers) is affected by physical-chemical factors such as:

Exposure time: Studies have suggested that the death of a microbial population follows a logarithmic pattern, indicating that if 90% of a population is killed in a unit of time, the next 90% of the remaining is destroyed in the next unit of time, leaving only 1% of the original number. Microbial load and the population of cells having varied susceptibility to the sanitizer due to age, spore formation, and other physiologic factors determine the time required for the sanitizer to be effective.

Temperature: The growth rate of the microorganisms and the death rate due to chemical application will increase as temperature elevates. A higher temperature generally lowers surface tension, increases pH, decreases viscosity, and creates other changes that may help bactericidal action. Generally, the degree of sanitation greatly exceeds the growth rate of the bacteria, so that the final effect of increasing temperature is to enhance the rate of destruction of the microorganisms.

Concentration: Increasing the concentration of the sanitizer increases the rate of destruction of the microorganisms.

pH: The activity of antimicrobial agents occurring as different species within a pH range may be dramatically influenced by relatively small changes in the pH of the medium. Chlorine and iodine compounds generally decrease in effectiveness with an increase in pH.

Equipment cleanliness: Hypochlorites, other chlorine compounds, iodine compounds, and other sanitizers can react with the organic materials of soil that have not been removed from equipment and other surfaces. Failure to clean surfaces properly can reduce the effectiveness of a sanitizer.

Water hardness: Quaternary ammonium compounds are incompatible with calcium and magnesium salts and should not be used with over 200 parts per million (ppm) of calcium in water or without a sequestering or chelating agent. As water hardness increases, the effectiveness of these sanitizers decreases.

Bacterial attachment: It has been demonstrated by Le Chevallier et al. (1988) that attachment of certain bacteria to a solid surface provides an increased resistance to chlorine. Other factors, such as nutrient limitation (stringent response), also do so, and with attachment, the resultant resistance to chlorine is increased.

Desired Sanitizer Properties

The ideal sanitizer should have the following properties:

- Microbial destruction properties of uniform, broad-spectrum activity against vegetative bacteria, yeasts, and molds to produce rapid kill
- Environmental resistance (effective in the presence of organic matter [soil load], detergent and soap residues, and water hardness and pH variability)
- Good cleaning properties
- Nontoxic and nonirritating properties
- Water solubility in all proportions
- Acceptability of odor or no odor
- Stability in concentrated and use dilution
- Ease of use
- Ready availability
- Inexpensive
- Ease of measurement in use solution

A standard chemical sanitizer cannot be effectively utilized for all sanitizing requirements. The chemical selected as a sanitizer should pass the Chambers test (also referred to as the *sanitizer efficiency test*): Sanitizers should produce 99.999% kill of 75 million to 125 million *Escherichia coli* and *Staphylococcus aureus* within 30 seconds after application at 20°C. The pH at which the compound is applied can influence the effectiveness of the sanitizer. Chemical sanitizers are normally divided according to the agent that kills the microorganisms.

Chlorine Compounds

Liquid chlorine, hypochlorites, inorganic chloramine and organic chloramines, and chlorine dioxide function as sanitizers. Their antimicrobial activity varies. Chlorine gas may be injected slowly into water to form the antimicrobial form, hypochlorous acid (HOCl). Liquid chlorine is a solution of sodium hypochlorite (NaOCl) in water. Hypochlororous acid is 80 times more effective as a sanitizing agent than an equivalent concentration of the hypochlorite ion. The activity of chlorine as an antimicrobial agent has not been fully determined. Hypochlorous acid, the most active of the chlorine compounds, appears to kill the microbial cell through inhibiting glucose oxidation by chlorine-oxidizing sulfhydryl groups of certain enzymes important in carbohydrate metabolism. Aldolase was considered to be the main site of action, owing to its essential nature in metabolism.

Other modes of chlorine action that have been proposed are: (1) disruption of protein synthesis; (2) oxidative decarboxylation of amino acids to nitrites and aldehydes; (3) reactions with nucleic acids, purines, and pyrimidines; (4) unbalanced metabolism after the destruction of key enzymes; (5) induction of deoxyribonucleic acid (DNA) lesions with the accompanying loss of DNA-transforming ability; (6) inhibition of oxygen uptake and oxidative phosphorylation, coupled with leakage of some macromolecules; (7) formation of toxic N-chlor derivatives of cytosine; and (8) creation of chromosomal aberrations.

Vegetative cells take up free chlorine but not combined chlorine. Formation of chloramines in the cell protoplasm does not cause initial destruction. Use of ^{32}P in the presence of chlorine has suggested that there is a destructive permeability change in the microbial cell membrane. Research by Camper and McFetters (1979) demonstrated that chlorine impairs cell membrane function, especially transport of extracellular nutrients, and that labeled carbohydrates and amino acids could not be taken by chlorine-treated cells. Research by Benarde et al. (1965), through use of ^{14}C-labeled amino acids, revealed that chlorine dioxide disrupted protein synthesis among *E. coli*, although the degree of disruption was not determined.

Chlorine-releasing compounds are known to stimulate spore germination and subsequently to inactivate the germinated spore. Research conducted by Kulikoosky et al. (1975) has demonstrated that chlorine alters the spore permeability through changes in the integument, with subsequent release of Ca^{2+}, dipicolinic acid (DPA), RNA, and DNA.

Granular chlorine sanitizers are based on the salts of an organic carrier that contains releasable ions. Chlorinated isocyanurate is a highly stable, rapidly dissolving chlorine carrier that releases one of its two chloride ions to form NaOCl in aqueous solution. Buffering agents, which are mixed with the dry chlorine carrier in these products, control the rate of antimicrobial activity, corrosion characteristics, and stability of solutions of the sanitizers by adjusting the solution to an optimal use pH.

The chemical properties of chlorine are such that when liquid chlorine (Cl_2), and hypochlorites are mixed with water, they hydroyze to form hypochlorous acid, which will dissociate in water to form a hydrogen ion (H^+) and a hypochlorite ion (OCl^-), according to the reactions shown below. If sodium is combined with hypochlorite to form sodium hypochlorite, the following reactions would apply.

$$Cl_2 + H_2O \rightarrow HOCl + H^+ + Cl^-$$
$$NaOCl + H_2O \rightarrow NaOH + HOCl$$
$$HOCl \rightleftharpoons H^+ OCl^-$$

Chlorine compounds are more effective antimicrobial agents at a lower pH where the presence of hypochlorous acid is dominant. As the pH increases, the hypochlorite ion, which is not as effective as a bactericide, predominates. Another chlorine compound, chlorine dioxide, does not hydrolyze in aqueous solutions. Therefore, the intact molecule appears to be the active agent.

Chlorine is known to be effective as a sanitizer for mechanically polished stainless steel, unabraded electropolished stainless steel, and the polycarbonate surfaces, reducing self-populations to less than 1.0 log CFU/cm^2. This sanitizer is less effective on abraded electropolished stainless steel and mineral resin surfaces, where populations exceed 1.0 log CFU/cm^2 (Frank and Chmielewski, 1997).

Hypochlorites, the most active of the chlorine compounds, are also the most widely used. Calcium hypochlorite and sodium hypochlorite are the major compounds of the hypochlorites. These sanitizers are effective in deactivating microbial cells in aqueous suspensions and require a contact time of approximately 1.5 to 100 seconds. A 90% reduction in cell population for most microorganisms can be attained in less than 10 seconds, with relatively low levels of free available chlorine (FAC). Bacterial spores are more resistant than vegetative cells to hypochlorites. The time required for a 90% reduction in cell population can range from approximately 7 seconds to more than 20 minutes (Odlaug, 1981). The concentration of free available chlorine needed for inactivation of bacterial spores is approximately 10 to 1,000 times as high (1,000 ppm, compared with approximately 0.6 to 13 ppm) for vegetative cells. *Clostridium* spores are less resistant to chlorine than *Bacillus* spores. These data suggest that in sanitizing applications, where the concentration of hypochlorous acid is low and the contact time is short, there is limited effect on bacterial spores. Although 200 ppm is effective for numerous surfaces, 800 ppm is suggested for porous areas.

The following example indicates how to formulate a 200-ppm solution of chlorine in a 200-L tank. This calculation assumes that the chlorine contains 8.5% NaOCl.

$$8.5\% \text{ NaOCl} = 85,000 \text{ ppm } (0.085 \times 1,000,000)$$
$$1 \text{ L} = 1,000 \text{ mL}$$
$$200 \text{ L} = 200,000 \text{ mL}$$
$$\frac{X}{200,000 \text{ mL}} = \frac{200 \text{ ppm}}{85,000 \text{ ppm}}$$
$$85,000 \ X = 40,000,000 \text{ mL}$$
$$X = 470 \text{ mL of } 8.5\% \text{ NaOCl}$$

Calcium hypochlorite, sodium hypochlorite, and brands of chlorinated trisodium phosphate may be applied as sanitizers after cleaning. The hypochlorites may also be added to cleaning compound solutions to provide a combination cleaner-sanitizer. Organic chlorine-releasing agents, such as sodium dichloroisocyanurate and dichlorodimethylhydantoin, can be formulated with cleaning compounds.

Molecular hypochlorous acid is present in highest concentration near pH 4, decreasing rapidly as pH increases. At a pH higher than 5, hypochlorite (OCL$^-$) increases; whereas, at pH<4, chlorine gas increases. Furthermore, the formation of Cl$_2$ is a safety issue. Because there are substantial amounts of hydrochlorous acid present when the pH exceeds 6.5, sanitizing operations are normally executed in the pH range of 6.5 to 7.0.

The reaction time of chlorine-based sanitizers is temperature-dependent. Up to 52°C, the reaction rate doubles for each 10°C increase in temperature. Although hypochlorites are relatively stable, Cl$_2$ solubility decreases rapidly above 50°C.

The efficacy of a buffered sodium hypochlorite solution in controlling bacterial contamination has been evaluated by Park et al. (1991). They found this sanitizing solution to be effective in reducing *Salmonella enteritidis*. Their research reflected no adverse effects on protein functionality, lipid oxidation, and starch degradation after exposure of food products to the sanitizing solution.

Active chlorine solutions are very effective sanitizers, especially as free chlorine and in slightly acid solutions. These compounds appear to act through protein denaturation and enzyme inactivation. Chlorine sanitizers are effective against gram-positive and gram-negative bacteria and conditionally against certain viruses and spores. However, the available chlorine from hypochlorite and other chlorine-releasing chemicals reacts with and is inactivated by residual organic matter. If the recommended volume of chlorine solution and sufficient concentration is applied, a sanitizing effect can still be achieved. Only freshly prepared solutions should be used. Storage of used solutions may result in a decline in strength and activity. Concentration of active chlorine can be easily measured by use of test kits to ensure application of the desired concentration.

Inorganic chloramines are compounds formed from the reaction of chlorine with ammonia nitrogen; *organic chloramines* are formed through the reaction of hypochlorous acid with amines, imines, and imides. Bacterial spores and vegetative cells are more resistant to chloramine than to the hypochlorites. Chloramine T apparently releases chlorine slowly, and, as a result, its lethal effects are slow when compared with the hypochlorites.

Other chloramine compounds are as effective as, or more effective than, the hypochlorites in deactivating microorganisms. Sodium dichloroisocyanurate is more active than sodium hypochlorite against *E. coli*, *S. aureus*, and other bacteria.

Less is known about the antimicrobial effects of *chlorine dioxide* than about the other chlorine compounds; however, interest in this compound has increased. New chemical formulations of this compound allow it to be shipped to areas of use (rather than being generated on site); consequently, it is being used more in the food industry. *Chlorine dioxide* (ClO_2) is known to have 2.5 times the oxidizing power of chlorine. This compound is not as effective as chlorine at pH 6.5, but at pH 8.5, ClO_2 is the most effective. This characteristic suggests that ClO_2 is less affected by alkaline conditions and organic matter, thus making it a viable agent for sewage treatment.

Examples of how chlorine dioxide sanitizers are produced are indicated by the reactions that follow:

$$5NaClO_2 + 4HCl \rightarrow 4\ ClO_2 + 5NaCl + 2H_2O$$

$$NaOCl + HCl \rightarrow NaCl + HOCl$$
$$HOCl + 2NaClO_2 \rightarrow ClO_2 + 2NaCl + H_2O$$

Meinhold (1991) reported that ClO_2 is being used in cleaning and sanitizing through foam generation. It can be produced by combining chlorine salt and chlorine or hypochlorite and acid, followed by the addition of chlorite. A biodegradable foam containing 1- to 5-ppm ClO_2 can be produced and is effective with a shorter contact time than the quats or hypochlorites. Chlorine dioxide is effective against a broad spectrum of microorganisms, including bacteria, viruses, and spore-formers. As a chemical oxidant, the residual activity significantly inhibits microbial redevelopment. It is active over the broad pH range normally encountered in food facilities. This compound is less corrosive than other chlorine sanitizers because of the low concentration necessary to be effective and produces less "undesirable" chlorinated organics.

The U.S. Food and Drug Administration (FDA) has approved the use of stabilized chlorine dioxide for sanitizing food processing equipment. Anthium dioxcide is a compound with 5% aqueous solution of stabilized chlorine dioxide supplied with a pH of 8.5 to 9.0. Free ClO_2 is the potential biocidal agent in the solution. Although anthium dioxcide does exhibit bacteriostatic properties, it is not nearly as effective as free ClO_2. The active biocide is free ClO_2, even though the stabilized ClO_2 at pH 8.5 is mildy bacteriostatic. The anthium dioxcide complex is a combination of oxygen and chlorine joined as ClO_2 in aqueous solution, which provides a longer residual effect than other chlorine sanitizers. Industrial applications include a no-rinse sanitizer at 100 ppm, poultry chill tanks at 3 to 5 ppm, and drinking water treatment.

Oxine has gained recent interest as a sanitizer. It differs from generated ClO_2 as it is formulated from scratch, using a proprietary process, as opposed to being converted from chlorite. Increased microbial kill is possible by adjusting the ratio of chlorite and chlorine dioxide, and of other oxychlorine species, through the formation of oxine. Oxine is stabilized by dissolving the gas into a proprietary aqueous solution and essentially converting it into its "salt" form (Flickinger, 1997). An activator, such as food-grade acid, is needed for this binary product to lower the pH and retrieve the gas. The major application of this compound is as a surface sanitizer that is effective against biofilms. Recent testing conducted with *E. coli* O157:H7 revealed that oxine destroys this pathogen at 6 ppm (Flickinger, 1997).

When chlorine compounds are used in solutions or on surfaces where available chlorine can react with cells, these sanitizers are bactericidal and sporicidal. Vegetative cells are more easily destroyed than are *Clostridium* spores, which are killed more easily than are *Bacillus* spores. Chlorine concentrations of less than 50 ppm lack antimicrobial activity against *Listeria monocytogenes*, but exposure to more than 50 ppm effectively destroys this pathogen. This lethal effect of most chlorine compounds is enhanced, with an increase in free available chlorine, a decrease in pH, and an increase in temperature. However, chlorine solubility in water decreases and corrosiveness increases with a higher temperature, and solutions with a high chlorine concentration and/or low pH can corrode metals. Chlorine compounds have the following advantages over other sanitizers:

- They are effective against a variety of bacteria, fungi, and viruses.
- They include fast-acting compounds that will pass the Chambers test at a concentration of 50 ppm in the required 30 seconds.
- They are the cheapest sanitizer (if inexpensive chlorine compounds are used).
- Equipment does not have to be rinsed if 200 ppm or less is applied.

- They are available in liquid or granular form.
- They are not affected by hard-water salts (except when slight variations, due to pH, exist).
- High levels of chlorine may soften gaskets and remove carbon from rubber parts of equipment.
- Toxic by-products are not produced.
- They are less corrosive than chlorine.

However, they have some disadvantages:

- They are unstable and drive off rather rapidly with heat or contamination with organic matter.
- Their effectiveness decreases with increased solution pH.
- They are very corrosive to stainless steel and other metals.
- They must be in contact with food handling equipment, especially on any type of dishes, for only a short time to prevent corrosion.
- They deteriorate during storage when exposed to light or to a temperature of above 60°C.
- Solutions at a lower pH can form toxic and corrosive chlorine gas (Cl_2).
- Concentrated in the liquid form, they may be explosive.

Iodine Compounds

The mode of antibacterial action of iodine has not been studied in detail. It appears that diatomic iodine is the major active antimicrobial agent, which disrupts bonds that hold cell protein together and inhibits protein synthesis (Anon., 1996). Generally, free elemental iodine and hypoiodous acid are the active agents in microbial destruction. The major iodine compounds used for sanitizing are iodophors, alcohol-iodine solutions, and aqueous iodine solutions. The two solutions are normally used as skin disinfectants. The iodophors have value for cleaning and disinfecting equipment and surfaces, and as a skin antiseptic. Iodophors are also used in water treatment.

The iodophor complex releases an intermediate triciodide ion, which, in the presence of acid is rapidly converted to hypoiodous acid and diatomic iodine. Both the hypoiodous acid and diatonic iodine are the active antimicrobial forms of an iodophor sanitizer.

Ionic surface active agents (surfactants) are compounds composed of two principal functional groups—a lipophilic portion and a hydrophilic portion. When placed in water, these molecules ionize, and the two groups induce a net charge to the molecule, which results in either a positive or a negative charge for the surfactant molecule. Cationic and anionic sanitizers have similar modes of action.

When elemental iodine is complexed with nonionic surface-active agents such as nonyl phenolethylene oxide condensates or a carrier such as polyvinylpyrrolidone, the water-soluble complexes known as *iodophors* are formed. Iodophors, the most popular forms of iodine compounds used today, have greater bactericidal activity under acidic conditions. Thus, these compounds are frequently modified with phosphoric acid. Complexing iodophors with surface-active agents and acids gives them detergent properties and qualifies them as detergent-sanitizers. These compounds are bactericidal and, when compared to aqueous and alcoholic suspensions of iodine, have greater solubility in water, are nonodorous, and are nonirritating to the skin.

To prepare the surfactant–iodine complex, the iodine is added to the nonionic surfactant and heated to 55°C to 65°C to enhance solution of the iodine and to stabilize the end product. The exothermic reaction between the iodine and surfactant produces a rise in temperature, dependent on the type of surfactant and the ratio of surfactant to iodine. If the iodine level does not exceed the solubilizing limit of the surfactant, the end product will be completely and infinitely soluble in water.

The behavior of surfactant–iodine complexes has been previously explained on the basis of equilibrium $R + I_2 \rightleftharpoons RI + HI$, where R represents the surfactant. Removal of the iodides formed by oxidation to iodine is responsible for further disposition of available iodine, presumably due to increased iodination of the surfactant.

The amount of free available iodine determines the activity of iodophors. The surfactant present does not determine the activity of iodophors but can affect the bactericidal properties of iodine. Spores are more resistant to iodine than are vegetative cells, and the lethal exposure times noted in Table 8–1 are approximately 10 to 1,000 times as long as for vegetative cells. Iodine is as effective in the deactivation of vegetative cells. It is as active in the deactivation of vegetative cells but not as effective as chlorine in spore inactivation.

Iodine-type sanitizers are somewhat more stable in the presence of organic matter than are the chlorine compounds. Because iodine complexes are stable at a very low pH, they can be used at a very low concentration of 6.25 ppm and are frequently used at 12.5 to 25 ppm. Iodine sanitizers are more effective than are other sanitizers on viruses. Only 6.25 ppm are required to pass the Chambers test in 30 seconds. Nonselective iodine compounds kill vegetative cells and many spores and viruses.

Iodophor sanitizers, used in the recommended concentration, usually provide 50 to 70 mg/L of free iodine and yield pH values of 3 or less in water of moderate alkaline hardness. Excessive dilution of iodophors with highly alkaline water can severely impair their efficiency because acidity is neutralized. Solutions of this sanitizer are most effective at a pH of 2.5 to 3.5.

Table 8–1 Inactivation of Bacterial Spores by Iodophors

| Organism | Concentration | | Time for a 90% Reduction (min) |
	pH	(ppm)	
Bacillus cereus	6.5	50	10
	6.5	25	30
	2.3	25	30
Bacillus subtilis	—	25	5
Clostridium botulinum type A	2.8	100	6

Note: All tests were done in distilled water at 15° to 25°C.
Source: Odlaug, 1981.

In a concentrated form, formulated iodophors have a long shelf life. In solution, however, iodine may be lost by vaporization. This loss is especially rapid when the solution temperature exceeds 50° C because iodine tends to sublime. Iodine can be absorbed by plastic materials and rubber gaskets of heat exchangers, with resultant staining and antiseptic tainting. Iodine stain can be advantageous because most organic and mineral soil stains yellow, thus indicating the location of inadequate cleaning. The amber color of iodine solutions provides visible evidence of the presence of the sanitizer, but color intensity is not a reliable guide to iodine concentration.

Because iodophor solutions are acidic, they are not affected by hard water and will prevent accumulation of minerals if used regularly. Yet, existing mineral deposits are not removed through the application of iodine sanitizers. Organic matter (especially milk) inactivates the iodine in iodophor solutions, with a subsequent fading of the amber color. Iodine loss from solutions is slight, unless excessive organic soils are present. Because iodine loss increases during storage, these solutions should be checked and adjusted to the required strength.

Iodine compounds cost a little more to use than does chlorine and may cause off-flavor in some products. Other disadvantages of iodine compounds are that they vaporize at approximately 50°C, are less effective against bacterial spores and bacteria phage than are chlorines, have poor low-temperature efficacy, and are very sensitive to pH changes. Iodine sanitizers are effective for sanitizing hands because they do not irritate the skin. They are recommended for hand-dipping operations in food plants and are used frequently on food handling equipment.

Bromine Compounds

Bromine has been used alone or in combination with other compounds, more in water treatment than as a sanitizer for processing equipment and utensils. At a slightly acid to normal pH, organic chloramine compounds are more effective in destroying spores (such as *Bacillus cereus*) than are organic bromine compounds, but chloramine with bromine tends to be less affected by an alkaline pH of 7.5 or higher. The addition of bromine to a chlorine compound solution can synergistically increase the effectiveness of bromine and chlorine.

Quaternary Ammonium Compounds

The quaternary ammonium compounds, frequently called the *quats,* are used most frequently on floors, walls, furnishings, and equipment. They are good penetrants and, thus, have value for porous surfaces. They are natural wetting agents with built-in detergent properties and are referred to as *synthetic surface-active agents*. The most common agents are the cationic detergents, which are poor detergents but excellent germicides. Quaternary ammonium compounds are very effective sanitizers for *L. monocytogenes* and are effective in reducing mold growth.

The quats are ammonium compounds in which four organic groups are linked to a nitrogen atom that produces a positively charged ion (cation). In these quaternary ammonium compounds, the organic radical is the cation, and chlorine is usually the anion. The mechanism of germicidal action is not fully understood but may be that the surface-active nature of the quat surrounds and covers the cell's outer membrane, causing a failure of the wall, which consequently causes leakage of the internal organs and enzyme inhibition. The general formula of the quaternary ammonium compound is:

$$R_1 \ :\overset{\overset{\displaystyle R_2}{\displaystyle \cdots}}{\underset{\displaystyle R_4}{N}}: \ R_3 \quad Cl^- \ or \ Br^-$$

The quats act against microorganisms differently than do chlorine and iodine compounds. They form a bacteriostatic film after being applied to surfaces. Although the film is bacteriostatic, these compounds are selective in the destruction of various microorganisms. The quats do not kill bacterial spores but can inhibit their growth. Quaternary ammonium compounds are more stable in the presence of organic matter than are chlorine and iodine sanitizers, although their bactericidal effectiveness is impaired by the presence of organic matter. Stainless steel and polycarbonate are more readily sanitized by the quats than are either the abraded polycarbonate or mineral resin surfaces (Frank and Chmielewski, 1997).

The quaternary ammonium compounds include alkyldimethylbenzylammonium chloride and alkyldimethylethylbenzylammonium chloride, both effective in water ranging from 500 to 1,000 ppm hardness without added sequestering agents. Haverland (1981) reported that diisobutylphenoxyethoxyethyl dimethyl benzyl ammonium chloride and methyldodecylbenzyltrimethyl ammonium chloride are compounds that require sodium tripolyphosphate to raise hard-water levels to a minimum of 500 ppm. These compounds require high dilution for germicidal or bacteriostatic action. As with other quats, these are nonconcorrosive and nonirritating to the skin, and have no taste or odor in use dilutions. The concentration of quat solutions is easy to measure. The quats are low in toxicity and can be neutralized or made ineffective by using any anionic detergent.

Quats have surfactants that cause them to foam (Carsberg, 1996). Quats can be "foamed" on, which provides a medium for them to cling to vertical and radial surfaces. When formulated with a specified detergent, they can be used as a cleaner-sanitizer. However, this application requires rinsing, although it is satisfactory for bathrooms, toilets, locker rooms, and other non–food-contact surfaces. These cleaner-sanitizers are not recommended for use in the food plant environment because there are insufficient detergent properties and pH or alkalinity levels to thoroughly clean. Because rinsing of this cleaner-sanitizer is required, there is no residual antibacterial activity on the surface.

The quaternary ammonium compounds should not be combined with cleaning compounds for subsequent cleaning and sanitizing because the quats may be inactivated by detergent ingredients such as anionic wetting agents (see Chapter 7) and, to a lesser extent, by others. However, an increase in alkalinity through formulation with compatible detergents may enhance the bactericidal activity of the quats.

The quats have the following major advantages (Anon., 1997):

- Colorless and odorless
- Stable against reaction with organic matter
- Resistant to corrosion of metals
- Stable against temperature fluctuation
- Nonirritating to the skin
- Effective at a high pH
- Effective against mold growth
- Nontoxic
- Good surfactants

They have these disadvantages:

- Limited effectiveness (including ineffectiveness against most gram-negative microorganisms except *Salmonella* and *E. coli*)
- Incompatibility with anionic-type synthetic detergents
- Film forming on food handling and food processing equipment

Acid Sanitizers

Acid sanitizers, which are considered to be toxicologically safe and biologically active, are frequently used to combine the rinsing and sanitizing steps. Organic acids, such as acetic, peroxyacetic, lactic, propionic, and formic acid, are most frequently used. The acid neutralizes excess alkalinity that remains from the cleaning compound, prevents formation of alkaline deposits, and sanitizes. Because bacteria have a positive surface charge, and negatively charged surfactants react with positively charged bacteria, their cell walls are penetrated, and cellular function is disrupted. These sanitizers destroy microbes by penetrating and disrupting the cell membranes, then dissociating the acid molecule and, consequently, acidifying the cell interior. Acid

treatment is dose-dependent for spoilage and pathogenic microorganisms. These compounds are especially effective on stainless steel surfaces or where contact time may be extended and has a high antimicrobial activity against psychrotrophic microorganisms.

The development of automated cleaning systems in food plants, where it is desirable to combine sanitizing with the final rinse, has made the use of acid sanitizers desirable. After the final rinse, the equipment may be closed to avoid contamination and held overnight with no danger of corrosion. Although these compounds are sensitive to pH change, they are less prone to be affected by hard water than are the iodines. In the past, the disadvantage of these synthetic detergents in automated cleaning systems was foam development, which made it difficult to get good drainage of the sanitizer from the equipment. Nonfoaming acid synthetic detergent sanitizers have become available, eliminating this problem and making these compounds even more valuable in the food industry. These sanitizers are less effective with an increase in pH or against thermoduric organisms. Acids are not as efficient as irradiation and, when applied at high concentrations, can cause slight discoloration and odor on food surfaces, such as meat. The cost-effectiveness of acid sanitizers has not been evaluated sufficiently, and experiments with acetic acid have revealed a lack of effectiveness in the reduction of *Salmonella* species contamination.

Acid sanitizers are fast-acting and effective against yeasts and viruses. The pH range of below 3 is the most ideal for the performance of acid sanitizers. Acid anionic sanitizers may be incorporated as an acid-rinse for equipment to leave it stainless, bright, and shiny. These sanitizers have very good wetting properties, are nonstaining and usually noncorrosive, permitting exposure to equipment overnight. Hard water and residual organic matter do not have a major effect on the ability of acid anionic sanitizers to destroy microorganisms, and they can be applied by cleaning-in-place (CIP) methods or by spray, or they can be foamed on if a foam additive is incorporated. Acid sanitizers can lose all of their effectiveness by the presence of alkaline residuals or by the presence of cationic surfactants. Thus, all cleaning compounds should be rinsed from surfaces before acid sanitizers are applied.

Carboxylic acid sanitizers are effective over a broad range of bactericidal activity. They are stable in dilutions, in the presence of organic matter, and at high temperatures. These sanitizers are noncorrosive to stainless steel, provide a good shelf life, are cost effective and act as a sanitizer and acid rinse. Carboxylic acid is less effective against yeasts and molds and not as effective above pH 3.4 to 4.0 as some chemical sanitizers. They are negatively affected by cationic surfactants, so thorough rinsing of detergents is essential. This sanitizer is corrosive to nonstainless steels, plastics, and some rubbers.

Carboxylic acid is known as a fatty acid sanitizer composed of free fatty acids, sulfonated fatty acids, and other organic acids. It is similar to acid-anionic sanitizers, except that significantly lower foam characteristics exist (Anon., 1997).

Acid Anionic Sanitizers

These sanitizers are formulated with:

- anionic surfactants (negatively charged)
- acids
  phosphoric acid
  organic acids

Acid anionic sanitizers act rapidly and kill a broad spectrum of bacteria. They have good stability, are effective in a wide temperature range, and are not affected by water hardness. An acidified rinse can be combined with the sanitizing step. These sanitizers can be corrosive to unprotected metals, may foam too much for CIP equipment, are less effective at a higher pH, and are more expensive than are the halogen sanitizers. The antimicrobial effect of acid anionics appears to be through reaction of the surfactant, with positively charged bacteria by ionic attraction to penetrate cell walls and disrupt cellular function.

Increased interest in *peroxyacetic acid* has developed in CIP sanitizing for dairy, beverage, and food processing plants. This sanitizer, which provides a rapid, broad-spectrum kill, works on the oxidation principle through the reaction with the components of cell membranes. It reduces pitting of equipment surfaces by being less corrosive than are iodine and chlorine sanitizers. Peroxyacetic acid can be applied during an acidified rinse cycle to reduce effluent discharge; it is also biodegradable. Because this sanitizer is effective against yeasts—such as *Candida, Saccharomyces,* and *Hansenula,* and molds—such as *Penicillium, Aspergillus, Mucor,* and *Geotrichum,* it has gained acceptance in the soft drink and brewing industry. Peroxyacetic acid is effective for sanitizing aluminum beer kegs. Increased use of this sanitizer in dairy and food processing plants is attributable to its efficacy against various strains of *Listeria* and *Salmonella.* The application rate of this sanitizer is 125 to 250 ppm. These sanitizers have the following advantages (Anon., 1997):

- Stable to heat and organic matter, have nonvolatile characteristics, and can be heated to any temperature below 100° C without loss of strength
- Generate low foam—suitable for CIP equipment
- Nonselective, permitting destruction of all vegetable cells
- Safe for use on most food handling surfaces (low toxicity—breaks down into water, oxygen, and acetic acid)

- Have rapid, broad-spectrum kill (bacteria, yeasts, and molds)
- Are pH-range tolerant
- Effective against biofilms
- Allow sanitizing and acid rinse steps to be combined

Disadvantages are high cost, odor, irritancy, tendency to corrode iron and other metals, lack of effectiveness in the presence of organic materials, and lower effectiveness against yeasts and molds than some sanitizers.

Acid-Quat Sanitizers

During the early portion of the 1990s, organic acid sanitizers formulated with quaternary ammonium compounds were marketed as acid-quat sanitizers. The brief track record of this type of sanitizer suggests that it is effective—especially against *L. monocytogenes*. A limitation of this type of sanitizer is that it is expensive when compared with the halogens.

Hydrogen Peroxide

A hydrogen peroxide-based powder in 3% and 6% solutions has been found to be effective against biofilms (Felix, 1991). This antibacterial agent may be used on all types of surfaces, equipment, floors and drains, walls, steel mesh gloves, belts, and other areas where contamination exists. This sanitizer has been demonstrated to be effective against *L. monocytogenes* when applied to latex gloves (McCarthy, 1996).

Use of hydrogen peroxide for the sterilization of food packaging material is in compliance if more than 0.1 ppm can be determined in distilled water packaged under production conditions. A hydrogen peroxide solution may be used by itself or in combination with other processes to treat food-contact surfaces prepared from ethylene-acrylic acid copolymers, isomeric resins, ethylene-methyl acrylate copolymer resins, ethylene-vinyl acetate copolymers, olefin polymers, polyethylene terephthalate polymers, and polystyrene and rubber-modified polystyrene polymers.

Ozone

Ozone, a molecule comprised of three oxygen atoms, is naturally occurring in the earth's upper atmosphere. It acts as an oxidant and disinfectant, and may be used as a control tool for microbial and chemical hazards. Common by-products of ozonation are acids, aldehydes, and ketones.

Ozone is being evaluated as a chlorine substitute. Like chlorine dioxide, ozone is unstable and should be generated as needed at the site of application. Because it oxidizes rapidly, it poses less environmental impact.

Ozone, a molecule comprised of three oxygen atoms, is naturally occurring in the earth's upper atmosphere. It acts as an oxidant and disinfectant and may be used as a control tool for microbial and chemical hazards. Common by-products of ozonation are acids, aldehydes, and ketones.

Glutaraldehyde

This sanitizer has been used to control the growth of common gram-negative and gram-positive bacteria, as well as species of yeasts and filamentous fungi found in conveyor lubricants used in the food industry. When added to selected lubricant formulations, glutaraldehyde reduces bacterial levels by 99.99% and fungal levels by 99.9% in 30 minutes.

Microbicides

The microbicide, 2-methyl-5-chloro-2-methyl isothiazolone, has been evaluated for the control of *L. monocytogenes* on product conveyors. This microbicide has been found to be effective against *L. monocytogenes* when it is incorporated in the use dilution of a conveyor lubricant at a continuous dosing rate of 10-ppm active ingredient. This biocide kills microorganisms quickly at a pH higher than 9.0, which is typical of most conveyor lubricants.

Table 8–2 summarizes the important characteristics of the commonly used sanitizers. Table 8–3 matches the recommended sanitizer with the specific area or condition.

Sanitizers and Applications for Pathogen Reduction of Carcasses

The potential presence of *E. coli* O157:H7 on carcasses (especially beef) has necessitated the need for intervention processes to reduce microbial load, including pathogens such as *E. coli* O157:H7. Potential barriers to microbial load on carcasses are chemical and thermal sanitizers.

Chemical sanitizers, such as chlorine and organic acids (acetic, citric, and lactic acids), have been investigated (Table 8–4). These sanitizers can reduce the microbial load but do not destroy all pathogens. Past results have been inconsistent, and some of the experimental design has been questionable. The use of phosphates, such as trisodium phosphate and sodium tripolyphosphate, can reduce the microbial load but do not destroy all pathogens. Overall effectiveness, due to the high pH, is similar to that achieved by organic acids (Fratamico et al., 1996).

Table 8–2 Characteristics of Commonly Used Sanitizers

Characteristics	Steam	Iodophors	Chlorine	Acid	Quats
Germicidal efficiency	Good	Vegetative cells	Good	Good	Somewhat selective
Yeast destruction	Good	Good	Good	Good	Good
Mold destruction	Good	Good	Good	Good	Good
Toxicity					
Use dilution	—	Depends on wetting agent	None	Depends on wetting agent	Moderate
Shelf strength	—	Yes	Yes	Yes	Yes
Stability					
Stock	—	Varies with temperature	Low	Excellent	Excellent
Use	—	Varies with temperature	Varies with temperature	Excellent	Excellent
Speed	Fast	Fast	Fast	Fast	Fast
Penetration	Poor	Good	Poor	Good	Excellent
Film forming	No	None to slight	None	None	Yes
Affected by organic matter	None	Moderate	High	Low	Low
Affected by other water constituents	No	High pH	Low pH and iron	High pH	Yes
Ease of measurement	Poor	Excellent	Excellent	Excellent	Excellent
Ease of use	Poor	Excellent	Excellent	High foam	High foam
Odor	None	Iodine	Chlorine	Some	None
Taste	None	Iodine	Chlorine	None	None
Effect on skin	Burns	None	Some	None	None
Corrosive	No	Not to stainless steel	Extensive on mild steel	Bad on mild steel	None
Cost	High	Moderate	Low	Moderate	Moderate

Source: Adapted from Lentsch, 1979.

Table 8–3 Specific Areas or Conditions Where Particular Sanitizers Are Recommended

Specific Area or Condition	Recommended Sanitizer	Concentration (ppm)
Aluminum equipment	Iodophor	25
Bacteriostatic film	Quat	200
	Acid-quat	Per manufacturer recommendations
	Acid-anionic	100
CIP cleaning	Acid sanitizer	130
	Active chlorine	Per manufacturer recommendations
	Iodophor	Per manufacturer recommendations
Concrete floors	Active chlorine	1,000–2,000
	Quat	500–800
Film formation, prevention of	Acid sanitizer	130
	Iodophor	Per manufacturer recommendations
Fogging, atmosphere	Active chlorine	800–1,000
Hand-dip (production)	Iodophor	25
Hand sanitizer (washroom)	Iodophor	25
	Quat	Per manufacturer recommendations
Hard water	Acid sanitizer	130
	Iodophor	25
High iron water	Iodophor	25
Long shelf life	Iodophor	
	Quat	
Low cost	Hypochlorite	
Noncorrosive	Iodophor	
	Quat	
Odor control	Quat	200
Organic matter, stable in presence of	Quat	200
Plastic crates	Iodophor	25
Porous surface	Active chlorine	200
Processing equipment (aluminum)	Quat	200
	Iodophor	25
Processing equipment (stainless steel)	Acid sanitizer	130
	Acid-quat	Per manufacturer recommendations
	Active chlorine	200
	Iodophor	25
Rubber belts	Iodophor	25
Tile walls	Iodophor	25
Visual control	Iodophor	25
Walls	Active chlorine	200
	Quat	200
	Acid-quat	Per manufacturer recommendations
Water treatment	Active chlorine	20
Wood crates	Active chlorine	1,000
Conveyor lubricant	Glutaraldehyde	Per manufacturer recommendations

Source: Adapted from Lentsch, 1979.

Table 8–4 Chemical Sanitizer Applications

Sanitizer	Application
Chlorine	All food contact surfaces, spray, CIP, fogging
Iodine	All food contact surfaces, approach as a hand dip
Peracetic acid	All food contact surfaces, CIP, especially cold temperature and carbon dioxide environments
Acid anionics	All food contact surfaces, spray, combines sanitizing and acid rinse into one operation
Quaternary ammonium compounds	All food contact surfaces, mostly used for environmental control; walls, drains, tiles

Carcass rinse methods lack effectiveness in killing microorganisms because of the ineffective penetration of water to all of the contaminated surfaces. Hair, feathers, and scale follicles are large enough to hide bacteria but too small to admit a liquid wash or spray. An unrealistic high-water pressure is needed to overcome the capillary pressure in a pore large enough to house a bacterium.

At this writing, it appears that thermal sanitizing may be more effective than chemical sanitizing in the destruction of pathogens on carcasses. Hot water at or above 82°C is effective, as is steam pasteurization.

Steam pasteurization involves passing carcasses through a tunnel that is approximately 12 m long, where large quantities of steam are applied to the carcass surface. A large percentage of bacteria on the carcass surface is destroyed, and the risk of enteric pathogens such as *E. coli* O157:H7 and *Salmonella* is reduced. This process involves immersion of the carcass in pressurized steam to envelop the entire surface area for 6 to 8 seconds to raise the temperature to approximately 82°C. Then the carcass is sprayed with chilled water to reduce the surface temperature to 20°C before storage in a chilled environment.

The steam-vacuum method was originally designed to take advantage of both hot water and steam, in combination with a physical removal of bacteria and contamination via vacuum. More recently, steam-only equipment has been designed and is used in beef processing plants for spot removal. The steam-vacuum method of pathogen reduction has resulted in a larger variation of reduction levels than other moist-heat interventions tested (Dorsa, 1997). This variation is attributable to repeated passes of the nozzle over the sampled surface of contaminated beef having possibly embedded bacteria, making them more difficult to remove by steam-vacuumizing. Plant studies have demonstrated that a commercial steam-vacuum system can consistently outperform knife trimming for the removal of bacterial contamination of beef carcasses.

The steam-vacuum system has been reported to achieve a 5-log cycle (100,000-fold) reduction of *E. coli* O157:H7 on inoculated beef surfaces (Dorsa et al., 1996). Use of low temperature steam retards the premature warming of meat and poultry surfaces. Air removal prior to treatment with steam increases effectiveness as air would otherwise retard the rate at which steam heats carcass surfaces.

The effects of *cetylpyridinium chloride* (CPC) on the inhibition and reduction of *Sal-*

monella have been demonstrated successfully as a pathogen intervention technique for poultry carcasses. This compound has been used safely for over 30 years as an oral hygiene product. Also, CPC is effective in preventing bacterial attachment and has potential in the reduction of cross-contamination. Treatment with CPC does not affect the physical appearance of poultry products. *Electrical stimulation* is another potential means of microbial load reduction on the surface of carcasses.

The fate of *E. coli* O157:H7 cells that have been removed from carcasses by rinses with sanitizing agents is not understood fully. It appears that the exposure times associated with carcass sanitizing are too short to achieve any significant direct inactivation. The primary effect of carcass rinses may be the physical removal of microorganisms (Buchanan and Doyle, 1997).

Tests for Sanitizer Strength

Sanitizers are effective only if the proper concentration is used. A number of tests has been devised to determine the concentration of the sanitizer being tested. The tests we examine here are recommended by the FDA.

Chlorine Sanitizers

The following methods can be used to determine chlorine concentration in the sanitizer being tested:

1. *Starch iodide method (iodometric).* This is a titration test in which chlorine displaces iodine from potassium iodide in an acid solution and forms a blue color with starch. Decolorization occurs by the addition of standard sodium thiosulfate. This test is generally used to measure high residuals.

2. *O-tolidine colorimetric comparison.* This is a test in which a colorless solution of o-tolidine is added to a chlorine solution. An orange-brown-colored compound proportional to its concentration is produced and is compared with a standardized color.

3. *Indicator paper test.* This is rapid test of limited accuracy in which test papers, usually impregnated with starch iodide, are immersed. The developed color is compared with a standard.

Iodophors

Although iodophors have a built-in color indicator that is relatively accurate, color comparative kits and other kits are available for testing.

Quaternary Compounds

There are several satisfactory tests for determining concentration of these compounds. Some reagents are available in tablets, and others use test papers by which a color comparison is made.

SUMMARY

Sanitizers are applied to reduce the pathogenic and spoilage microorganisms of food facilities and equipment. Soils must be completely removed for sanitizers to function properly.

The major types of sanitizers are thermal, radiation, and chemical. Thermal and radiation techniques are less practical for food production facilities than is chemical sanitizing. Of the chemical sanitizers, the chlorine compounds tend to be the most effective and the least expensive, although they tend to be

more irritating and corrosive than are the iodine compounds or the quaternary ammonium compounds. Bromine compounds are
more beneficial for wastewater treatment
than for sanitizing cleaned surfaces, although
bromine and chlorine are synergistic when
combined. The quats are more restrictive in
their activities but are effective against mold
growth and have residual properties. They do
not kill bacterial spores but can limit their
growth. Acid-quat and chlorine dioxide
sanitizers may be the "wave of the future" for
the control of *L. monocytogenes*, and ozone is
being evaluated as a chlorine substitute. Glutaraldehyde can be incorporated as a sanitizer
for conveyor lubricants used for food operations. Various tests are available to determine
the concentration of sanitizing solutions.

STUDY QUESTIONS

1. What are the advantages and disadvantages of hot water as a sanitizer?
2. What factors contribute to the effectiveness of a sanitizer?
3. How is chlorine dioxide produced for
   use in a food facility?
4. What are the advantages and disadvantages of chlorine as a sanitizer?
5. What are the advantages and disadvantages of iodine as a sanitizer?
6. What are the advantages and disadvantages of the "quats" as a sanitizer?
7. What are the advantages and disadvantages of acid sanitizers?
8. What sanitizers are frequently added to
   lubricants?

REFERENCES

Anon. 1996. *Sanitizers for meat plants.* Wyandotte, MI:
Diversey Corp.

Anon. 1997. Guide to sanitizers. *Prepared Foods* March, 81.

Benarde, M.A., Israel, B.M., Olivieri, V.P., and Granstrom,
M.L. 1965. Efficiency of chlorine dioxide as a bactericide.
Appl Microbiol 13:776.

Buchanan, R.L., and Doyle, M.P. 1997. Foodborne disease
significance of *Escherichia coli* O157:H7 and other
entero-hemorrhagic *E. coli. Food Technol* 51, no. 10:69.

Camper, A.K., and McFetters, G.A. 1979. Chlorine injury and
the enumeration of water-borne coliform bacteria. *Appl
Environ Microbiol* 37:633.

Carsberg, H. 1996. Selecting your sanitizers. *Food Quality* 2,
no. 11:35.

Dorsa, W.J., Cutter, C.N., and Siragusa, G.R. 1996. Effectiveness of a steam-vacuum sanitizer for reducing *Escherichia
coli* O157:H7 inoculated to beef carcass surface tissue.
Lett Appl Microbiol 23:61.

Dorsa, W.J. 1997. New and established carcass decontamination procedures commonly used in the beef processing industry. *J Food Prot* 60:1146.

Felix, C.W. 1991. Sanitizers fail to kill bacteria in biofilms.
Food Prot. Rep., no. 5:6.

Flickinger, B. 1997. Automated cleaning equipment. *Food
Quality* III 23:30.

Frank, J.F., and Chmielewski, R.A.N. 1997. Effectiveness of
sanitation with quaternary ammonium compound vs.
chlorine on stainless steel and other food preparation surfaces. *J Food Prot* 60:43.

Fratamico, P.M., Schultz, F.J., Benedict, R.C., Buchanan, R.L,
and Cooke, P.H. 1996. Factors influencing attachment of
Eschericia coli O157:H7 to beef tissues and removal using selected sanitizing rinses. *J Food Prot* 59:453.

Haverland, H. 1981. Cleaning and sanitizing operations. *Dairy
Food Environ Sanit* 1:331.

Kulikoosky, A., Pankratz, H.S., and Sandoff, H.L. 1975. Ultra-
structural and chemical changes in spores of *Bacillus
cereus* after action of disinfectants. *J Appl Bacteriol* 38:39.

Le Chevallier, M.W., Cawthon, C.D., and Lee, R.G. 1988.
Factors promoting survival of bacteria in chlorinated water supplies. *Appl Environ Micribiol* 54:649.

Lentsch, S. 1979. Sanitizers for an effective cleaning program.
In *Sanitation Notebook for the Seafood Industry,* ed. G.J.
Flick et al., 77. Blacksburg: Department of Food Science
and Technology, Virginia Polytechnic Institute and State
University.

McCarthy, S.A. 1996. Effect of sanitizers on *Listeria mono-cytogenes* attached to latex gloves. *J Food Safety* 16:231.

Meinhold, N.M. 1991. Chlorine dioxide foam effective against Listeria—less contact time needed. *Food Proc* 52, no. 2:86.

Odlaug, T.E. 1981. Antimicrobial activity of halogens. *J Food Prot* 44:608.

Park, D.L., Rua, S.M., Jr., and Acker, R.F. 1991. Direct application of a new chlorite sanitizer for reducing bacterial contamination on foods. *J Food Prot* 54:960.

SUGGESTED READINGS

Guthrie, R.K. 1988. *Food sanitation.* 3rd ed. New York: Van Nostrand Reinhold.

Giese, J.H. 1991. Sanitation: The key to food safety and public health. *Food Technol* 45, no. 12:74.

Haverland, H. 1980. Cleaning and sanitizing operations. In *Current concepts in food protection,* 57. Cincinnati: U.S.

Department of Health, Education and Welfare, Public Health Service, Food and Drug Administration.

Jowitt, R. 1980. *Hygienic design and operation of food plant.* Westport, CT: AVI Publishing Co., Inc.

Marriott, N.G. 1990. *Meat sanitation guide II.* Blacksburg: American Association of Meat Processors and Virginia Polytechnic Institute and State University.

CHAPTER 9

Sanitation Equipment

In Chapters 7 and 8, we examined characteristics of cleaning compounds and sanitizers and suggestions for potential applications. This chapter provides information on cleaning and sanitizing equipment and discusses a systems approach to cleaning and sanitizing. A variety of cleaning equipment, cleaning compounds, and sanitizers is available, making selection of the optimal cleaning technique confusing. There are no cleaning compounds, sanitizers, or cleaning units available that are truly all-purpose, because such products would need to possess too many chemical and physical requirements. Therefore, we will examine various types of cleaning equipment and explore their uses.

Mechanical cleaning and sanitizing equipment merits serious consideration because it can reduce cleaning time and improve efficiency. An efficient system can reduce labor costs by up to 50% and should have a pay-out period of fewer than 30 months. In addition to labor savings and increased efficiency, a mechanized cleaning unit can more effectively remove soil from surfaces than can the hand method.

Management frequently fails to recognize that there is a technology of cleaning that should be applied for effective performance. The well-managed firm should not make large expenditures for effective cleaning and sanitizing equipment without hiring skilled employees to operate the equipment and qualified management to supervise the operation. Although many technical representatives of chemical companies that manufacture cleaning compounds and sanitizers are qualified to recommend cleaning equipment for various applications, people who manage the sanitation program should not rely on the recommendations of an enthusiastic sales representative who may not have adequate technical expertise. It is important to approach cleaning and sanitizing problems on a technological basis. The observation of a plant during cleanup to evaluate the operation of cleaning equipment can be used to determine whether the operation is satisfactory.

SANITATION COSTS

A typical cleaning operation has the following breakdown of costs:

Cost	%
Labor	46.5
Water/sewage	19.0
Energy	8.0
Cleaning compounds and sanitizers	6.0
Corrosion damage	1.5
Miscellaneous	19.0

The largest cost of cleaning is *labor*. Approximately 46.5% of the sanitation dollar is spent for cleaning, sanitizing, and quality assurance personnel and supervision. This expense, however, can be reduced more than other costs, through the use of mechanized cleaning systems.

Water and sewage have the next highest costs. Food plants use large quantities of water for the application of cleaning compounds. In addition, this category encompasses sewage discharge costs and surcharges. Energy requirements and sewage treatment costs are major because sewage from food plants can be high in biochemical oxygen demand (BOD) and chemical oxygen demand (COD).

The cost and availability of *energy* for generating hot water and steam are important factors. Most cleaning systems, cleaning compounds, and sanitizers are effective when the water temperature is below 55°C. A lower temperature will conserve energy, reduce protein denaturation on surfaces to be cleaned (thus increasing ease of soil removal), and decrease injury to employees.

Although *cleaning compounds* and *sanitizers* are expensive, this cost is reasonable if one considers that sanitizers destroy residual microorganisms and that these compounds contribute to more thorough cleaning with less labor. The optimal cleaning system combines the most effective cleaning compounds, sanitizers, and equipment to perform the cleaning tasks economically and effectively. Chemical costs can be reduced by using the correct amounts of cleaning solution to perform tasks.

Improper use of cleaning compounds and sanitizers on processing equipment constructed of stainless steel, galvanized metal, and aluminum costs the industry millions of dollars through *corrosion* damage. This cost can be reduced through use of appropriate construction materials and the proper cleaning system, including noncorroding cleaning compounds and sanitizers.

An accumulation of *miscellaneous sanitation costs* includes the cost of water and sewage treatment. Miscellaneous costs encompass equipment depreciation, returned goods, general and administrative expenses, and other operating costs. The general nature of these costs makes it more difficult to identify a specific approach for their reduction. The most effective course is careful management.

EQUIPMENT SELECTION

At least three sources are available to the industry to provide information related to the optimal sanitation system: a planning division (or similar group) of the food company, a consulting organization (internal or external), and/or a supplier of cleaning and sanitizing compounds and equipment. Regardless of which source is used, a basic plan should be followed to guide the selection and installation of equipment.

Sanitation Study

A sanitation study should start with a plant survey. A study team or individual specialist should identify cleaning procedures in use (or procedures recommended for a new operation), labor requirements, chemical requirements, and utility costs. This information is needed to determine recommended cleaning procedures, cleaning and sanitizing supplies, and cleaning equipment. The survey data should reflect required expenses and projected annual savings from the proposed sanitation system. A report of this study should be distributed to key management personnel.

Sanitation Equipment Implementation

After the appropriate equipment has been recommended and acquired, the vendor or a designated expert should supervise the installation and startup of the new operation. Personnel training should be provided by the vendor or by the organization responsible for manufacture of the system. After startup, regular inspections and reviews should be conducted jointly with the organization performing the sanitation study and a management team designated by the food company. In addition to daily inspections, reviews should be conducted every 6 months. Both inspections and reviews should be documented so that records are available.

Reports should contain information related to the effectiveness of the program, periodic inventory data, and cleaning equipment condition. Information related to labor, cleaning compounds, sanitizers, and maintenance costs provided by reports should be compared with costs projected in the sanitation study. This approach provides a way to pinpoint trouble spots and to verify that actual costs approximate projected costs. This technique will contribute to savings of up to 50% when compared with an unmonitored system.

The HACCP Approach to Cleaning

The Hazard Analysis Critical Control Point (HACCP) approach should be applied when considering the evaluation of a cleaning system. A sanitation survey will permit application of the HACCP concept. This survey should designate areas that require cleaning as highly critical, critical, or subcritical for physical and microbial contamination. These areas can be grouped according to required cleaning frequency as demanding attention:

1. Continuously
2. Every 2 hours (during each break period)
3. Every 4 hours (during lunch break and at the end of the shift)
4. Every 8 hours (end of shift)
5. Daily
6. Weekly

For verification purposes, the microbial methods should be appropriate for the task. Sampling should be accomplished where the information will most accurately reflect the cleaning effectiveness. Examples are:

Flow sheet sampling is the measurement of microbial load on food samples collected after each step in the preparation sequence. When samples are collected from the first food coming into contact with cleaned equipment, the contribution of microorganisms from each piece of equipment that the food contacts can be measured.

Environmental samples taken from the food processing environment are important in the control of pathogens such as species of *Salmonella* and *Listeria*. Examples are air intakes, ceilings, walls, floors, drains, air, water, and equipment.

CLEANING EQUIPMENT

Cleaning is generally accomplished by manual labor with basic supplies and equipment or by the use of mechanized equipment that applies the cleaning medium (usually water), cleaning compound, and sanitizer. The cleaning crew should be provided with the tools and equipment needed to accomplish the cleanup with minimal effort and time. Storage space should be provided for chemicals, tools, and portable equipment.

Mechanical Abrasives

Although abrasives such as steel wool and copper chore balls can effectively remove soil when manual labor is used, these cleaning aids should not be used on any surface that has direct contact with food. Small pieces of these scouring pads may become embedded in the construction material of the equipment and cause pit corrosion (especially on stainless steel) or may be picked up by the food, resulting in consumer complaints and even consumer damage suits. Wiping cloths should not be used as a substitute for abrasives or for general purposes because they spread molds and bacteria. If cloths are necessary, they should be boiled and sanitized before use.

Water Hoses

Hoses should be long enough to reach all areas to be cleaned, but should be no longer than required. For rapid and effective cleanup, it is important to have hoses equipped with nozzles designed to produce a spray that will cover the areas being cleaned. Nozzles with rapid-type connectors should be provided for each hose. Fan-type nozzles give better coverage for large surfaces in a minimum amount of time. Debris lodged in deep cracks or crevices is best dislodged by small, straight jets. Bent-type nozzles are beneficial for cleaning around and under equipment. For a combination of washing and brushing, a spray-head brush is needed. Cleanup hoses, unless connected to steam lines, should have an automatic shutoff valve on the operator's end to conserve water, reduce splashing, and facilitate exchange of nozzles. Hoses should be removed from food production areas after cleanup, and it is necessary to clean, sanitize, and store them on hooks off of the floor. This precaution is especially important in the control of *Listeria monocytogenes*.

Brushes

Brushes used for manual or mechanical cleaning should fit the contour of the surface being cleaned. Those equipped with spray heads between the bristles are satisfactory for cleaning screens and other surfaces in small operations where a combination of water spray and brushing is necessary. Bristles should be as harsh as possible without creating surface damage. Rotary hydraulic and power-driven brushes for cleaning pipes aid in cleaning lines that transport liquids and heat exchanger tubes.

Brushes are manufactured from a variety of materials—horsehair, hog bristles, fiber, and nylon—but are usually nylon. Bassine, a coarse-textured fiber, is suitable for heavy-duty scrubbing. Palmetto fiber brushes are less coarse and are effective for scrubbing with medium soil, such as metal equipment and walls. Tampico brushes are fine fibered and well adapted for cleaning light soil that requires only gentle brushing pressure. All nylon brushes have strong, flexible fibers that are uniform in diameter, durable, and do not absorb water. Most power-driven brushes are equipped with nylon bristles. Brushes made of absorbent materials should not be used.

Scrapers, Sponges, and Squeegees

Sometimes scrapers are needed to remove tenacious deposits, especially in small operations. Sponges and squeegees are most effectively used for cleaning product storage tanks when the operation has insufficient volume to justify mechanized cleaning.

High-Pressure Water Pumps

High-pressure water pumps may be portable or stationary, depending on the volume and needs of the individual plant. Portable units are usually smaller than centralized installations. The capacity of portable units is from 40 to 75 L/min, with operating pressures of up to 41.5 kg/cm^2. Portable units may include solution tanks for mixing of cleaning compounds and sanitizers. Stationary units have capacities ranging from 55 to 475 L/min. Piston-type pumps deliver up to 300 L/min, and multistage turbines have capacities of up to 475 L/min, with operating pressures of up to 61.5 kg/cm². The capacity and pressure of these units vary from one manufacturer to another.

In a centralized unit, the high-pressure water is piped throughout the plant, and outlets are placed for convenient access to areas to be cleaned. The pipes, fittings, and hoses must be capable of withstanding the water pressure, and all of the equipment should be made of corrosion-resistant materials. The choice of a stationary or portable unit depends on the desired volume of high-pressure water and the ease with which a portable unit can be moved close to areas being cleaned. Other uses of high-pressure water in the plant can also determine whether a stationary unit is warranted.

High-pressure, high-volume water pumps have been used primarily when supplementary hot, high-pressure water is desired. Because this equipment uses a large volume of water and cleaning compounds, it is frequently considered inefficient. This concept has been applied to portable and centralized high-pressure, low-volume equipment that blends cleaning compounds for dispensing in areas to be cleaned. With a lower volume and lower water temperature, it becomes a more efficient approach that can effectively clean areas that are difficult to reach and penetrate.

Low-Pressure, High-Temperature Spray Units

This equipment may be portable or stationary. The portable units generally consist of a lightweight hose, adjustable nozzles, steam-heated detergent tank, and pump. Operating pressures are generally less than 35 kg/cm². Stationary units may operate at the main hot-water supply pressure or may use a pump. These units are used because no free steam or environment fogging is present, splashing during the cleaning operation is minimal, soaking operations are impractical and hand brushing is difficult and time-consuming, and the detergent stream is easily directed onto the soiled surface.

High-Pressure Hot-Water Units

This equipment utilizes steam at 3.5 to 8.5 kg/cm² and unheated water at any pressure above 1 kg/cm². These units convert the high-velocity energy of steam into pressure in the delivery line. The cleaning compound is simultaneously drawn from the tank and mixed in desired proportions with hot water. Pressure at the nozzle is a function of the steam pressure in the line; for example, at 40 kg of steam pressure, the jet pressure is approximately 14 kg/cm². This equipment is easy to operate and maintain but has the same inefficiency as the high-pressure, high-volume water pumps.

Steam Guns

Various brands of steam guns are available that mix steam with water and/or cleaning compounds by aspiration. The most satisfactory units are those that use sufficient water and are properly adjusted to prevent a steam fog around the nozzle. Although this equipment has applications, it is a high-energy-consuming method of cleaning. It also reduces safety through fog formation and increases moisture condensation, sometimes resulting in mold growth on walls and ceilings, and increased potential for the growth of *L. monocytogenes*. High-pressure, low-volume equipment is generally as effective as steam guns if appropriate cleaning compounds are used.

High-Pressure Steam

High-pressure steam may be used to remove certain debris and to blow water off processing equipment after it has been cleaned. Generally, this is not an effective method of cleaning because of fogging and condensation, and it does not sanitize the cleaned area. Nozzles for high-pressure steam and other high-pressure, high-volume equipment should be quickly interchangeable and have a maximum capacity below that of the pump. An orifice of approximately 3.5 mm is considered satisfactory for an operating pressure of approximately 28 kg/cm^2.

Hot-Water Wash

This technique should be considered a method instead of a kind of equipment or a cleaning system. Because only a hose, nozzle, and hot water are required, this method of cleaning is frequently used. Sugars, certain other carbohydrates, and monovalent compounds are relatively soluble in water and can be cleaned more effectively with water than can fats and proteins. Investment and maintenance costs are low, but the hot-water wash is not considered a satisfactory cleaning method. Although hot water can loosen and melt fat deposits, proteins are denatured; removal from the surface to be cleaned is complicated because these coagulated deposits are more tightly bound to the surface. Without high pressure, penetration of areas of poor accessibility is difficult, and labor is increased if a cleaning compound is not applied. As with the other equipment that uses hot water, this method increases both energy costs and condensation.

Portable High-Pressure, Low-Volume Cleaning Equipment

A portable high-pressure, low-volume unit contains an air- or motor-driven high-pressure pump, a storage container for the cleaning compound, and a high-pressure delivery line and nozzle (Figure 9–1). The self-contained pump provides the required pressure to the delivery line, and the nozzle regulates pressure and volume. This portable unit simultaneously meters the predetermined amount of cleaning compound from the storage container and mixes it in the desired proportion of water as the pump delivers the desired pressure. The ideal high-pressure, low-volume unit delivers the cleaning solution at approximately 55°C with 20 to 85 kg/cm^2 pressure and 8 to 12 L/min, depending on equipment specifications and nozzle design.

The high-pressure cleaning principle is based on automation of the cleaning compound through a high-pressure spray nozzle. The high-pressure spray provides the cleaning medium for application of the cleaning compound. The velocity, or force, of the cleaning solution against the surface is the major factor that contributes to cleaning effectiveness. High-pressure, low-volume equipment is necessary to reduce water and cleaning compound consumption. This equipment conserves water and cleaning compounds, and it is less hazardous than high-pressure, high-volume equipment because the low volume results in reduced force as distance from the nozzle increases.

Portable high-pressure, low-volume equipment is relatively inexpensive and quickly connected to existing utilities. Some suppliers of cleaning compounds provide these units at little or no rent to customers who agree to purchase their products exclusively. These units do require more labor than does

Figure 9–1 A portable high-pressure, low-volume cleaning unit that can be used where a centralized system does not exist. This unit is equipped with racks for hoses, foamer, and cleaning compound storage, and provides two rinse stations and a sanitizer unit. Two workers can simultaneously prerinse, clean, postrinse, and sanitize. This equipment can also apply foam if the spray wand is replaced with a foam wand accessory. Courtesy of Ecolab, Inc., Mendota Heights, Minnesota.

centralized equipment because transportation throughout the cleaning operation is necessary and because less automation can be provided without a centralized system. Portable equipment is not as durable and can require an excessive amount of maintenance. High-temperature sprays tend to bake the soil to the surface being cleaned, providing the optimum temperature for microbial growth.

This hydraulic cleaning equipment is beneficial for small plants because the portable units can be moved through the facility. Portable equipment can be utilized for cleaning parts of equipment and building surfaces, and is especially effective for conveyors and processing equipment where soaking operations are impractical and hand brushing is difficult and time-consuming. It appears that this method of cleaning may receive more attention in the future because it may be more effective in the removal of *L. monocytogenes* from areas that are difficult to clean with less labor-intensive equipment such as foam-dispensing units. A trend exists toward centrally installed equipment because of the potential labor savings and reduced maintenance.

Centralized High-Pressure, Low-Volume Systems

This system, which uses the same principles as does the portable high-pressure, low-volume equipment, is another example of mechanical energy being harnessed and used as chemical energy. Centralized systems utilize piston-type or multistage turbine pumps to generate desired pressure and volume. Like the portable equipment, the cleaning action of high-pressure spray units is primarily due to the impact energy of water on the soil and surface. The pump(s), hoses, valves, and nozzle parts of the ideal centralized high-pressure cleaning system should be resistant to attack by acid or alkaline cleaning products. Automatic, slow-acting shutoff valves should be provided to prevent hose jumping, indiscriminate spraying, and wasting of water. The centralized system is more flexible, efficient, safe, and convenient because there is no live steam to block vision or injure personnel.

If improperly used, this cleaning system can be counterproductive by blasting loose dirt in all directions. Therefore, a low-pressure rinse-down should precede high-pressure cleaning. Most suppliers of these systems provide customers with technical assistance and match cleaning product and cleaning equipment to obtain maximum value.

The penetrating and cleaning action of a centralized high-pressure, low-volume system is similar to that of a commercial dishwashing machine. The system automatically injects either a detergent disinfectant or a solvent solution into a water line so that the hydraulic scouring action of the spray cleans exposed surfaces and reaches into inaccessible or difficult-to-reach areas. Cracks and crevices where soil has accumulated can be flushed out to reduce bacterial contamination. Cutting and scouring action is applied to all surfaces by the jet, and chemical cleaning action is improved by the water spray, which is automatically charged with a detergent or detergent-disinfectant solution. An example of equipment components of a high-pressure cleaning system is given in Figure 9–2.

The flexibility and major benefits of the centralized high-pressure cleaning system are realized if there are quick-connection outlets available in all areas requiring cleaning. Several detergents—acid, alkaline, or neutral cleaners and sanitizers—can be dispersed through the system, and mechanized spray heads can be mounted on belt conveyors with automatic washing, rinsing, and cutoff.

Centralized systems are far more expensive than portable units because they are generally custom built. The cost varies according to facility size and system flexibility. Initial investment may range from $15,000 to over $200,000.

Factors Determining Selection of Centralized High-Pressure Equipment

Generally, two types of central equipment are most common: medium pressure (10 kg/cm^2 boost to 20 kg/cm^2) and high pressure (40 kg/cm^2 boost to 55 kg/cm^2). Medium-pressure systems are normally used in processing plants where heavy soils dominate. High-pressure equipment is found mostly in beverage and snack food plants where soils are light and cutting action is needed to clean processing equipment. However, several factors must be considered to determine which equipment will provide the best long-term results for each specific plant.

There is normally an inverse relationship between the rinse nozzle flow rates and pressure. Each cleaning task requires a specific impingement force to dislodge the soil and flush it off the equipment. At high pressures of 40 to 55 kg/cm^2, the nozzle flow rates can average approximately 5 L/min; however, lower pressure requires high flow rates at the nozzles to achieve the same impingement. For example, if a plant has a medium-pressure 20 kg/cm^2 system with 30 to 40 L/min rinse nozzles and management wants to conserve water usage, the method to

Figure 9–2 A centralized high-pressure, low-volume system for a large cleaning operation. Courtesy of Diversey Lever, 255 E. 5th Street, Suite 1200, Cincinnati, Ohio, 800-233-1000.

accomplish this is to increase the pressure to 40 or 50 kg/cm^2 and reduce the nozzle flow rates to 10 or 15 L/min. The result is the same impingement force with a 50% reduction in rinse-water usage.

Water conservation, in addition to being responsible management, carries additional benefits that are not always obvious. Reduced nozzle flow means less sewage and less energy used to heat the water. Paybacks of less than 6 months are not unusual and often run as low as 3 months.

During the past, a trend has existed to use lower pressure in plants with a heavy soil because high pressure tends to dislodge particles with such force as to move them to another undesired location (splatter). Heavier soils require heavier impingement. Most processors with less heavy soil use medium pressure.

For long-term water conservation purposes, 40 to 50 kg/cm^2 with an average rinse hose nozzle flow rate of 10 to 20 L/min is suggested. The usual exception is if the plant has an unusually short time period with which to prepare for production. If only 4 or 5 hours are available for cleanup, higher flow rates will be required. This condition is usually temporary but must be planned for, i.e., the central system must have the flow capacity.

The price of central equipment is usually the main determinant in the purchasing decision. High-pressure equipment requires the largest investment to buy and maintain. The pumps are more expensive than medium-pressure pumps, and all of the piping, valves, and other components cost more because of the high-pressure ratings required. Usually, the benefits of low water usage outweigh the initial cost and operating expenses over the long term.

Medium-pressure equipment requires less investment to purchase and operate. The pumps are mechanically less sophisticated, and none of the piping, valves, or related components requires a high-pressure rating; therefore, maintenance is usually lower than on high-pressure systems.

If water usage is not a critical factor in overall plant operations, a medium-pressure system should be considered. Water conservation cannot be accomplished with medium pressure. Many processors utilize 20 kg/cm^2 with 20 to 30 L/min nozzles in most areas of the plant. Proper utilization of the equipment and training in sanitation procedures are the key elements.

Portable Foam Cleaning

Because of the ease and speed of foam application, this cleaning technique has been popular during the past two decades. With this method, foam is the medium for application of the cleaning compound. The cleaning compound is mixed with water and air to form the foam. Clinging foam is readily visible and allows the worker to see where the cleaning compound has been applied, thus reducing the chance of job duplication.

Foam cleaning is particularly beneficial for cleaning large surface areas because of its ability to cling, increasing the contact time of the cleaning compound. This technique is frequently used to clean the interior and exterior of transportation equipment, ceilings, walls, piping, belts, and storage containers. Equipment is similar in size and cost to portable high-pressure units. Portable cleaning units that may be used to apply cleaning compounds by foam are illustrated in Figures 9–3 and 9–4. This equipment requires a foam-charging operation to blend the cleaning compound, water, and air prior to use.

Centralized Foam Cleaning

This equipment applies cleaning compounds with the same technique used by portable foam equipment except that drop stations for quick connection to a foam gun are strategically located throughout the plant. Centralized equipment provides desirable features similar to the centralized high-pressure system. As with portable foam cleaning, the cleaning compound is automatically mixed with water and air to form a foam. This equipment does not require the foam-charging operation that portable foam units require. Equipment components of a centralized foam cleaning system are illustrated in Figures 9–5 and 9–6.

The compact wall-mounted foam generation unit shown in Figure 9–5 is designed to blend and dispense cleaning compounds from reservoirs or original shipping containers. Wall-mounted units can be located in convenient areas where cleaning is concentrated. The equipment shown in Figure 9–5 can blend and dispense cleaning compounds through an adjustable air regulator and water-metering valve. The easily accessible chemical-metering pump and other controls are in the latching stainless steel cabinet. This equipment contains a built-in vacuum breaker and check valves in the air and water lines.

The equipment in Figures 9–6 a, b, and c features drop stations that can be used to dispense foam, high-pressure water for rinsing, and a sanitizer. The foam station provides adjustable air and detergent regulators to create proper foam consistency. The rinse unit can provide up to 69 kg/cm^2 of pressure.

Portable Gel Cleaning

This system is similar to portable high-pressure units except that the cleaning com-

Figure 9–3 A portable air-operated foam unit that applies the cleaning compound as a blanket of foam. Courtesy of Ecolab, Inc., Mendota Heights, Minnesota.

pound is applied as a gel (due to air restriction) rather than as foam slurry or high-pressure spray. This medium is especially effective in cleaning food-packaging equipment because the gel clings to the moving parts for subsequent soil removal. Equipment costs and arrangement are comparable to those for portable foam and high-pressure units.

Centralized or Portable Slurry Cleaning

This method is identical to foam cleaning except that less air is mixed with the cleaning compounds. A slurry is formed that is more fluid than foam and penetrates uneven surfaces more effectively. Exposure time of cleaning compounds applied as a slurry is

less than with foam, as the foam has superior clinging ability.

Combination Centralized High-Pressure and Foam Cleaning

This arrangement is the same as centralized high-pressure cleaning, except that foam can also be applied. This method offers more flexibility than most cleaning equipment because foam can be used on large surface areas with high pressure applied to belts, stainless steel conveyors, and hard-to-reach areas. A system with these capabilities is expensive—anywhere from $15,000 to over $200,000, depending on size—because most must be custom designed and built.

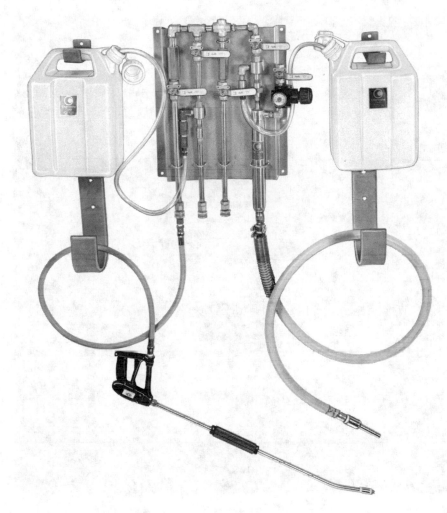

Figure 9–4 Combination high-pressure rinse/foam/flood sanitize medium-pressure rinse station. Courtesy of Ecolab, Inc., Mendota Heights, Minnesota.

Cleaning-in-Place

As labor rates continue to increase and hygienic standards are raised, cleaning-in-place (CIP) systems become more valuable. Dairies and breweries have used CIP for many years. It has been adapted sparingly in other plants because of equipment and installation costs and the difficulty of cleaning certain processing equipment. Because of these limitations, CIP is considered a solution for specific cleaning applications and is custom designed. CIP equipment is best used for cleaning pipelines, vats, heat exchangers, centrifugal machines, and homogenizers.

Custom-designed CIP equipment can vary in the amount of automation, according to cleaning requirements—from simple cam timers to fully automated computer-controlled systems. The choice depends on capital availability, labor costs, and type of soil. It should be designed by a reliable consulting firm and/or a reputable equipment and detergent supplier. These organizations can pro-

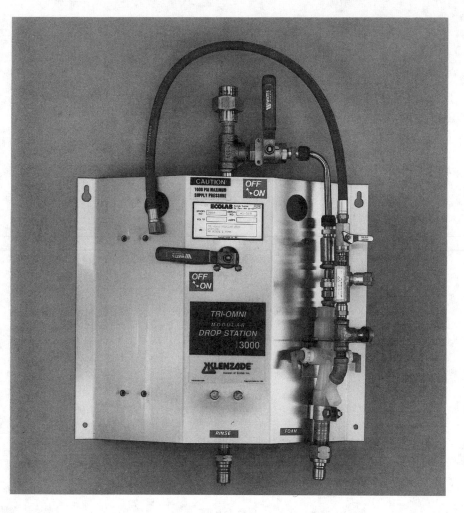

Figure 9–5 A wall-mounted foam and rinse station that can provide foam application at convenient locations in a food plant through automatic metering and mixing of the cleaning compound. Courtesy of Ecolab, Inc., Mendota Heights, Minnesota.

vide site surveys and confidential reports on the hygienic status of existing equipment and cleaning techniques.

Small-volume plants cannot always justify full automation. With reduced automation, the required circuits can be set manually by means of a flow-selector plate. Pipelines can be brought to a back plate with required connections made by a U-bend inserted in the appropriate parts. Microswitch logic can be interlocked to the CIP set. With full automa-

tion, the entire process and CIP operations can be automatically controlled. Electric interlocks negate the possibility of an error in valve operation.

The CIP principle is to combine the benefits of the chemical activity of the cleaning compounds and the mechanical effects of soil removal. The cleaning solution is dispensed to contact the soiled surface, and the proper time, temperature, detergency, and force are applied. For this system to be effective, a relatively high volume of solution has to be applied to soiled surfaces for at least 5 minutes and up to 1 hour. Therefore, recirculation of the cleaning solution is necessary for repeated exposure and to conserve water, energy, and cleaning compounds.

For optimal use of water and reduced effluent discharge, CIP systems are being designed to permit the final rinse to be utilized as makeup water for the next cleaning cycle. The dairy industry has attempted to recover a spent cleaning solution for further use by concentration through ultrafiltration or through use of an evaporator. Various installations have incorporated systems that integrate the advantages of single-use systems of known reliability and flexibility with water and solution recovery procedures that aid in reducing the total amount of water required for a specific cleaning operation. These installations combine the spent cleaning solution and past rinsings for temporary storage and use as a prerinse for the next cleaning cycle. Thus, the requirements of water, cleaning compounds, and required energy are reduced.

Properly designed CIP systems are capable of cleaning certain equipment in food plants as effectively as dismantling and cleaning by hand. In many food plants, CIP equipment has completely or partially replaced hand cleaning.

The simplified flow chart in Figure 9–7 illustrates how a CIP system operates. The arrangement illustrates how to provide mixing and detergent tank(s), pipelines, heat exchanger(s), and storage tank(s). These features permit cleaning of storage tanks, vats, and other storage containers by use of spray balls. Pipelines and various plant items can be cleaned by a high-velocity cleaning solution of water and designated cleaning compounds, which are recirculated. A typical cleaning cycle for the CIP system is outline in Table 9–1.

The plant layout for a CIP system is important because dismantling of equipment is un-necessary. Jowitt (1980) recommends the use of crevice-free joints of pipe work and the design of all tanks so that they can be enclosed with smooth walls that can be cleaned by liquid spray. Spray balls, whether fixed or rotating, should produce a high-velocity jet of liquid in a 360° pattern to cover the interior of the tank and thoroughly remove residual soil or other contamination. This cleaning concept is illustrated in Figure 9–8.

Development of circuits is important. The circuits must be flexible. The location of every pipe should be permanent and based on its possible function during cleaning. Large processing operations may be separated into several major circuits for separate cleaning. Circuit design should be based on soil characteristics. Circuit development can permit a limited cleanup force to proceed through the plant in an orderly sequence as process operations are completed.

Use of a drain selector valve facilitates the direction of flush water, cleaning compounds, and rinse water directly to a sewer instead of discharging onto the floor, with subsequent splashing and chemical damage. The selector valves and auxiliary tank in the spray cleaning circuit permit flushing with clear water from the supply tank, discharge to the sewer, recirculation of the cleaning solution, and rinsing with clear water metered continuously from the supply tank with subsequent discharge to the sewer.

There are two basic CIP designs: single-use and reuse systems. Another approach has been to incorporate combined systems, which provides the best characteristics of the single-use and reuse equipment. This type of unit is referred to as a *multiuse system*.

Single-Use Systems

Single-use systems use the cleaning solution only once. They are generally small units, frequently located adjacent to the equipment to be

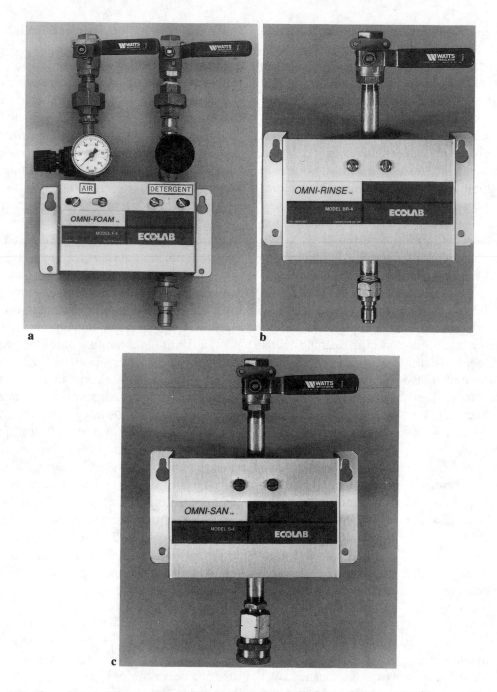

Figure 9–6 These drop stations provide a quick connection for the (a) foam detergent, (b) rinse, and (c) sanitizer applications. Courtesy of Ecolab, Inc., Mendota Heights, Minnesota.

cleaned and sanitized. Because the units are located in the area where cleaning is accomplished, the quantity of chemicals and rinse water can be relatively small. Heavily soiled equipment makes a single-use system more desirable than the others because reuse of the solution is less feasible. Some single-use systems are designed to recover the cleaning solution and rinse water from a previous cycle for use as a prerinse cycle in the subsequent cleaning cycle.

When compared to other CIP systems, single-use units are more compact and have a lower capital cost. These units are less complex and may be purchased as preassembled parts for easier installation. Figures 9–9 and 9–10 illustrate typical single-use systems. The unit

shown consists of a tank with level probes and pneumatically controlled valves to inject steam, introduce water, and regulate the circuit, inclusive of discharge, overflow, and through-flow. Discharging is normally accomplished at the end of the rinse cycle. Also included as part of a single-use system is a centrifugal pump and control panel; and a program cabinet with temperature controller, solenoid valves, and pressure and temperature instrumentation.

A typical sequence for cleaning equipment such as storage tanks or other storage containers takes about 20 minutes, with the following procedures:

1. Three prerinses of 20 seconds with intervals of 40 seconds each to remove

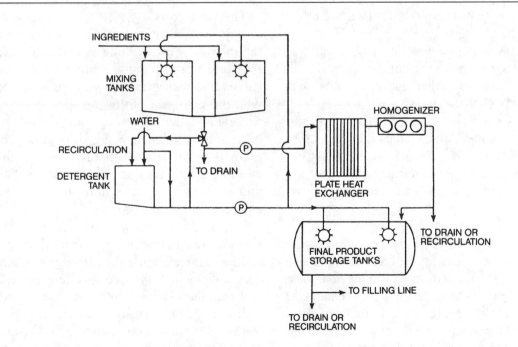

Figure 9–7 Flow arrangement illustrating the operation sequence of a simplified CIP system. *Source:* From Jowitt, 1980.

Table 9–1 Typical Cycle for CIP System

Operation	Function
1. Preliminary rinse (hot or cold water)	Remove gross soil
2. Detergent wash	Remove residual soil
3. Rinse	Remove cleaning compounds
4. Sanitization	Destroy residual microorganisms
5. Final rinse (optional, according to sanitizer use)	Remove CIP solutions and sanitizers

the gross soil deposits are initially applied. The water is subsequently pumped by a CIP return pump for discharge to the drain.

2. The cleaning medium is mixed with injected steam (if used) to provide the preadjusted temperature directly into the circuit. This status is maintained for 10 to 12 minutes prior to discharge of the spent chemicals to the drain or recovery tank.

3. Two intermediate rinses with cold water for an interval of 40 seconds each are followed by transfer to water recovery or to the drain.

4. Another rinse and recirculation is established and may include the injection of acid to lower the pH value to 4.5. Cold circulation is continued for about 3 minutes, with subsequent drainage.

Reuse Systems

Reuse CIP systems are important to the food industry because they recover and reuse cleaning compounds and cleaning solutions. It is important to understand that contamination of cleaning solutions is minimal because most of the soil has been removed during the prerinse cycle, enabling cleaning solutions to be used more than once. For this system to be effective, the proper concentration of the cleaning solution is essential. The concentration can be determined by following the guidelines recommended by the chemical

supplier and the equipment vendor. Sequencing versatility permits the timing and sequencing of operations (acid/alkaline or alkaline/acid) to be varied.

A tank for each chemical is provided with reuse CIP systems. A hot-water tank or bypass loop is necessary to save energy and water if a hotwater rinse is used. The cleaning solution is frequently heated by a coil.

The basic parts of a CIP reuse system are an acid tank, alkaline tank, fresh-water tank, return-water tank, heating system, and CIP feed and return pumps. Remote-controlled valves and measuring devices are provided with the piping layout of this cleaning system. The predetermined cleaning operations have automatic sequencing through a program control unit. With this system, the cleaning solution is transported from the CIP unit through the production plant and the equipment to be cleaned.

Two-tank systems for reuse of the wash water consist of one tank for rinse water and another for reclaiming the cleaning solution. CIP equipment with three tanks contain one tank for the cleaning solution, one for reclaiming the prerinse solution, and one for a fresh-water final rinse. Both single-use and reuse systems require careful design and monitoring to avoid the danger of unwanted mixing of food products with cleaning solutions (Giese, 1991).

Two tanks for alkaline cleaning compounds are frequently provided for solutions

Figure 9–8 A portable hand-size unit that mechanically cleans large and small tanks. It consists of a stainless steel head hydraulically propelled on a self-cleaning alkali- and acid-resistant ball race. The spherical head is fitted with a wide-angle nozzle and slotted to provide saturation spray in every direction for surface coverage. An offset jet, adjustable for speed, provides a vertical spinning motion. Courtesy of Oakite Products, Inc.

of differing concentrations. The less concentrated solution can be used for cleaning tanks, other storage facilities, and pipelines. The stronger solution is available for cleaning the plate heat exchanger. Pumps that feed the cleaning compounds into the tanks are used to automatically adjust the use of neutralization tanks with automatically adjusted acid concentrations.

Two CIP circuits can be cleaned simultaneously by the addition of extra CIP feed pumps. The tank capacity is determined by circuit volumes, temperature requirements, and desired cleaning programs. In mechanized plants, a central control console uses remote-controlled valves to switch the cleaning circuits on and off. Through use of a return water tank, water consumption of a reuse

Figure 9–9 A CIP single-use solution recovery unit that is part of a system containing a water supply tank and CIP circulating unit. Courtesy of Ecolab, Inc., Mendota Heights, Minnesota.

system can be optimized. Recirculation of the cleaning solution is usually necessary for best results; thus, reuse equipment has a higher initial cost but permits operational expense savings.

The ideal CIP reuse system has the ability to fill, empty, recirculate, heat, and dispense contents automatically. A typical operation of this system with a program for storage tank and pipeline cleaning with recovery of the cleaning solution is described in Table 9–2.

Multiuse Systems

These units, which combine the features of both single-use and reuse systems, are designed for cleaning pipelines, tanks, and other storage equipment that can be cleaned effectively by the CIP principle. These systems function through automatically controlled programs that entail various combinations of cleaning sequences involving circulation of water, alkaline cleaners, acid

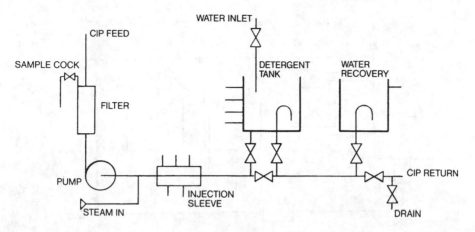

Figure 9–10 In this single-use CIP system with limited recovery, an additional tank with high-level probe is mounted so that the wash and rinse water can be collected for the next prewash cycle. *Source:* From Jowitt, 1980.

cleaners, and acidified rinses through the cleaning circuits for differing time periods at varying temperatures.

A simplified flow chart of a typical multiuse CIP system is presented in Figure 9–11.

This versatile modular unit accommodates differing CIP sequencing, chemical strength, and thermal performance. The multiuse CIP system contains tanks for chemical and water recovery, with an associated single pump, re-

Table 9–2 Operation of an Ideal CIP Reuse System

Operation*	Time (min)	Temperature
Prerinse: Application of cold water from the recovery tank with subsequent draining	5	Ambient
Detergent wash: A 1% alkaline-based cleaning compound purges the remaining rinse water to the drain with subsequent diversion by a conductivity probe to the cleaning compound tank for circulation and recovery	10	Ambient to 85°C, depending on equipment to be cleaned and type of soil
Intermediate water rinse: Softened cold water from the rinse forces out the remaining cleaning solution to the cleaning solution tank; water is then diverted to the water recovery tank	3	Ambient
Acid wash: An acid solution of 0.5–1.0% forces out residual water to the drain; then this solution is diverted through a conductivity probe to the acid tank for recirculation and recovery	10	Ambient to 85°C, depending on equipment to be cleaned and type of soil
Final water rinse: Cold water purges out the residual acid solution with collection of water in the water recovery tank; overflow is diverted to the drain	3	Ambient

*Pasteurizing equipment tanks and pipelines may also be subjected to a final flush of hot water at 85°C.
Source: Jowitt, 1980.

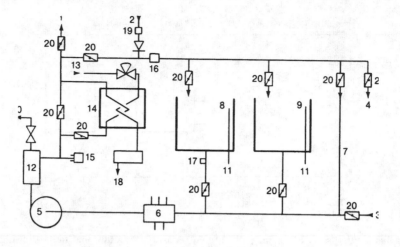

Figure 9–11 Typical multiuse CIP system, simplified:

1.	CIP feed	11.	Overflow
2.	CIP return	12.	Filter
3.	Water inlet	13.	Steam in
4.	Drain	14.	"Paraflow" heat exchanger
5.	Puma pump	15.	Temperature probe
6.	Injection sleeve	16.	Soluvisor
7.	Recirculating loop	17.	Conductivity probe
8.	Detergent tank	18.	Condensate
9.	Water recovery	19.	No-flow probe
10.	Sample cock	20.	Butterfly valve

Source: From Jowitt, 1980.

circulating pipe work, and heat exchanger. The plate heat exchanger heats the incoming water and cleaning liquid to the required temperature. Flexibility of temperature control, optimal utilization of tank capacity, and flexibility in heating of water or cleaning solutions can be realized through the use of a heat exchanger.

An automatic multiuse CIP system follows the following operation sequence:

1. *Prerinse.* This step occurs from water recovery or the water supply provided at the desired temperature. The solution from this operation can be directed to the drain or diverted by a recirculatory loop for a timed period, then transferred to the drain.

2. *Cleaning solution recirculation.* The recirculation step occurs by the cleaning compound vessel or the bypass loop. A desired combination of cleaning chemicals can be used for variable recirculation times, and the chemical injection can boost the strength or use of the solution. The plate heat exchanger or its bypass loop can contribute to the cleaning solution recirculation. With a bypass loop, variable-temperature programming permits total detergent-tank heating. Cleaning solutions can be recovered or drained.

3. *Intermediate rinse.* This operation is similar to the prerinse, except that it is important to remove residual cleaning chemicals from the previous operation.

4. *Acid recirculation.* This optional operation, which is similar to the cleaning recirculation operation, may occur with or without an acid tank. With an acid tank, the recirculatory loop is established on water, either through the plate heat exchanger or via the plate heat exchanger bypass loop. The acid is injected to a preset strength based on timing for a specific circuit volume.

5. *Sanitizer recirculation.* This operation, designed to reduce microbial contamination, is similar to the acid injection operation except that heating is not normally required.

6. *Hot water sterilization.* Variable times and temperatures are available for this operation, which involves use of a recirculation loop on fresh water via the plate heat exchanger. The spent water can be either returned to water recovery or drained.

7. *Final water rinse.* Water is pumped via the CIP route and sent to water recovery. Water rinse times and temperatures are variable.

The desirable features of CIP equipment are:

- *Reduced labor.* Manual cleaning is reduced because the CIP system automatically cleans equipment and storage utensils. This feature becomes increasingly important as wages increase and it becomes more difficult to locate dependable workers.
- *Improved sanitation.* Automated operation cleans and sanitizes more effectively and consistently. Through timed or computer-controlled equipment, cleaning and sanitizing operations are more precisely controlled.
- *Conservation of cleaning solution.* Optimal use of water, cleaning compounds, and sanitizers is possible through automatic metering and reuse.
- *Improved equipment and storage utilization.* With automated cleaning, equipment, tanks, and pipelines can be cleaned as soon as use is discontinued so that immediate reuse is possible.
- *Improved safety.* Workers are not required to enter vessels that are cleaned with CIP equipment. The risk of accidents from slippery internal surfaces is eliminated.

The disadvantages of CIP systems are:

- *Cost.* Because most CIP systems are custom designed, design and installation costs add to the high price of the equipment.
- *Maintenance.* More sophisticated equipment and systems tend to require more maintenance.
- *Inflexibility.* These cleaning systems can effectively clean in only those areas where equipment is installed, whereas portable cleaning equipment can cover more area. Heavily soiled equipment is not as effectively cleaned by CIP systems, and it is difficult to design units that can clean all processing equipment.

Microprocessor Control Unit

It is now possible to control CIP equipment more precisely. Microprocessor-based monitoring systems, interfaced with CIP controllers, permit tracking of cleaning parameters to provide for troubleshooting and process control. Improved product protection, labor and cost reduction, and improved efficiency may be achieved.

Sophisticated CIP equipment includes a microprocessor control unit (Figure 9–12). Programming flexibility enables this unit to be used for the operation of a wide variety of CIP systems, single or reuse, using either time or low measurement as the base. The principle involved with this innovation is that a graphic recording of the CIP unit is provided to monitor operating parameters, including supply and return temperature, pressure, flow, pH, and conductivity. This equipment can spread the data out in detailed chronological graphic form for temperature, return velocity, and solution concentration. Items such as valve pulsings, pump cavitation, and program steppings can be charted and documented, which are not clearly visible in normally chronological CIP recording charts. Therefore, this feature is a valuable tool for CIP monitoring, documentation, and maintenance.

Computer-based CIP monitoring equipment contains a control panel equipped with a display for the operator. However, primary monitoring is performed through printed records of CIP performance. The printer can produce a series of strip charts. On each chart, a graph of time is plotted against a corresponding CIP parameter. The unit is programmed to account for variation of cleaning parameters for different cleaning circuits. The same printer can record CIP cycles for various equipment, storage vessels, tank trucks, or transfer lines. Some equipment contains alarms that warn of performance outside the limits of programmed set points.

Figure 9–13 illustrates a microprocessor and distribution system that includes a pump-and-fill station for dispensing cleaning compounds into rugged, capped allocation containers. The service station provides air, water, electrical, and thermostat control connections. Designated employees can access the system using a swipe card, thus simplifying production selection by application. Management personnel can access the unit remotely via modem to track chemical usage by application. This system can lower chemical costs by 15% to 20% and reduce the cleaning

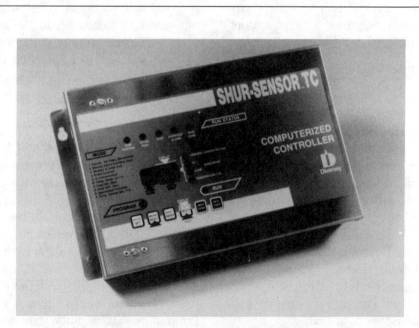

Figure 9–12 A microprocessor controller that regulates the monitoring capabilities of this equipment. Courtesy of Diversey Lever, 255 E. 5th Street, Suite 1200, Cincinnati, Ohio, 800-233-1000.

Figure 9–13 Use of a microprocessor for programmed distribution of sanitation compounds. Courtesy of Ecolab, Inc., Mendota Heights, Minnesota.

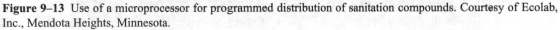

cycle time by 10% as a result of more efficient chemical allocation.

The microprocessor control unit enhances cleaning effectiveness and reduces cleaning costs through precise control of the variables associated with mechanized cleaning. One of these units can be designed with the capacity for as many as 200 separate and variable programs that can provide product recovery, rinse and/or cleaning compound recovery, manual rinsing, sanitizing cycle, concentration of chemical strength, extended wash duration, and many other options. The microprocessor control unit can be designed with self-contained, on-line programming while running via an integral key pad or an off-line programming package available for use on personal computers.

Cleaning Out of Place

Systems designed for cleaning out of place (COP) require cleaning by disassembly and/or removal from the normal location. Fluid flow is utilized in the application of force for cleaning. Regulatory agencies have previously used velocity as a means of measuring the fluid flow force, employing the rule of thumb of 1.5 m/sec. This guideline should no longer be emphasized because COP equipment can effectively clean with less velocity. Velocity and turbulence, the actual cleaning force, are not equally related under all conditions of flow.

Many small parts of equipment and utensils, as well as small containers, can be washed effectively in a recirculating-parts washer, also called a *COP unit*. These units, like sanitary pipe washers, contain a recirculating pump and distribution headers that agitate the cleaning solution. Also, a COP unit can serve as the recirculating unit for CIP operation. The normal wash recirculation time is approximately 30 to 40 minutes, with an additional 5 to 10 minutes for a cold acid or sanitizing rinse.

A COP unit is frequently constructed with a double-compartment stainless steel sink equipped with motor-driven brushes. The same motor also pumps a cleaning solution through a preformatted pipe onto the brushes. Desired temperature of the cleaning solution (45°C to 55°C) is maintained by a thermostatically controlled heater. The first compartment is allocated to use of the cleaning solution. The cleaned parts or utensils are rinsed with a spray nozzle in the second compartment. Drying is normally accomplished by air within the COP unit or on a suitable drain board or rack.

Equipment that functions as a COP unit contains a brush assembly and a rinse assembly. A tank is included for the cleaning solution. Many COP units contain rotary brushes for cleaning both the inside and outside of parts and utensils, with the cleaning solution being introduced through the brushes that clean the inside.

The major appeal of COP equipment is that it can effectively clean parts that are disassembled as well as small equipment and utensils. This equipment can also reduce labor requirements and improve hygiene. COP units are considered reasonable in cost to buy and maintain. Their major limitations for small-volume operations are initial cost, maintenance, and labor requirements for loading and unloading these washers.

The COP concept is frequently used to clean equipment and utensils for the food preparation and food service industry. Stainless steel bins may be cleaned and sanitized in an enclosed stainless steel cabinet washer through the use of a computer-controlled cycle. A programmable logic controller governs the timing of each sequential step in the cleaning operation. Further discussion of COP equipment in the dairy and food service industries is given in Chapters 14 and 19.

SANITIZING EQUIPMENT

Equipment for the application of sanitizing compounds can vary from hand sprayers, such as units used to apply insecticides and herbicides, to wall-mounted units and headers mounted on processing equipment. Many mechanized cleaning units may contain sanitizing features as part of the system (see Figure 9–14).

Centralized high-pressure, low-volume cleaning and foam cleaning equipment include sanitizing lines with stations for application of the sanitizer by hose and wand or by spray headers on processing equipment, especially moving belts or conveyors. A benefit of the latter feature is that sanitizing is mechanized and can be uniformly administered through the use of timer switches. Metering of sanitizing compounds provides a more accurate and precise application of the sanitizer.

Figure 9–15 illustrates a wall-mounted sanitizing unit. The high-pressure rinse water passes through flow control with the orifice size necessary to achieve a prespecified flow rate. To sanitize, the high-pressure water passes through the sanitizer injector, which meters a specific amount of sanitizer into the water stream. A flood sanitizing nozzle may be incorporated to spread the solution effectively without automation.

Sanitation Application Methods

The application methods that are available provide acceptable methods for transport of the sanitizer to the desired area. The optimal method depends upon individual operations.

Chemical sanitizers are normally applied by one of the following methods.

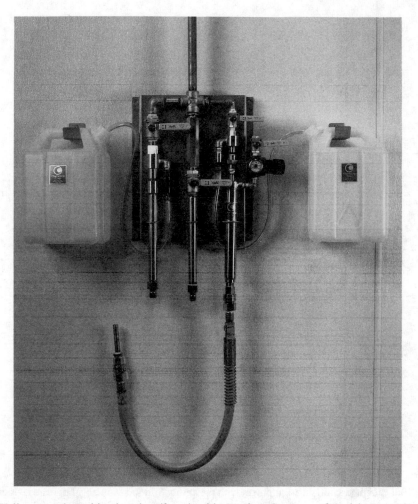

Figure 9–14 Wall-mounted combination rinse/foam/sanitize station. Courtesy of Ecolab, Inc., Mendota Heights, Minnesota.

- *Spray Sanitizing*. This method involves use a sanitizer dissolved in water and a spray device to transport the sanitizer to the area to be sanitized.
- *Fogging*. Fogging involves application of the sanitizer as a fine mist to sanitize the air and surfaces in a room.
- *Flood Sanitizing*. This method involves the application of a sanitizer dissolved in water and applied in a large quantity to ensure extensive exposure. The use of flood sanitizing has been increased to combat the proliferation of *L. monocytogenes*. The disadvantages of this method are the cost of the sanitizer and water, and the wet condition created.
- *Immersion/COP Sanitizing*. This technique involves the submersion of equip-

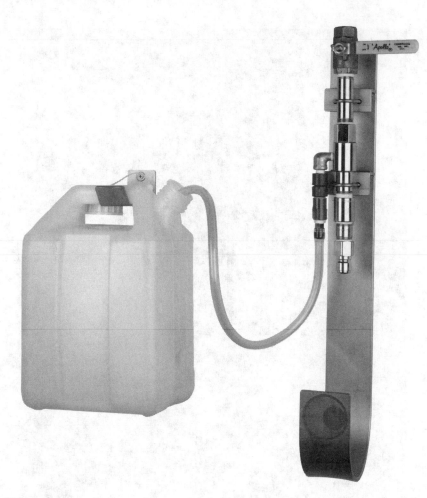

Figure 9–15 Wall-mounted sanitizing equipment with the ability to flood sanitize a large area. Courtesy of Ecolab, Inc., Mendota Heights, Minnesota.

ment, utensils, and parts in a tank that contains a sanitizing solution.

- *CIP Sanitizing*. CIP sanitizing involves sanitizing by circulation of the sanitizer inside pipes, lines, and equipment.

LUBRICATION EQUIPMENT

Figure 9–16 illustrates typical equipment for effective maintenance of high-speed bottling and canning conveyors used in the beverage industry and on shackle chains and conveyors, smokehouse chain drives, and other applications requiring precise continuous conveyor and/or chain lubrication. This prin-

ciple involves water under pressure during a reciprocating piston in the chemical pump. This piston subsequently drives a chemical-concentrate-metering piston that draws the lubricant from the drum and injects it into the water from the cylinder.

SUMMARY

A major function of cleaning equipment is to dispense the cleaning compound and sanitizer to facilitate cleaning and sanitizing and reduce the microbial flora. An efficient cleaning system can reduce cleaning labor by up to 50%.

Figure 9–16 Lubrication equipment for high-speed conveyors, drives, and shackle chains. Courtesy of Ecolab, Inc., Mendota Heights, Minnesota.

High-pressure, low-volume cleaning equipment generally is among the most effective cleaning equipment in the removal of soil deposits with penetration ability in difficult-to-reach areas. Foam, which has gained wider acceptance because it is easily applied and has the ability to cling to surface areas, tends to be more effective for large surface areas. A slurry provides a medium similar to foam, except that less air is present and it has reduced clinging ability. A gel medium is most effective for cleaning equipment with small moving parts.

A portion of the equipment used in food processing plants for fluid processing, such as for dairy products and beverages, can be cleaned effectively with CIP units, which reduce cleaning labor. However, this equipment is expensive and is less effective where heavy soil and a variety of processing systems exist. Sophisticated CIP equipment includes a microprocessor control unit to monitor operating parameters. Parts and small utensils can be cleaned effectively with COP equipment. More sanitary lubrication of high-speed conveyors and other equipment is possible through the use of mechanized lubrication equipment.

STUDY QUESTIONS

1. What is the largest cost factor for sanitation?
2. How does high-pressure, low-volume cleaning equipment function?
3. What are the advantages and disadvantages of high-pressure, low-volume cleaning equipment?
4. What is the difference between centralized and portable cleaning equipment?
5. Why is foam cleaning a popular and accepted method of cleaning?
6. What is gel cleaning?
7. What is slurry cleaning?
8. What is CIP equipment and how does it function?
9. What is COP equipment and how does it function?
10. What is a CIP reuse system?
11. What are the advantages and disadvantages of CIP equipment?
12. What is the typical cycle for a CIP system?

REFERENCES

Giese, J.H. 1991. Sanitation: The key to food safety and public health. *Food Technol* 15, no. 12: 74.

Jowitt, R. 1980. *Hygienic design and operation of food plant.* Westport, CT: AVI Publishing Co.

SUGGESTED READINGS

Guthrie, R.K. 1988. *Food sanitation.* 3rd ed. New York: Van Nostrand Reinhold.

Marriott, N.G. 1990. *Meat sanitation guide II.* Blacksburg, VA: American Association of Meat Processors and Virginia Polytechnic Institute and State University.

CHAPTER 10

Waste Product Disposal

Waste materials generated from food processing and food service facilities can present difficult treatment problems because they contain large amounts of carbohydrates, proteins, fats, and mineral salts. For example, the wastes from dairy plants; food freezing and dehydration plants; and processing plants for red meats, poultry, and seafood can produce distinct odors and heavy pollution of water if the discharge is not properly treated. Organic matter of these wastes must be treated by biological stabilization processes before it is discharged into a body of water. Improper waste disposal is a hazard to humans and to aquatic forms of life.

Increasingly, federal, state, and local regulatory agencies, as well as the public, are demanding better waste treatment by the industry. Processors and regulatory agencies are responsible for the disposal of waste materials promptly and completely. Accumulation of wastes, even for short periods of time, can attract insects and rodents, produce odors, and become a public nuisance or an unsightly condition inside or outside the plant.

The major problem with these wastes is that the organic matter provides a food source for microbial growth. With an abundant food supply, microorganisms multiply rapidly, reducing the dissolved oxygen contained in the water. Water normally contains approximately 8 parts per million (ppm) of dissolved oxygen. A minimum standard for fish life is 5 ppm of dissolved oxygen. If values are below this level, fish can suffocate.

If dissolved oxygen is eliminated from water through high organic matter content, a septic condition with foul odors and darkening of water occurs. Septic conditions with sulfur-containing proteins or water with a high natural content of sulfates can produce hydrogen sulfide, which has a foul odor and can blacken buildings.

Waste disposal from food processing and food service facilities can present a hazard if the wastes are not properly handled because of the high content of organic matter, which is measured as biochemical oxygen demand (BOD). Most facilities that discharge a large quantity of effluent with a high BOD into a municipal treatment system have to pay a surcharge because of the increased wastewater treatment load. Because of this burden on small municipal treatment facilities, many large firms elect to treat effluent discharge partially or completely.

The large volume of wastewater produced in food plants contains vast quantities of organic residues. The intermittent production schedule of many plants places greater demands on wastewater treatment systems. During processing, water is essential to help

187

cleanse the product and to serve as a cleaning medium in conveying unwanted materials to the sewage system. This water becomes the problem during wastewater treatment because it contains suspended and dissolved organic matter.

STRATEGY FOR WASTE DISPOSAL

A waste disposal survey is needed to identify the quantity of waste materials and the waste characteristics that will be discussed in this chapter.

Planning the Survey

The first step in a waste disposal survey is an operations study, which identifies sources of wastes. Construction drawings showing the plot plan, piping plans, and equipment layouts should be studied to determine all sources of incoming and outgoing water. The piping plans should show water lines, storm sewer lines, sanitary sewer lines, and processing waste drains and lines. The pipe sizes, locations and types of connections to processing equipment, and the flow direction should be included on the drawings.

The operating schedule of the food plant—the number of shifts and types and volumes of products produced in a single day and over a week, a month, a season, and the entire year—is important to this survey. Production records for several preceding years can provide this information. Water consumption records should also be examined.

An initial waste survey is conducted to ensure that a plant can comply with federal, state, and local effluent requirements in order to obtain or sustain a National Pollution Discharge Elimination System (NPDES) permit. The NPDES permit places the burden of monitoring the waste effluent stream on the firm that creates the discharge. The Environmental Protection Agency (EPA) periodically monitors waste discharges to check the accuracy of reports submitted by applicants and permit holders.

An initial survey is also beneficial to determine locations and types of required monitoring equipment to establish a continuous monitoring program. Another advantage of an initial survey is to determine whether waste treatment is needed to meet discharge regulations and, if needed, the most ideal waste treatment approach.

Conducting the Survey

Information obtained from the operations study should determine what to include in the survey. It may be necessary to conduct individual surveys in each season if the types and volumes of products processed in the plant vary widely with the season of year, as is typical in many food plants, especially fruit and vegetable processing plants. These steps must be part of the survey: determination of the water balance, sampling of wastewater, and determination of extent of pollution.

Determination of Water Balance

The volume of wastewater and flow rates from all sources must be measured by meters placed on all incoming water lines. Suitable measuring devices are Parshall flumes, rectangular and triangular weirs, and Venturi tubes and orifices. Through calculation of the water balance for an entire plant, the quantity of water in the waste effluent, together with the quantities lost through steam leaks, evaporation, and other losses, and the amounts used in the products of the plant in a given period of time can be determined. All of these quantities should equal the amount of water supplied to the plant during a given period. This calculation can be used to identify hidden water losses or major leaks, which can affect the sanitation program and cause increased waste, additional effluent discharge, and reduced profits.

Sampling of Wastewater

Samples of the wastes should be obtained in proportion to flow rates. Random "grab" samples—taken by catching a given quantity of the effluent discharge in a container without consideration of variations in volumes and flow rates and changes in plant operations—are of limited value for determining the true characteristics of wastes and can provide misleading results. Statistical sampling at planned times during the operating and nonoperating periods, and in proportion to flow, can provide valid data related to the characteristics of the wastewater effluent from a plant.

The sampling device should be located in the wastewater discharge system to obtain a representative sample. Samples should be collected where wastes are homogeneous—perhaps below a weir or flume. Caution should be exercised to avoid sampling errors resulting from a deposition of solids upstream from a weir or from accumulation of grease immediately downstream. The sample should be collected near the center of the channel and at 20% to 30% of the depth below the surface, where the velocity is sufficient to prevent deposition of solids. Sewers and deep, narrow channels should be sampled at 33% of the water depth from the bottom to the surface, with the collection point rotated across wide channels. During sample collection and handling, agitation should be avoided for dissolved oxygen determination. Food plant wastes readily decompose at room temperature; thus, it is important to chill samples promptly to 0°C to 5°C if they are not analyzed immediately after sampling.

Determination of Extent of Pollution

A large percentage of the waste discharged in fruit and vegetable waters, wash water from animal slaughter, and cleanup water discharge are product pieces (larger pieces can be removed by screening). Finer solids, which pass through a screen, and organic matter in colloidal and true solution usually have an oxygen demand in excess of the dissolved oxygen content of the water.

Biochemical Oxygen Demand. A frequently used method of measuring pollution strength is the 5-day BOD test. The BOD of sewage, sewage effluents, and waters of industrial wastes is the oxygen (in ppm) required during stabilization of the decomposable organic matter by action of aerobic microorganisms. The sample is stored in an airtight container for a specified period of time and temperature. Complete stabilization can require more than 100 days at 20°C. Because such long periods of incubation are impractical for routine determinations, the procedure recommended and adopted by the Association of Official Analytical Chemists (AOAC) is a 5-day incubation period and is referred to as the *5-day BOD*, or *BOD$_5$*. This value is only an index of the amount of biodegradable organic matter, not an actual measure of organic waste.

Domestic sewage that contains no industrial waste has a BOD of approximately 200 ppm. Food processing wastes are normally higher and frequently exceed 1,000 ppm. Table 10–1 gives the typical amount of BOD$_5$ and suspended solids from food and related industries. Note that the BOD data and values for suspended solids generally show a parallel relationship. However, BOD$_5$ is not as closely related to dissolved solids.

Although BOD is a common measurement of pollution of water and the test is relatively easy to conduct, it is time-consuming and lacks reproducibility. Tests such as chemical oxygen demand (COD) and total organic carbon (TOC) are quicker, more reliable, and more reproducible.

Chemical Oxygen Demand. The COD test for measuring pollution strength oxidizes

Table 10–1 Typical Composition of Wastes from Food and Related Industries

Type of Waste	BOD₅ (ppm)	Suspended Solids (ppm)
Dairy and milk products	670	390
Food products	790	500
Glue and gelatin	430	300
Meat products	1,140	820
Packing house and stockyard waste	590	600
Rendered products	1,180	630
Vegetable oils	530	475

compounds chemically rather than biologically by a dichromate ($K_2Cr_2O_7$) acid reflux method. Because it is a chemical analysis, this test also measures nondegradable materials, which are not measured by BOD testing. When a plant monitors effluent to be discharged for municipal treatment, daily COD measurements can serve as a guide to determine whether and when a biological or chemical effluent could create a treatment problem at the wastewater treatment plant. This test, however, gives no indication as to whether the organic matter can be degraded biologically and, if so, at what rate. According to Forster (1985), not all molecules are oxidized by this type of treatment. Although overlapping occurs, this test does not duplicate the BOD₅. Data from the COD test closely relate to dissolved organic solids. Unless a ratio has been established for COD/BOD, regulatory agencies have not accepted COD data as a substitute for BOD data in the past.

Dissolved Oxygen. Dissolved oxygen (DO) concentration is of major concern for both wastewater and receiving water because it affects aquatic life and is important in treatment systems such as aerated lagoons. Determination of dissolved oxygen can be accomplished by an iodometric titration procedure using the azide and permanganate procedures to remove interfering nitrite and ferrous ions,

even though this method is not considered to be very reliable. Alternatively, electrode probes can be used for this measurement. They are faster and more convenient than the iodometric tritrimetric method and more adaptable for use in most industrial wastewaters. However, certain metal ions, gaseous oxidants stronger than molecular oxygen, and high concentrations of cleaning compounds will interfere with the electrode probes used to measure dissolved oxygen.

Total Organic Carbon. Total organic carbon determines all materials that are organic. It measures the amount of CO_2 produced from the catalytic oxidation at 900°C of solid matter in wastewater. This method of pollution measurement is rapid and reproducible, and correlates highly with standard BOD₅ and COD tests, but it is difficult to conduct and requires sophisticated laboratory equipment. This test can be effectively conducted where total solid matter is mostly organic and if the operation has a large volume. However, the cost of performing TOC analysis is frequently prohibitive for smaller and/or seasonal processing plants.

Residue in Wastewater. Residue can be considered pollution because it affects the measurements that have been previously discussed. Residues of evaporation (total solids) and the volatile (organic) and fixed (ash) fractions are routinely recognized.

Settleable solids (SS) settle to the bottom in 1 hour. They are usually measured in a graduated Imhoff cone and reported as mL/L SS. Settleable solids are an indication of the amount of waste solids that will settle out in clarifiers and settling ponds. This examination technique is easy to perform and can be conducted at field sites.

Total suspended solids, sometimes referred to as *nonfilterable residue*, are determined by filtration of a measured volume of wastewater through a tared membrane filter (or glass fiber mat) in a Gouch crucible. The dry weight of the total suspended solids (TSS) is obtained after 1 hour at 103°C to 105°C.

Total dissolved solids (TDS), or filterable residue, is determined by the weight of the evaporated filtered sample or as the difference between the weight of the residue on evaporation and the weight of TSS. These pollutants are difficult to remove from wastewater, so knowledge of them is essential. Treatment requires microorganisms, which are normally present, for conversion to particulate matter, i.e., microbial cells.

Fats, oil and grease (FOG) are detrimental to biota and are unaesthetic. Interchange of air and water is reduced by the thin film created by FOG, which is detrimental to fish and other marine life. Water fowl are also affected by heavy oil films. These compounds increase oxygen demand for complete oxidation.

Although *turbidity* is not a pollutant, it is caused by the presence of suspended matter (organic matter, microorganisms, and other soil particles). Turbidity is an optical property of the sample, which causes light to be scattered and/or absorbed, rather than transmitted. It is measured by a candle turbidimeter. This measurement is not an accurate indication of suspended matter that has been determined gravimetrically because the latter method involves particle weight, and the former relates to optical properties.

In waste material, *nitrogen* can exist in forms ranging from reduced ammonium to oxidized nitrate compounds. High concentrations of the nitrogen forms can be toxic to certain plant life. The most common forms of nitrogen found in wastewater are ammonia, proteins, nitrites, and nitrates. The reduced forms, i.e., organic nitrogen and ammonia, can be measured by the total Kjeldahl nitrogen (TKN) method. Other tests are necessary to measure the oxidized forms, i.e., nitrate and nitrite.

Phosphorus occurs in wastewater as phosphate in the forms of orthophosphate and polyphosphate. This element is present as either mineral or organic compounds. Although trace amounts of soluble phosphates occur in natural waters, too much is detrimental to marine life. Routine analyses measure only soluble orthophosphate. Analyses for total phosphates, polyphosphates, and precipitated phosphates are accomplished by converting the polyphosphates and precipitated phosphates to orthophosphate by acid hydrolysis, with subsequent testing for orthophosphate by colorimetric methods. These methods can be performed by a trained technician, with the required chemical reagents and a colorimeter or spectrophotometer.

Use of sulfur dioxide in pretreatment of fruits or sodium bisulfide in processing may cause the *sulfur* content of wastewater to be high enough to cause pollution problems. These pollutants exist primarily as sulfite and sulfate ions or precipitates. Also, sulfides require more available oxygen if present in water. Sulfide ions combine with various multivalent metal ions to form insoluble precipitates, which can settle out and be removed with the sludge. Sulfate and sulfide determinations are possible with a trained technician and minimal equipment. Sulfides contribute to an undesirable odor and taste in drinking water. Thus, it is important to test for these compounds if the wastewater is discharged into a stream that supplies drinking water.

SOLID WASTE DISPOSAL

Solid waste disposal is a major challenge for the food industry. In food industries such as canneries, up to 65% of the raw materials received must be disposed of as solid waste. The most common method for disposal has been to truck the wastes to municipal garbage dumps. If a dump is not nearby and the wastes are disposed on the plant site, odor and insect problems will be created. Some processing firms handle solid wastes by composting, and the finished compost can then be applied to the soil as fertilizer. A typical analysis of composted material is 1.25% nitrogen, 0.4% phosphates, and 0.3% potash. Some municipal waste treatment facilities manufacture and sell solid waste materials for agricultural application.

If composting is used, the organic matter in waste material must be stabilized through microbial action. Humus, which results from stabilization of waste material, improves fertility and tillage properties. The basic composting procedures has four steps:

1. Solid waste material should be comminuted (pulverized) to expose the organic matter to microbial attack.
2. The comminuted waste should be stacked in windrows approximately 2 m high and 3 m wide.
3. Aeration should be provided.
4. After extensive aeration, the compost should be comminuted again.

Addition of an inoculum will accelerate the composting process. A compost can be produced through those aerobic thermophilic microorganisms present in the waste material

in 10 to 20 days, depending on temperature and waste composition.

In addition to compost, various food product wastes can be dehydrated and ground for feed use. An example is the press liquors of tomato processing wastes. The residue from alcohol manufacture can be dried and fed to livestock. Citrus wastes, including activated sludge from treatment, can also be dried and used as animal feed because they contain B vitamins and protein. Processed whey and rendered animal by-products are also valuable foods for animal consumption.

LIQUID WASTE DISPOSAL

Whenever food is handled, processed, packaged, and stored, wastewater is generated. Quantity, pollutant strength, and nature of constituents of processing wastewater have both economic and environmental consequences concerning treatability and disposal. Economics of treatment are affected by the amount of product loss from the processing operations and by the treatment costs of this waste material. Significant characteristics that determine the cost for wastewater treatment are the relative strength of the wastewater and the daily volume of discharge.

Wastewater can be salvaged through recycling, reuse, and the recovery of solids. The degree of conservation and salvage value of wastewater is based on factors such as wastewater treatment facilities for recoverable materials, operating costs of independent treatment, market value of the recoverable materials, local regulations regarding effluent quality, surcharge cost for plants discharging into public sewers, and anticipated discharge volume in the future. The economics of disposal of solids, concentrates, blood, and concentrated stick (in wet rendering) determine how much of these polluting solids are kept out of the sewer. A wastewater control plan must be able to remove and convey organic solids using "dry" methods, without discharging those solids to the sewer and by using a minimal amount of water in the cleaning operation.

Spent cleaning compounds and sanitizers are discharged into waste treatment facilities. The toxicity of these materials causes concern because sanitizers, which destroy microorganisms, are toxic by definition. However, they meet requirements of the Food and Drug Administration (FDA) as an indirect food additive because these organic compounds are diluted in water and their toxic properties become reduced to a safe level. Many of the ingredients used in cleaning compounds and lubricants are generally recognized as safe as food additives (Bakka, 1992). It appears that the major concerns for wastewater treatment of this effluent are pH fluctuation and possible long-term exposure to trace heavy metals. However, these effects can be controlled and waste minimized through appropriate plant design and optimal concentration use of cleaning compounds and sanitizers.

Cleaning compounds and sanitizers increase BOD/COD because they utilize surfactants, chelators, and polymers in addition to organic acids and alkalis. Conveyor lubricants utilize similar materials that increase the BOD/COD of the effluent. However, these compounds account for less than 10% of the BOD/COD contributions from a food processing plant. Water volumes associated with sanitation from a food processing plant can account for up to 30% of the total water discharge. Because of the low BOD/COD contributions, pH of wastewater is a major concern.

A eutrophic condition can develop from the discharge of biodegradable, oxygen-consuming compounds if inadequately treated wastewater is discharged to a stream or other body of water. If this condition continues, the ecological balance of the receiving body of water will be harmed.

Frequently it is more economical to invest in waste prevention techniques and utilization of waste products than in waste treatment facilities. Yet many food plants generate waste effluents that pollute. Insufficient treatment capacity of many municipal waste treatment plants necessitates special waste facilities for a large percentage of food plants. Wastewater treatment is still a developing technology, and one that is going to need the cooperation of the EPA, suppliers, and processors.

Pretreatment

The pretreatment of food processing wastewater is frequently required prior to discharge into a municipal waste treatment system. A sewer use ordinance defines specified municipal discharge limitations that determine the degree of pretreatment required. The EPA has previously concluded that many wastewaters from processing plants are compatible and biodegradable.

Municipal sewage plants normally place certain restrictions on wastewater discharge from food processing plants. Although toxic substances are not frequently associated with food process waste streams, certain wastes that are present cannot be treated and can cause obstruction and required additional maintenance. Troublesome wastes include oils and fats, plant and animal tissues, and waste materials. Therefore, some form of isolation and pretreatment of the waste stream is essential prior to discharge in a municipal waste treatment facility.

If increased wasteload reduces the ability of the municipal waste treatment system to treat the additional waste adequately, the food processor usually has to accept more responsibility related to pretreatment or support of a municipal waste treatment plant modification or expansion program. The processor should calculate the cost of the added sewage treatment load and determine that the projected cost should be handled by pretreatment or by paying a surcharge to a municipal expansion program keyed to specific wastewater parameters.

Surcharge calculations start with a flow base rate and utilize multipliers for concentrations of such ingredients as BOD_5, suspended solids, and grease. An example would be to charge the flow base rate to all sewer users as 50% of the water bill, including flow from private supplies. Treatment costs chargeable to BOD and suspended solids frequently include surcharges for concentrated wastes when above an established minimum based on normal load criteria.

Small plants frequently determine that it is advantageous to provide only enough pretreatment of wastewater to ensure compliance with municipal regulations. Yet, larger processors, in contrast, have discovered that providing pretreatment beyond the level required by the ordinance can be advantageous. Some plants provide enough pretreatment to reduce the surcharge for discharging untreated wastewater. Many large-volume processors treat all of their wastewater to avoid high surcharges or because the municipal plant lacks the capacity to handle the addition effluent.

The following advantages of pretreatment of wastewater beyond the level required by the local ordinance should be considered:

- Grease and solid materials from plant and animal products frequently have a good market value. Demand from soap plants, feed plants, and other industries can make a recovery of waste solids a profitable operation. Such operations also reduce the amount of wastewater treatment.

- If municipal charges and surcharges are high, additional pretreatment can be economically advantageous because better pretreatment will reduce these charges.
- Municipality complaints can be reduced through additional treatment responsibilities assumed by the food processor.

The following disadvantages can discourage pretreatment of wastewater:

- Pretreatment facilities are expensive and increase the complexity of the processing operation.
- Maintenance costs, monitoring costs, and record keeping of a wastewater treatment operation can be expensive.
- Pretreatment facilities are placed on the property tax roll unless state regulations permit tax-free waste treatment.

If pretreatment is conducted, this process should be based on facts revealed from the waste disposal survey. Results from the plant survey and review of viable waste conservation and water reuse systems are essential for identification, design, and cost estimates of a pretreatment system. Cost estimates should include those parts of the pretreatment attributable to flow, such as dissolved air flotation and grease basins. Thus, major in-plant expenses for waste conservation and water recycling can be determined based on the estimated reduction in flow, BOD, suspended solids, and grease.

Most common pretreatment processes include flow equalization and the separation of floatable matter and SS. Separation is frequently increased by the addition of lime and alum, ferric chloride ($FeCl_3$), or a selected polymer. Paddle flocculation may follow alum and lime, and lime or ferric chloride additions to assist in coagulation of the suspended solids. Separation is usually accomplished by gravity or by air flotation. Screening by vibrating, rotary, or static-type screens is a step that precedes the separation process and concentrates the separated floatables and settled solids.

Flow Equalization

Flow equalization and neutralization are used to reduce hydraulic loading in the waste stream. Facilities required are a holding device and pumping equipment designed to reduce the fluctuation of effluent discharge. This operation can be economically advantageous, whether processing firms treat their own wastewater or discharge into a municipal sewage treatment facility after pretreatment. An equalizing tank has the capacity to store wastewater for recycling or reuse, or to feed the flow uniformly to the treatment facility day and night. This unit is characterized by a varying flow into and a constant flow from the tank. Equalizing tanks can be lagoons, steel construction tanks, or concrete tanks, often without a cover.

Screening

The most frequently used process for pretreatment is screening, which normally employs vibrating screens, static screens, or a rotary screen. Vibrating and rotary screens are more frequently used because they can permit pretreatment of a larger quantity of wastewater that contains more organic matter. These screening devices are well adapted to a flow-away (water in forward flow and passing through with solids constantly removed from the screen) mode of operation and can vary widely in mechanical action and in mesh size. Mesh sizes used in pretreatment range from approximately 12.5 mm in diameter for a static screen to approximately 0.15 mm in diameter for high-speed circular vibratory polishing screens. Screens are sometimes used in combination (e.g., prescreen-

polish screen) to attain the desired efficiency of solids removal.

Skimming

This process is frequently incorporated if large, floatable solids are present. These solids are collected and transferred into some disposal unit or preceding equipment. Lime and $FeCl_3$, or a selected polymer may be added to enhance separation of solids, and paddle flocculation may follow to assist with the coagulation of these solids.

Primary Treatment

The principal purpose of primary treatment is to remove particles from the wastewater. Sedimentation and flotation techniques are used.

Sedimentation

Sedimentation is the most common primary treatment technique used to remove solids from wastewater influent because most sewage contains a substantial amount of readily settleable solid material. As much as 40% to 60% of the solids, or approximately 25% to 35% of the BOD_5 load, can be removed by pretreatment screening and primary sedimentation. Some of the solids removed are refractory (inert) and are not measured by the BOD test.

A rectangular settling tank or a circular tank clarifier is most frequently used in primary treatment. Many settling tanks incorporate slowly rotating collectors with attached flights (paddles) that scrape settled sludge from the bottom of the tank and skim floating scum from the surface.

Design of a sedimentation system should incorporate sizing of the detention vessel and provide a quiescent state for the raw wastewater. Temperature variation of the wastewater also affects sedimentation because of the development of heat convection currents and the potential interference with marginal setting participles. Grease removal is accomplished during this pretreatment process through elimination of the surface scum.

Flotation

In this treatment process, oil, grease, and other suspended matter are removed from wastewater. A primary reason that flotation is used in the food industry is that it is effective in removing oil from wastewater.

Dissolved air flotation (DAF) removes suspended matter from wastewater by using small air bubbles. When discrete particles attach to tiny air bubbles, the specific gravity of the aggregate particle becomes less than that of water. The particle separates from the carrying liquid in an upward movement by attaching to the air bubble. The particles are then floated for removal from the wastewater. Also, this pretreatment process involves contact of the raw wastewater with a recycled, clarified effluent that has been pressurized through air injection in a pressure tank. The combined flow stream enters the clarification vessel, and the release of pressure causes tiny air bubbles to form, which move up to the surface of the water, carrying the suspended particles with them.

Air bubbles, which incorporate the flotation principle by removal of oil and suspended particles, can be created in the wastewater by (1) use of rotating impellers or air diffusers to form air bubbles at atmospheric pressure; (2) saturation of the liquid medium with air and subsequent combination of the mixture to a vacuum to create bubbles; and (3) saturation of air with liquid under high pressure and subsequent release to form bubbles.

Flocculating agents are commonly used to pretreat wastewater prior to treatment by a DAF unit. Treatment by DAF is widespread because of the relatively fast passage and be-

cause solids of nearly the same as or lower density than water can be removed. This treatment technique requires high investment and operating costs, especially for chemical additives and sludge handling.

DAF systems maintain a concentration of bacteria that are kept alive within the system to biodegrade pollutants in the effluent. A dewatering device, such as a belt filter, can be incorporated with DAF. After floatable oils and grease are captured, they can be chemically treated and the material conditioned, similar to a liquid-solid separation process.

Flotation technology has also been adapted to sludge-handling and to secondary and tertiary treatments. Food processors with substantial quantities of grease and oil in their wastewater use this technique as part of their waste treatment systems. In the past, one problem of flotation has been the presence of a turbulent flow; however, commercial high-rate flotation devices that eliminate turbulent flow are now available. The installation of lamellas (vertical baffles) can prevent unfavorable currents and short-circuiting and, with a properly designed feed well, can improve solid/liquid separation, producing higher underflow solid concentration in gravity thickeners and better effluent quality in gravity clarifiers.

Collected sludge from primary treatment contains approximately 2% to 6% solids, which should be concentrated before final disposal. Sludge treatment and disposal costs are the major expenses of sewage treatment if this product is not used as a fertilizer or for some other practical application. Some treatment systems biodegrade most of the organic matter and create little sludge. These systems can reduce treatment and disposal costs. If sludge is recovered as a by-product, disposal costs can be reduced, and the value of the salvaged material can provide enough profit to defray other treatment costs. Recovered solids (sludges) can also be treated by biological oxidation methods as a means of ultimate disposal.

A method developed in the past (Sofranec, 1991) utilizes a series of coagulants formed from corn starch to separate oil, grease, and suspended solids from wastewater prior to its discharge. The resultant grease and solids recovered from the DAF can be rendered. These starch-based coagulants are normally added to an equalization tank prior to the DAF system, where they can reduce the surface charge on the solids and grease, allowing the materials to coalesce and be removed by DAF.

In the past, wastewater treatment has normally involved the removal of solids from liquids. New equipment that utilizes a water loop principle can filter water from behind a chiller and flow it through a series of filters before returning it to the chiller. In this process, organic matter is filtered out so that the water can be recycled. Furthermore, water concentrates of as little as 3% of organic matter can be recycled through rendering equipment such as a disk dryer to concentrate a product into dry powder, with the vapors directed back into the evaporative system to be used as an energy source. The evaporative system provides free energy (Sofranec, 1991).

Secondary Treatment

Treatment through biological (or bacterial) degradation of dissolved organic matter through biological oxidation is the most common technique for secondary treatment. However, secondary treatment can range from the use of lagoons to sophisticated activated sludge processes and may also include chemical treatment to remove phosphorous and nitrogen or to aid in the flocculation of solids.

Most lagoons are earthen basins that contain a mixture of water and waste. The mixture in the lagoon is removed continuously without emptying the lagoon (Safley et al.,

1993). The design of most lagoons is similar. A dike or berm usually surrounds a lagoon as a lip of the basin that prevents spills and overflows. The depth of an impoundment (lagoon) depends on the volume of waste expected to be handled, with increased depth necessary to contain unforeseeable events, such as weather.

To accommodate such unforeseeable environmental changes, there is usually a storm event space left free of water. This is usually the amount of precipitation determined to have accumulated in 24 hours during the worst storm in the previous 100 years or the amount of precipitation from the wettest month in 25 years. Additional space reserved for safety measures includes wind set-up and wave run-up spaces to prevent overflows.

Circular or square lagoons enhance mixing and are usually less expensive to construct. If rectangular lagoons are used, a length:width ratio of 3:1 or less is recommended. Narrow areas isolated from the main body of water should be avoided because they may encourage mosquito proliferation. Although most lagoons are approximately 3 m deep, a greater depth requires less land, enhances mixing, and minimizes odors.

Lagoons must be sealed to prevent seepage that causes groundwater contamination. A lagoon can be sealed with hard-packed clay soil or with an industrial liner. A lagoon is considered sealed in most states if its lower boundaries (bottom and sides) have a maximum hydraulic conductivity of 10^{-7} cm/sec (Safley et al., 1993). A minimum of 30 cm of clay seal on the bottom and sides is required for most locations, but local ordinances may vary in their regulations. As lagoon depth increases, a thicker seal is required. Soil type, depth to water table, and depth to bedrock should be considered when locating a lagoon.

Although primary treatment removes screenable and readily settleable solid material, dissolved solids remain. The primary purpose of secondary treatment is to continue the removal of organic matter and to produce an effluent low in BOD and suspended solids. Microorganisms most frequently involved in biological oxidation of existing solids are those that naturally occur in water and soil environments. Microbial flora involved in biological oxidations can assimilate some of the dissolved solids and convert them into terminal oxidation products, such as carbon dioxide and water, or into cellular material that can be removed as particulate matter. Microbial cellular matter and assimilated organic matter continue to undergo aerobic degradation via the following endogenous respiratory reaction:

$$C_2H_9O_3N + 4O_2 \rightarrow 0.2C_3H_9O_3N + 4CO_2 + 0.8NH_3 + 2.4H_2O$$

Oxygen is required for these reactions. After treatment, the microbial suspended solids are separated from the water by gravity sedimentation. Some of the dissolved solids and small suspended solid matter in the form of colloidal and supracolloidal particles escape secondary clarification. If the effluent concentration is too high, the flow should be filtered before discharge, or clarification can be improved by the addition of flocculating chemicals.

Anaerobic Lagoons

Anaerobic lagoons can be designed with either a single stage or multiple stages. The disadvantages of multiple-stage lagoons are increased construction and land costs. Advantages are:

- There is less floating debris on the second and third stages, with a reduction in clogging of the flushing system or irrigation pump.

- The first lagoon, containing a higher concentration of waste, will not overflow.
- An adequate amount of bacteria will be available for waste treatment.
- The resulting effluent will be treated more thoroughly.

According to Safley et al. (1993), the lagoon start-up should be planned to minimize the amount of biological stress. Time is required for the appropriate bacteria to become established. Because anaerobic bacteria are slow growers, it may require a year or more for a lagoon to become fully mature. Lagoons should be started up in late spring or summer to permit bacterial establishment during the warmer weather. The amount of waste added should be increased gradually over 2 to 3 months.

Lagoons will accumulate fluid over time, due to precipitation, and should have fluid removed periodically. Typically, 40% to 50% of the active lagoon volume should remain, and fluid removal should be done only during warmer months to ensure that the bacteria can replenish themselves and will not decline below an effective level. In multiple-stage lagoons, the effluent should be removed from the last stage.

After 10 to 20 years, a lagoon will build up sludge that should be removed to prevent biological overloading. Three techniques are used for sludge removal. The first technique involves agitation equipment to resuspend the sludge and pump it out while the contents are thoroughly mixed. The remaining sludge will resettle once the agitation is stopped. The second technique involves the use of a floating dredge to move across the lagoon while a pump located on the dredge pumps the sludge over to another pump located on the shore. The second pump either sends the sludge to a holding tank or applies it to the land. The third technique is to pump the liquid to a lagoon and permit the remaining sludge to dry naturally. This long process may require several months.

Waste sludges may be produced by both primary and secondary treatment. These sludges typically require further stabilization before final disposal. Anaerobic lagoons and aerobic lagoons, frequently referred to as *stabilization ponds*, have been used for wastewater treatment and sludge stabilization. Use of this treatment technique has increased since the 1950s because of the relatively low capital investment, low operating costs, and ease of operation. Anaerobic and aerobic lagoons are not well suited where land costs are extremely high or for extremely large waste loads.

The treatment principle underlying lagoons is biological oxidation and solids sedimentation. Dissolved, suspended, and settled solids are converted to volatile gases, such as oxygen, carbon dioxide, and nitrogen; water; and biomass, such as microflora, macroflora, and fauna. Anaerobic and other lagoons equalize the discharge flow to further treatment facilities or receiving waters.

The depth of anaerobic lagoons varies from 2.5 m to 3.0 m. Surface area:volume ratios should be minimal. Anaerobic conditions are created throughout the entire lagoon, through heavy organic loads. Under anaerobic conditions, anaerobes digest the organic matter. Loading rates are expressed as BOD_5, COD, SS, and other measurements per unit volume of the lagoon. BOD_5 loadings range from 225 to 1,120 kg/ha/day. Operating temperatures of 22°C or higher are needed, with 4 to 20 days of detention. BOD reduction efficiency is typically 60% to 80% but is a fraction of the influent BOD and the determination time. Anaerobic lagoons are used as primary or secondary treatment of primary effluents containing high organic loads or as sludge treatment systems. Anaerobic lagoons are normally followed by aerobic lagoons or by trickling filters because their effluents remain high in organic matter (i.e., more than 100 mg of BOD_5).

Some treatment processes incorporate a combination of anaerobic and aerobic treatment. A completely mixed anaerobic tank reactor provides an environment for breaking down complex organic compounds into CO_2, CH_4, and simple organic compounds. The anaerobic tank reduces BOD_5 by 85% to 95%. The gases separate from the water and contain approximately 65% to 70% CH_4. The effluent flows on to an aerobic reactor for further treatment.

The previously described process involves the flow of anaerobically treated water to a degasification and flocculation tank, followed by a lamella clarifier, where the anaerobic microorganisms are separated and returned to the anaerobic tank. The supernatant flows by gravity to an aeration basin, where oxygen is supplied by mechanical aerators. Because the aeration step of the process has to remove only 5% to 15% of the original BOD_5, aerobic energy requirements are reduced. This process further involves settling out of aerobic sludge in the final clarifier, with a return to the aeration basin. Surplus sludge is recirculated into the anaerobic tank, where it enhances the bacterial activity and undergoes decomposition.

A combination of anaerobic and aerobic treatment can handle wide effluent variations. Anaerobic treatment responds slowly to flow variation because of the slow growth rate of the anaerobic microorganisms, but the faster growing aerobic microorganisms can generally treat the higher loads in the anaerobic effluent. (Note: It is then no longer anaerobic.)

Aerobic Lagoons

Aerobic lagoons use mechanical aerators to supply atmospheric oxygen for aiding biological oxidation. Mechanical agitators, designed to pull air under water and circulate it horizontally, can maintain a dissolved oxygen concentration of 1 to 3 mg/L at a BOD loading rate of up to 450 kg/ha/day. Because oxygen transfer occurs under water, neither freezing nor clogging occur. Aerated lagoons are classified as either aerated facultative lagoons—which have enough mixing to dispense dissolved oxygen but not enough to keep all the solids suspended—or as completely mixed aerated lagoons—which are mixed enough to keep all solids suspended. Approximately 20% of the BOD sent to an aerobic lagoon is converted to sludge solids, and the BOD influent is reduced by 70% to 90%. The solids produced will partially decompose in anaerobic sludge banks in facultative lagoons, but the completely mixed effluents usually require additional treatment, such as clarification or polishing pond treatment.

Trickling Filters

Trickling filters reduce BOD and SS by bacterial action and biological oxidation as wastewater passes in a thin layer over stationary media (usually rocks) arranged above an overdrain. Biological degradation occurs almost exactly as in the activated sludge process, except that the filter is a three-phase system in which the biofilm is fixed on the solid medium (stones or plastic). Aeration is accomplished by exposing large surface areas of wastewater to the atmosphere. Layers of zoogloea (filter sludge) grow on and attach to the medium surface. Primary treatment should precede this process if the wastewater suspended solids concentration exceeds 100 mg/L.

The efficiency of trickling filters is affected by temperature, waste characteristics, hydraulic loading rate, characteristics of the filter media, and depth of the filter. Media characteristics such as size, void space, and surface area, as well as hydraulic loading rates, tend to affect the performance of trickling filters more than do other factors. Removal efficiency is relatively independent of surface organic loading rate within broad ranges. Incorporation of plastic media with more surface area and void space than rock filter media has permitted improvements in design and efficiency. This treatment method is considered more rugged in operation and easier to maintain than activated sludge plants.

Activated Sludge

The activated sludge process is widely used for wastewater treatment. It requires a reactor that is an aeration tank or basin, a clarifier, and a pumping arrangement for returning a portion of the settled sludge to the reactor and discharging the balance to waste disposal. Primary treatment is optional. A portion of the clarifier-settled sludge is returned to be mixed with wastewater entering the reactor. The resulting biological solids concentration is much higher than what could be maintained without the recycle. The term *activated sludge* applies because this returned sludge has microorganisms that actively decompose the waste being treated. This mixture of influent wastewater and returned biological suspended solids is termed the *mixed liquor*. The activated sludge process is frequently called the *fluid-bed* biological oxidation system, whereas the trickling filter is referred to as a *fixed-bed* system.

The conventional activated sludge system has been designed for continuous secondary treatment of domestic sewage. It is not effective in treating inorganic dissolved solids but is very effective for the removal of all organic matter in the wastewater. This process may incorporate either surface aerators or air diffusers to achieve mixing. The influent organics are mixed with the activated sludge and undergo biological decomposition as they pass from the influent end of the reactor to the discharge end. The detention time in the reactor can vary from 6 hours to 3 days or more, depending on the strength of the wastewater and the method of operation selected. When the activated sludge contacts the influent waste, there is a short period—less than 30 minutes—when influent particulate matter is rapidly absorbed onto the gelatinous matrix of the returned sludge. Absorption removes a large portion of the influent BOD. The aeration mechanical and electrical equipment components of an activated sludge system are relatively expensive, and the energy costs are relatively high. This process can be operated for a very high treatment efficiency (95% to 98%) and can be modified to remove nitrogen and phosphorus without the use of chemicals.

The *extended aeration process* is a modification of the activated sludge plant. A typical application is the Pasveer and Carrousel type of oxidation ditches used in Europe and in other countries that serve a large population. The term *extended aeration* was given to this process because it is operated to minimize waste sludge production. This results in a lengthening of the aeration time to maintain the mixed liquor suspended solids at a concentration that will still settle efficiently in the clarifier. This sludge is sufficiently mineralized, and the excess quantity does not require any further treatment in a digester before dewatering. However, more power is consumed in extended aeration systems because all organics are stabilized aerobically. The major advantage of this process is that it is generally capable of giving high BOD removal efficiency (95% to 98%) while minimizing waste sludge handling. This process is operated without primary treatment.

The aerobic digestion of sludge achieves volatile solids stabilization similar to that of aerobic digestion if mechanical or pneumatic aeration is provided. This approach is sometimes used to stabilize surplus biological sludges generated in the activated sludge process and in its modifications, or in trickling filtration. It can also be used to stabilize primary sludges generated by settling prior to biological treatment.

The *contact stabilization process* is another modification of the activated sludge process, where advantage is taken of the fact that substrate removal occurs in two stages. The first stage, which lasts 0.5 to 1.0 hour, involves rapid adsorption of the colloidal, finely suspended and dissolved organic compounds in the sewage by the activated sludge solids. In the second phase, the adsorbed organic material is separated by gravity sedimentation, and the concentrated mixed liquor is oxidized in 3 to 6 hours. The first step occurs in the contact tank and the second in the stabilization tank. Therefore, the adsorption phase is separated from the oxidation decay phases.

Oxidation Ditch

This treatment technique has been developed as an efficient, easy-to-operate, and economical process for treating wastewater. The process maintains waste materials in contact with the sludge biomass for 20 to 30 hours under constant mixing and aeration. After the biological reactor step, the stabilized suspended solids enter a clarification step, which removes them from the water by settling. An oxidation ditch can accommodate BOD loadings of from 200 to 500 g/day applied for each cubic meter of available aeration space. Sludge solids should have a 16- to 20-day turnover (i.e., solids retention time or sludge age). For each kg of BOD applied, approximately 200 to 300 g of new sludge solids can be produced, with an expected BOD reduction of 90% to 95%. Temperature can have a significant influence on the waste removal performance of the oxidation ditch. Pinpoint biological flocs may develop and be discharged with the clarifier effluent, decreasing the performance efficiency under cold-weather operating conditions.

The typical oxidation ditch aeration basin design is either a single closed-loop channel or multiple closed-loop channels with serial flow. An attractive feature of oxidation ditches is that a minimum of operation attention is required once a proper operation is established. Several food processing firms use oxidation ditches for wastewater treatment.

There is a current interest in the use of the total barrier oxidation ditch (TBOD) design for treating municipal food processing and industrial wastewater. TBOD biologically purifies water as it mixes oxygen with waste particles and permits the bacteria to feed on these pollutants. The system can achieve high oxygen transfer efficiency at a single point along the ditch, which allows for effective process control and design flexibility. A constant, powerful flow of wastewater is then maintained, preventing settling of the biomass at the bottom of the ditch reactor. The aeration and pumping system consists of submerged, turbine draft tube aerators that transfer oxygen into the mixed liquid (Sofranec, 1991).

Land Application

The two types of land application techniques that are the most efficient are *infiltration* and *overland flow*. With land application techniques, the pollutants can harm vegetation, soil, and surface and ground waters if not properly operated. However, both of these treatment techniques can effectively remove organic carbon from high-strength wastewater. Pollutant removal efficiencies of approximately 98% for the infiltration flow system and 84% for the overland flow system can be attained. The advantage of higher efficiency obtained with an infiltration system is offset by its more expensive and complex distribution system. Less pollution of potable ground water supplies is usually experienced with the overland flow system.

Although land application has been a standby in the past for discharge of some food processing wastes, this approach is now limited. Hydraulic loads that are high may necessitate an unreasonably large amount of land. Runoff and proper utilization of nutrients can restrict the vegetation. Buildup of minerals and other materials in the soil has the potential for long-term liability for residues possibly as yet undiscovered (Rushing, 1992).

Rotating Biological Contactor

The rotating biological contactor (RBC) is an attached growth type of biological treatment system similar in concept to the trickling filters. Initial costs of this equipment are high, but operating costs and space requirements are moderate. This system consists of a number of large-diameter (approximately 3 m) and lightweight discs that are mounted 2 to 3 cm apart (to prevent bridging between the growths) on a horizontal shaft (in groups or packs, with baffles between each group to minimize surging or short-circuiting) to form an RBC unit (Figure 10–1). The discs are partially (30% to 40%) immersed and rotate slowly (0.5 to 10 rpm) as wastewater passes through a horizontal open tank, which usually has a semicircular bottom to fit the contour of the discs.

The RBC unit functions by attachment of microorganisms to the surface of the discs and grows by assimilating nutrients from the wastewater. Aeration is achieved through direct exposure of microorganisms to air when the surface of the disc is rotated above the water and by a thin film of water, which is

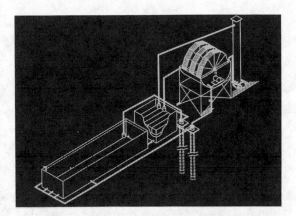

Figure 10–1 Rotating biological contacters used to remove ammonia from water of an aquaculture production operation. Courtesy of George Libey, Virginia Polytechnic Institute and State University, Blacksburg, Virginia.

aerated as it adheres to the disc's surface and rises out of the water. An acceleration in rotational speed increases the dissolved oxygen in the tank. The biofilm undergoes sloughing as in trickling filters, and these solids must be settled and removed. Although this treatment process is considered to be secondary treatment, primary sedimentation may be eliminated if the wastewater suspended solids are not unusually high (greater than 240 mg/L).

Magnetic Separation

This secondary physical treatment method has applications for tertiary use. The organic waste solids in suspension are chemically treated with magnetite (Fe_3O_4). Alum or ferric chloride coagulation flocculation is performed, and the coagulated particles subsequently contain magnetite. This process consists of a chamber containing a stainless steel wool matrix located in a magnetic field. The magnetized coagulated particles in wastewater suspension are passed through the chamber and adhere to the stainless steel wool in the magnetic field. The collected organic waste is removed by reducing the magnetic field to

zero and washing out the waste solids. This process was developed in Australia and has seen very few applications in North America.

Tertiary Treatment

Tertiary treatment processes for wastewater, which are collectively known as *advanced wastewater treatment*, are incorporated to improve the quality of waste treatment effluents to meet NPDES guidelines. Tertiary waste treatment is applied to food processing wastewaters to remove pollutants of food processing, such as colors, odors, brines, and flavoring compounds. Some of the processes for tertiary treatment of municipal waste treatment are frequently used as a primary waste treatment for certain food processors.

Physical Separation

Sand filters and microstrainers have been developed for tertiary wastewater treatment and purification. Both of these physical separation methods remove suspended solids down to the micrometer particle range.

The *microstrainer* is a rotating cylinder covered by a screening material (usually fine mesh nylon or metal fabric) housed in a horizontal position in an open tank. Wastewater enters the inside of the cylinder and is filtered through the screen. As the cylinder rotates slowly, an exposed section above the wastewater surface is backwashed to clean the screen and to collect the solids into a separate channel. Particle removal by microstaining is a function of screen pore size, which normally ranges from 20 to 65 μm. This is a relatively low-cost method of tertiary treatment because the screens are self-cleaning, and operating and maintenance costs are low. The effectiveness of this treatment method is limited by partial screen clogging, with a resultant decrease in the life of the screen. Also, microorganisms can grow in secondary water inside the cylinder, causing slime formation on the screen. Ultraviolet light or chlorination treatment has been used to reduce slime formation.

The *rapid sand filter* and *mixed media and continuous countercurrent filtration* are frequently used in tertiary wastewater treatment. This treatment method requires underdrains for removal of clarified liquids and a system for recovering collected solids. Automatic backwash mechanisms are available to enable self-cleaning of filters.

Physical-Chemical Separation

Food processing wastewaters contain a substantial amount of dissolved solids that can be removed effectively by various physical-chemical separation methods. One of the least costly tertiary treatments for removing refractory organics is *activated carbon adsorption*. The affinity of the organic solute for the carbon depends on the type of carbon and the solubility coefficient of the solute to water.

Ion exchange processes remove minerals, either cations or anions, by replacing them with other ions through the medium of charged resins. Multivalent ions are usually replaced by monovalents, such as Na^+ or H^+, and anions are replaced by OH^- or Cl^-. The principle purpose of this technique is to remove minerals considered harmful to the water supply or to recover valuable minerals from industrial processing wastewater.

The ion exchange resin usually consists of a network of cross-linked organic molecules known as *polymers*, which contain reactive functional groups that are usually strongly acidic, weakly acidic, or strongly basic. The resin is charged with ions such as H^+ or Na^+, which are replaced by multivalent ions from the wastewater passing through. Periodic recharging of the resin is necessary. This can be accomplished using strong acid or base solution. Ion exchange is especially beneficial for demineralizing water and whey. With the development of pulse-type ion exchange units, this method of treatment is becoming economically feasible.

Electrodialysis is used to remove minerals from brines and to demineralize whey. This process functions through the principle of alternately located cation- and anion-selective membranes placed in a current path. As ionic solutions pass through as a function of electric current, cations are transported through the cation-selective membrane and anions through the anion-selective membrane. Portions of the solution within the electrodialysis unit become concentrated with ions while the remainder is demineralized. Because of problems related to precipitation of salts, mineral fouling of anion plates, and membrane clogging by organic components in the water, electrodialysis as a tertiary treatment method has limited utility.

Tertiary Lagoons

These maturation lagoons, which are frequently called *polishing ponds*, are used for tertiary treatment of secondary effluents from activated sludge or trickling filter systems. This type of lagoon is usually from 0.3 to 1.5 m deep. Natural aeration, mechanical aeration, or photosynthesis serve as the oxygen source. BOD_5 loading rates normally range from 17 to 34 kg/ha/day, with a reduction range for BOD and SS of 80% to 90%. The waste removal efficiency of this system is influenced by temperature. This simple method of treatment requires practically no equipment or power, and minimal attention is required for the day-to-day operation. However, the land requirement of this process is the highest of the treatment methods.

Chemical Oxidations

Chemical oxidations through various chemicals are used for further oxidizing wastewater components in the tertiary treatment process. *Ozone* is a viable chemical oxidation treatment process. Ozone-generation equipment has made the process economically feasible. Ozone is a strong-oxidant that breaks down in water to form oxygen and nascent oxygen, which rapidly reacts with organic matter. This process also disinfects, removes taste and odor, and bleaches. Other chemicals used in chemical oxidations are chlorine, chlorine dioxide, oxygen, and permanganate.

Disinfection

For public health reasons, treated wastewaters should be disinfected before final discharge. Less disinfection is required as a result of the removal of microbes by primary and secondary wastewater treatment and by death of pathogenic microorganisms from extended exposure to natural environments. Because of the potential reaction of disinfectants with organic matter, it is more practical to disinfect at the end of wastewater treatment. Table 10–2, which relates the typical microbial population and load in domestic wastewater, illustrates the amount of contamination that can occur from wastewater of food processing operations.

Chemical oxidants; ultraviolet, gamma, and microwave irradiation; and physical methods, such as ultrasonic disruption and thermal application, are used as disinfectants. Chlorination has received less emphasis in recent years because of potentially carcinogenic organohalides in chlorinated waters. In addition, overchlorination of wastewater effluents can be toxic to fish. Chlorination and other chemical treatments do not kill all microorganisms. Certain algae, spore formers, and viruses (including pathogenic viruses) survive chlorination treatment.

It is practical to disinfect moderate volumes of effluent with ultraviolet irradiation equipment, an effective method with no residual effects that harm flora or fauna in receiving water. Thermal treatment is effective but is impractical for large volumes of effluent.

SUMMARY

To determine the optimal waste treatment systems, it is necessary to conduct a survey to ascertain waste volume and characteristics and water consumption records. The extent of waste pollution is measured by BOD, COD, DO, TOC, SS, TSS, TDS, and FOG.

Wastewater can be salvaged through recycling and reuse and recovery of solids. The basic phases of wastewater treatment are pretreatment by flow equalization, screening, and skimming; primary treatment by sedimentation and flotation; secondary treatment by anaerobic lagoons, aerobic lagoons, trickling filters, activated sludge, oxidation ditch processes, land application, RBCs, and ter-

Table 10–2 Microbial Characteristics of Domestic Wastewater

Microorganism	Quantity per 100 mL Wastewater
Total bacteria	10^9–10^{10}
Coliforms	10^6–10^9
Fecal streptococci	10^5–10^6
Salmonella typhosa	10^1–10^4
Viruses (plaque-forming units)	10^2–10^4

Source: Arceivala, 1981.

tiary treatment by physical separation, tertiary lagoons, and chemical oxidations. Disinfection of treated wastewater should follow other treatment phases to reduce the reaction of organic matter with the disinfectant.

STUDY QUESTIONS

1. What is biochemical oxygen demand?
2. What are the advantages and disadvantages of pretreatment of wastewater?
3. What are three methods of wastewater pretreatment?
4. What are two methods of primary treatment of wastewater?
5. Why are anaerobic lagoons used as a method of secondary treatment of wastewater?
6. How do aerobic lagoons function?
7. What is activated sludge?
8. What is the function of sand filters and microstrainers?

REFERENCES

Arceivala, S.J. 1981. *Wastewater treatment and disposal.* New York: Marcel Dekker, Inc.

Bakka, R.L. 1992. Wastewater issues associated with cleaning and sanitizing chemicals. *Dairy Food Sci Environ Sanit* 12: 274.

Forster, C.F. 1985. *Biotechnology and wastewater treatment.* London: Cambridge University Press.

Rushing, J.E. 1992. Water issues in food processing. *Dairy Food Sci Environ Sanit* 12: 280.

Safley, L.M., et al. 1993. *Lagoon management. Pork industry handbook.* West Lafayette, IN: Purdue University, Coop. Ext. Serv.

Sofranec, D. 1991. Wastewater woes. *Meat Proc* 46, November.

SUGGESTED READING

AOAC. 1990. *Official methods of analysis.* 15th ed. Washington, DC: Association of Official Analytical Chemists.

CHAPTER 11

Pest Control

The intent of this chapter is not to train pest control experts but to provide the sanitarian additional understanding of the impact of insects, rodents, and birds on the contamination of food supplies. The purpose of discussing pest control is to acquaint readers with the major pests that can contaminate the food supply and how the presence of these unwanted guests can be controlled. The food sanitarian has to contend with relatively few species of insects, rodents, and birds, but those encountered can cost the food industry billions of dollars every year.

An effective program against pests begins with a basic understanding of the characteristics of pest contamination sources and a comprehensive knowledge of safe and effective extermination and control procedures. If a pest control operator is not used to control pests, one or more employees (depending on the size of the organization) should be trained and held responsible for maintaining effective pest control.

Thorough housekeeping is an effective practice in ridding the premises of pests. A tidy operation facilitates the extermination of pests within the building(s) and complicates entry of pests from the outside. In addition to more difficult entry, pests have more difficulty finding suitable shelter where they can thrive and reproduce. Elimination of shelters, rubbish, decaying material, discarded supplies, and equipment will discourage the presence of insects and rodents. Pests may be found in enclosed areas under shelves, platforms, chutes, and ducts, especially if debris is allowed to accumulate in these areas. The same is true for breaks in walls and insulation. Discussion of pests and their control will follow.

INSECT INFESTATION

Cockroaches

The most common pests among food processing plants and food service facilities throughout the world are the cockroaches. Control of these pests is essential because they carry and spread various disease organisms. Many carry approximately 50 different microorganisms (such as *Salmonella* and *Shigella*), poliomyelitis, and *Vibrio cholerae*, the causative agent of cholera.

Cockroaches spread undesirable organisms through contact with food, especially through biting and chewing. Although they prefer foods that contain a large amount of carbohydrates, they will feed on any substance that humans will consume, as well as on human waste, decaying materials, dead insects (including other cockroaches), shoe linings, and paper and wood materials. Cockroaches are most active in dark areas and at night, when there is less disturbance from human activities.

These pests multiply rapidly by monthly production of small egg cases that may contain 15 to 40 eggs. The egg case is deposited in a hiding place for added protection. Young cockroaches begin feeding on the same material as the adults shortly after they hatch. Immature cockroaches look like adults except that they are smaller and do not have wings. They develop wings after growing larger and shedding their skin several times. Cockroaches live up to over a year and mate several times.

Identification of the specific kind of cockroach infesting an establishment can aid in the determination of the control technique. Three cockroach species most commonly invade commercial establishments in the United States. However, other cockroaches, such as the field cockroach (*Blatella vaga*), a pest in field crops, is gradually spreading to domestic premises in parts of the southern United States (Hill, 1990).

Species

German Cockroach (Blatella germanica)

The German cockroach is 13 to 20 mm long and pale brown, with two dark-brown stripes behind the head. Adults of both sexes have well-developed wings. The female carries the egg case protruding from the tip of the abdomen until hatching occurs. During the approximate lifetime of 9 months, each adult female produces approximately 130 offspring.

In food establishments, German cockroaches can infest the main processing or preparation rooms in addition to storage areas, offices, and welfare facilities. They prefer to inhabit warm crevices near heat sources and are not usually found in storage areas below ground level. German cockroaches are especially common in restaurants and may be found from floor to ceiling levels in rooms.

American Cockroach (Periplaneta americana)

This species is approximately 40 to 60 mm long and is the largest cockroach in the United States. Adults are reddish-brown to brown, and the young are pale brown. The female hides egg cases as soon as they are produced. This species produces more young than does the German cockroach because the adult female lives for 12 to 18 months, lays as many as 33 egg cases, and produces approximately 430 offspring.

American cockroaches tend to inhabit open, wet areas, such as basements, sewers, drainage areas, and garbage areas, although this species may be found in storage rooms. They tend to stay in places that are slightly cooler and have larger cracks and crevices than does the German cockroach. This species is most frequently found in large storage areas below ground level, on loading docks, or in basements of food processing plants.

Oriental Cockroach (Blatta orientalis)

The Oriental cockroach grows to approximately 25 mm long and is a shiny, dark brown to black. The wings are very short in the male and absent in the female. Young cockroaches of this species are pale brown. Egg cases from the females of this species are hidden soon after their formation. Females live 5 to 6 months and can produce one egg case per month for an approximate production of 80 cockroaches.

This species prefers a habitat similar to that of the American cockroach. In food plants, they normally inhabit below ground storage areas or those areas with a moist environment.

Detection

Cockroaches may be found in any location where food is being processed, stored, prepared, or served. These insects tend to hide and lay eggs in dark, warm, hard-to-clean areas. Their favorite harborages are small spaces in and between equipment and shelves and under shelf liners. When cockroaches need food that is not in these areas or when they are forced out by other cockroaches, they will come out into the light.

One of the easiest methods of checking for cockroach infestation is to enter a darkened production or storage area and turn on the lights. Also, a strong, oily odor that arises from a substance given off by certain glands of this insect can indicate the presence of cockroaches. Cockroaches deposit their feces almost everywhere they have visited. These droppings are small, black or brown, and almost spherical.

Control

Cockroaches are a year-round pest in food establishments. Thus, control of them should be a continuous operation through effective sanitation and use of chemicals. The most important form of control is effective sanitation. These pests require food, water, and a sheltered hiding place. Because these insects will eat practically anything, elimination of debris and maintenance of a tidy operation, including welfare facilities, through an ongoing sanitation program is the foundation for cockroach control.

Infestation can be reduced by filling cracks in floors and walls with caulking or other sealants. It is especially important to seal spaces where large pieces of equipment are improperly fitted to their bases or to the floor. These spaces provide an ideal habitat for these pests. Infestation can be reduced by deprivation of easy access via other sources. These hitchhikers can enter food establishments as cockroaches or as eggs in boxes, bags, raw foodstuffs, or other supplies. Incoming materials should be thoroughly examined and any insects or eggs removed. Cartons and boxes should be removed from the premises as soon as the supplies have been unpacked.

Use of chemical control should follow sanitary practices. Chemical control can be handled by a pest control operator, but integrated chemical control and sanitary practices can be more effective and more economical. Because insects such as cockroaches become inactive at approximately 5°C, refrigerated storage and refrigeration of other areas will reduce infestation. Cockroach control is usually based on the use of baits and bait stations, fungi, and possible nematodes.

Diazinon offers potential for the control of cockroaches. Amidinohydrozone (Dursban) has been developed and sold as a bait, and can be effective against cockroaches that resist other poisonous compounds, but the use of this insecticide indoors is not acceptable. A residual insecticide such as diazinon sprayed in hiding places is considered effective if these pests have not developed a resistance to this compound. This compound is sometimes supplemented with a pyrethrin-based nonresidual insecticide to force the insects from the hidden areas to the sprayed region, where improved contact with the insecticide can occur. Other compounds, such as flowable microencapsulated diazinon, are available for the control of cockroaches and other insects through spot, crack, or crevice treatment, but not for application in food-handling areas. Any compound applied as an insecticide for the control of cockroaches or other pests should be used according to the directions on the label.

Other Insects

The most common of the seasonal insects in food service and food processing plants are flies. The most populous varieties of flies associated with these establishments are the housefly and the fruit fly.

The housefly *(Musca domestica)*, which is found throughout the world, is an even greater pest than the cockroach. It is a pest to all segments of the community, transmitting a variety of pathogenic organisms to humans and their food. Examples are human disease such as typhoid, dysentery, infantile diarrhea, and streptococcal and staphylococcal infections.

Flies transmit diseases primarily because they feed on animal and human wastes and collect these pathogenic microorganisms on the feet, mouth, wings, and gut. These pathogens are deposited when the fly crawls on food or in the fly excrement. Because flies must take nourishment in liquid form, they secrete saliva on solid food and let the food dissolve before consumption. Fly spittle, or vomitus, is loaded with bacteria that contaminate food, equipment, supplies, and utensils.

Control of flies can be a challenge because these pests may enter a building that has openings only slightly larger than the head of a pin. Flies normally remain close to the area where they emerged as adults, even though they are lured to locations with odors and decaying materials. Air currents frequently carry flies a much greater distance than they normally travel. Flies are most likely to reside in warm locations protected from the wind, such as electric wires and garbage can rims. Houseflies lay an average of 120 eggs within a week of mating and can produce thousands of offspring during a single breeding season.

Warm, moist, decaying material that is protected from the sunlight provides an ideal environment for housefly eggs to hatch, with subsequent growth of fly larvae or maggots.

Houseflies are more abundant in the late summer and fall because the population has been building rapidly during the warm weather. When adult flies enter buildings for food and shelter, these pests generally remain. Flies are most active in a 12°C to 35°C environment. Below 6°C they are inactive, and below -5°C death can occur within a few hours. Heat paralysis sets in at approximately 40°C, and death can occur at 49°C.

It is difficult to control the size of a housefly population because they frequently breed in areas away from food establishments where decaying material exists. Therefore, the most effective means of controlling the fly population is to prevent them from entering processing, storage, preparation, and serving areas and reducing their population size within these areas.

Prevention of entry into food establishments can be accomplished by prompt and thorough removal of waste materials from food areas. Air screens, mesh screens (at least 16 mesh, recommended by the U.S. Public Health Service), and double doors discourage fly entry. Doors should be opened for receiving and/or shipping for a minimal amount of time, and air screens should be operational. Self-closing doors should remain open for a minimal amount of time.

To reduce attraction of flies around a food establishment, outdoor garbage storage should be as far away from doors as possible. If garbage is stored inside, this area should be separated by a wall from other locations and refrigerated to reduce decay and fly activity. Garbage should be stored in closed containers.

If flies have entered a facility, they can be controlled by the use of an electric flytrap or by other commercial traps, which attract adult flies to blue lights, killing them in electric grids. Electric fly traps should be used all day, and the catch basin should be cleaned daily. Chemical control through aerosols, sprays, or fogs, using chemicals such as pyrethrins can aid in fly control. The limited results are temporary, and use of chemicals is restricted in food facilities. Therefore, one should try control by exclusion and by the use of flytraps. At this writing, flytraps that contain the insecticide nithiazine appear to be effective against fly control outside of buildings.

Fruit flies *(Drosophila melanogaster)*, which are smaller than the housefly, are also considered seasonal and are most abundant in late summer and fall. Adult fruit flies are approximately 2 to 3 mm long, with red eyes and light-brown bodies. They are attracted to fruit, especially decaying fruit. These pests are not attracted to sewage or animal waste; thus, they carry less harmful bacteria.

The life cycle and feeding habits of fruit flies are similar to those of houseflies, except that these insects are attracted specifically to fruits. These pests proliferate most rapidly in late summer and early fall, when rotting plants and fruits are more abundant. The life span of a fruit fly is approximately 1 month.

Total eradication of the fruit fly is difficult. Use of mesh screens and air screens will decrease entry into food establishments. When entry occurs, electric traps are somewhat effective. One of the most effective methods of controlling these pests is to avoid accumulation of rotting fruits and fermenting foods.

Miscellaneous insect pests that plague food processing and food service operations are ants, beetles, and moths. The last two are generally found in dry storage areas. These pests can be identified through their webbing and holes in food and packaging materials. These insects can be kept in check through a tidy environment, good ventilation, cool and dry storage areas, and stock rotation.

Ants frequently nest in walls, especially around heat sources, such as hot-water pipes.

If infestation is suspected, sponges saturated with syrup should be placed in a number of locations to serve as bait in determining where the insecticides should be applied. Because ants, beetles, and moths can thrive on very small amounts of food, good housekeeping and proper storage of food and supplies are essential safeguards against these pests.

Silverfish and firebrats can reside in cracks, baseboards, window and doorframes, and between layers of pipe insulation. Because these pests thrive in undisturbed areas, their presence suggests inadequate and/or infrequent cleaning. Silverfish prefer a moist environment, e.g., basements and drains. The firebrat is more likely to be found in warmer environments, such as around steam pipes and furnaces.

INSECT DESTRUCTION

Pesticides

Pests should be destroyed without chemicals, if possible, because of the controversy and potential danger of pesticides. However, if these techniques are ineffective, it is necessary to use pesticides. To ensure proper and effective application of pesticides, use of a professional pest control firm should be considered. Restricted pesticides should be applied by a commercial applicator. Even if an exterminating firm is contracted, supervisory personnel from the food establishment should have a basic knowledge of these pests, insecticides, and regulations affecting use of these chemicals.

Residual insecticides are applied to obtain insecticidal effects for an extended period of time. In residual treatment, the chemicals are normally applied in spots or cracks and crevices. Some residual insecticides cannot be legally used in food areas. Therefore, extreme caution should be taken to avoid contamination of food, equipment, utensils, supplies, and other objects that come in contact with workers. People who use these chemicals should be familiar with the terms on the product labels, which describe the types of authorized applications and potential effects.

Another method of residual application of insecticides is *crack and crevice treatment*. Small amounts of insecticides are applied to cracks and crevices where insects hide or in areas where these pests may enter buildings—for example, expansion joints between the various elements of construction and between equipment and floors. Treatment at these locations is critical because these openings frequently lead to voids, such as hollow walls or equipment legs and bases. Other important areas where treatment is essential are conduits, junction or switch boxes, and motor housings.

Nonresidual insecticides are applied for the control of insects only during the time of treatment and are applied either as contact or as space treatments. *Contact treatment* is the application of a liquid spray for an immediate insecticidal effect. *Contact* refers to actual touching of the pests. This treatment method should be used only when there is a high probability that the spray will touch the pests. In *space treatment*, foggers, vapor dispensers, or aerosol devices are used to disperse insecticides into the air. This technique can control flying insects and crawling insects in the exposed area. Space spraying should be done to control the insect population.

Nonresidual insecticides may be dispensed by fogging as aerosols in food production areas when food is not exposed. This technique is used to apply pyrethrins, which are usually synergized with piperonyl butoxide. Other common insecticides are pyrethroids. Aerosol applications, which effectively kill flying and exposed insects, are frequently dispensed on a timed-release basis at a prearranged convenient time when food production and contact does not occur.

Fumigants are used in the food industry primarily to control insects that attack stored products. Their primary feature is the ability to reach hidden pests. These compounds are normally used for space treatment, typically on weekends, when processing operations are ceased for safety precautions. To ensure adequate dispersion, fumigants are often applied with air-moving equipment, such as ventilation machinery or fans. The major mode of fumigant action is through the activation of respiratory enzymes within the pest. Oxygen assimilation is blocked or delayed by most fumigants. The following chemicals are common fumigants for insects:

- *Phosphine*: The principal active compound in this fumigant is aluminum phosphide, which is usually contained in a permeable package or in pellets. This method of use permits controlled contact of the phosphine with moisture in the air to release hydrogen phosphide (phosphine), the active ingredient. This gas is very flammable. Instructions provided for use and storage provided by the supplier should be followed.
- *Methyl bromide*: This nonflammable fumigant is widely used. Methyl bromide penetrates effectively and acts as a respiratory toxin, apparently absorbed through the insect's cuticle. This fumigant has been evaluated critically by regulators, and it appears that it will be phased out in the future.
- *Ethylene oxide*: This nonresidual fumigant is normally mixed with carbon dioxide in a ratio of 1:9 (by weight) to reduce flammability and explosiveness. This insecticide, most frequently used for stored commodities, should be applied by a professional pest control operator.
- *Carbonyl sulfide*: This compound has been found to be toxic to a large number of species of stored-product insects. It has been patented as a fumigant for control of insects and mites in postharvest commodities. According to Brunner

(1994), carbonyl sulfide has many characteristics indicating that it could replace methyl bromide or phosphine, or both, under some circumstances. It is environmentally friendly, with good penetration and aeration characteristics. It is versatile, being toxic in short exposure periods or for a longer exposure time. This fumigant shows no adverse effects on seed germination and is an effective fumigant for other commodities.

Other Chemical Methods of Insect Control

Other potential methods of insect control include the use of *baits*. Baits are a combination of insect-attracting foods, such as sugar, and an insecticide.

Although baits are not always as convenient to use as other methods, they can be effective in controlling inaccessible areas of ant and cockroach infestations and in reducing outside fly populations. Because baits are a poisonous food, special precautions should be exercised in their use and storage. Commercial dry granular baits should be scattered thinly over feeding surfaces daily, or as needed, to provide initial knockdown and control of populations. Granular fly baits are satisfactory for outdoor use only. Liquid baits consist of an insecticide in water with an attractant such as sugar, corn syrup, or molasses. They may be applied using a sprayer or sprinkling can to walls, ceilings, or floors frequented by flies. Fly bait should be used regularly during the summer months to control population growth.

Mechanical Methods

None of the conventional devices to control insects mechanically is especially effective. Fly swatters are contaminated and spread insect carcasses and parts when being used, so they should not be permitted in food processing, storage, preparation, or sales areas. A viable mechanical device for the control of insects is the air curtain, which not only reduces cold air loss in a refrigerated facility but also protects against insect and dust entry into food establishments. Air curtains can be used for personnel doors and entrances large enough for loading trucks or for the passage of large equipment. An air curtain supplies a downward-directed fan that sweeps air across the door opening at rates of up to 125 m^3/min. Air curtains are most effective if the area being protected is under positive air pressure. This equipment is normally mounted outside and above the opening to be protected.

Insect Light Traps

One of the safest and most effective methods of fly control is the use of insect light traps. This technique does not have the potential hazard of toxic sprays.

Insect light traps use a high-voltage, low-amperage current on a conducting grid (Figure 11–1) placed in front of a quasi-ultraviolet (UV) irradiation source. This light source attracts the flies toward the light source, where they are electrocuted. Some light traps contain a "black light," which is effective at night, and a "blue-light," which is effective in the daytime.

Insect light traps in food processing plants and warehouses should be installed in stages, as follows:

- *Stage 1, interior perimeter:* These units should be placed near shipping and receiving doors, employee entrances, and personnel doors that provide access to the outside or anywhere else that flying insects may enter. Units should be placed 3 to 8 m inside the doors, away from strong air currents and out of traffic areas, where forklifts or other equipment may damage the units.
- *Stage 2, interior:* These units should be placed along the path that insects may

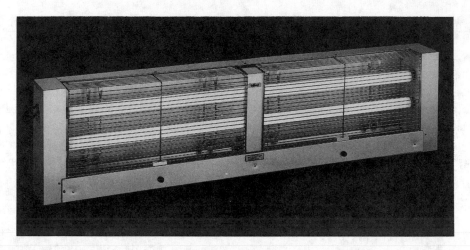

Figure 11–1 An insect light trap that attracts flies to the light source, subsequently electrocuting them. Courtesy of Dan Gilbert Industries, Inc.

follow to the processing areas. Within the processing areas, units with wings should be used to prevent dead insects from falling on the floor or on processing equipment.

• *Stage 3, exterior perimeter:* Covered docks, especially if refuse is being staged, should be protected. The units should be installed between the insects and the entrances, but not directly at the entrances.

Although this method of control can be effective, some precautions should be considered. The UV light source should be replaced yearly to attain optimal effectiveness. The trap should be strategically located to obtain optimal exposure and not to attract insects from the outside. The pan that collects the electrocuted insects should be emptied regularly to prevent infestation by dermestid beetles and pests that feed on dead insects.

Sticky Traps

These traps can consist of sticky flypaper, pieces of waterproofed cord, or flat pieces of plastic covered with a slow-drying adhesive.

Yellow plastic strips with a sticky covering will catch a wide variety of flying insects. Some sticky traps contain pheromones so that a specific insect species can be caught (Hill, 1990).

Biological Control

Use of biological control is frequently incorporated into integrated pest management (IPM) programs (discussed near the end of this chapter). One of the most widely used biological control schemes for the control of phytophagous insects is the development and incorporation of host-plant resistance. Resistance is attained through the use of plant species that are known to be refractory to attack. One of the promising techniques is the incorporation of gene splicing and recombinant DNA manipulation, which is being investigated universally. Other possibilities are the use of viruses, fungi, and bacteria to produce diseases in specific pests and of growth regulators, hormones, and pheromones that can influence sexual activity, primarily those that sterilize male pests. Equally important are growth regulators that interrupt the life cycle of insects and prevent their reproduction, usually in the pupal stage of development.

Growth regulators have been evaluated experimentally to control mosquitoes, fleas, and other insects. Insects can be potentially controlled by the use of milled diatomaceous earth. The milling process fragments the diatom shell into sharp microscopic particles, which penetrate the insects' wax coating whenever contact is made, causing moisture depletion and death. If particles of the shell enter the body cavity, they interfere with digestion, reproduction, and respiration.

Pheromone Traps

Some of these traps are based on the use of a specific sex pheromone and have a trapping chamber where the insects are caught. Some are constructed with a plastic funnel leading into the reception chamber, which contains an insecticide strip. Recently developed products containing microencapsulated pheromones provide a slow chemical release over a long period of time. Chemical attractants are now available for various species, and some are being used to control fruit flies. Hill (1990) indicated that food aroma attractants are usually more effective than are pheromones.

Hydroprene, a nonpesticide insect growth regulator (IGR), is appropriate for cockroach control in sensitive environments because of its margin of safety and toxicity. It has been approved by the Environmental Protection Agency for use in areas where food is present. An IGR can be destructive through disruption of the normal growth and development of immature cockroaches. Growth and development abnormalities include deformed wings and the inability to reproduce.

Monitoring of Infestants

A systematic inspection or surveillance and the recording of the species of pests present, their quantity, and origin should be established. Monitoring should include raw materials, adjuncts, and production and storage premises. Laboratory testing of samples should be performed using a filth test method. These methods can be found in the fifteenth edition of *Official Methods of Analysis* (Association of Official Analytical Chemists, 1990) or in other specialized analytical publications. Insects, insect fragments, eggs, larvae, and chrysalises should be identified, counted, and recorded to permit immediate pinpointing of dangerous infections or the appearance of abnormal variations. The same should be done for rodent hairs and excrement.

RODENTS

Rodents such as rats and mice are difficult to control because they have highly developed senses of hearing, touch, and smell. These pests can also effectively identify new or unfamiliar objects in their environment and protect themselves against these changes in the surroundings.

Rats

Rats can force their entry through openings as small as a quarter, can climb vertical brick walls, and can jump up to a meter vertically and 1.2 meters horizontally. These rodents are strong swimmers and are known for their ability to swim up through toilet bowl traps and floor drains.

Rats are dangerous and destructive. The National Restaurant Association has estimated that the loss from rodent damage could be as high as $10 billion per year. This includes consumption and contamination of food and structural damage to property, including damage from fires caused by rats' gnawing on electrical wiring. Of greater importance than economic losses from rat infestation is the serious health hazard from contamination of food, equipment, and utensils.

Rats directly or indirectly transmit diseases such as leptospirosis, murine typhus, and salmonellosis. Several million harmful microorganisms can be found in one rat dropping. When droppings dry and fall apart or are crushed, the particles can be carried into food by air movement within a room.

The most abundant kind of rat in the United States is the *Rattus norvegicus* (Norway rat), a red-brown to gray-brown rodent, sometimes known as the *sewer rat, barn rat, brown rat*, or *wharf rat*. Norway rats are normally brown, and are 18 to 25 cm long, excluding the tail, weigh 280 to 480 g, have a rather blunt nose and a thick-set body, and tend to live in burrows. A rat generally found in the South and along the Pacific coast and Hawaii is called the *Rattus rattus* (roof rat). This rat, which seeks an elevated location for its habitat, has more coordination than does the Norway rat and is smaller. It is black to slate-gray, 16.5 to 20 cm long, excluding the tail, and weighs 220 to 340 g. Roof rats will burrow or create nests in trees, vines, and other locations above the ground.

The female rat becomes fertile within 6 to 8 weeks after birth and can produce 6 to 8 young per litter, 4 to 7 times per year, if conditions are optimal for reproduction and survival. The typical female weans an average of 20 offspring per year.

Rats that receive an adequate amount of food will usually not move more than 50 m from their nest if mates are available. However, rat populations will adjust as food becomes scarce in one location or as a portion of the population starts to die from eradication methods. Rats and mice instinctively avoid uninterrupted expanses, especially if this potential barrier is lightly colored. Therefore, a potential rodent deterrent can be created by the construction of a 1.5-m-wide band of white gravel or granite chips around the outside perimeter of a building.

Mice

Mice, found frequently as the *Mus musculus domesticus* and *M. musculus brevirostris* varieties, are almost as cunning as rats. They are known to enter a building through a hole as small as a nickel. They are skilled swimmers that can swim through floor drains and toilet bowl traps; and they have an excellent sense of balance. Like rats, mice are filthy rodents and can spread diseases similar to those spread by rats. The house mouse, which is found everywhere in the United States, has a body length of 6 to 9 cm and weighs approximately 14 to 21 g. It has a small head and feet and large prominent ears.

Mice attain sexual maturity in approximately 1.5 months. Female mice produce 5 to 6 offspring per litter, up to 8 times per year. The typical female weans 30 to 35 young per year. Mice do not need a source of water because they can survive on water that they metabolize from food sources. However, they will drink liquids if available.

Mice are easily carried into food premises in crates and cartons. They are easier to trap than rats because they are less wary. Metal and wood-base snap traps are normally effective. Several traps may be spaced about 1 m apart. Hill (1990) stated that mice will usually accept a new object, such as a trap, often after about 10 minutes. Sodium fluorosilicate and the anticoagulant chlorophacinone are poisonous tracking powders that are effective in mice control. Except red squill, mice are destroyed by the same poisons as are rats.

Determination of Infestation

Rats and mice are nocturnal animals. Because they tend to be inactive during daylight hours, their presence is not always immediately detected. The presence of fecal droppings is one of the obvious signs of rodent infestation. Rat droppings range from 13 to 19 mm in length and up to 6 mm in diameter. Fecal material from the house mouse is approximately 3 mm long and 1 mm in diameter. Fresh droppings are black and shiny, with a pasty consistency. Older fecal material is brown and falls apart when touched.

Rats and mice generally follow the same path or runway between their nests and sources of food. In time, grease and dirt from their bodies form visible streaks on floors and other surfaces. Because rodents tend to keep in contact with vertical surfaces when they travel, runways along walls, rafters, steps, and inner sides of pipes are frequently visible. Rat and mouse tracks can be seen on dusty surfaces with light shining from an acute angle. Rodent tracks can be identified by spreading talc in areas of suspected rodent activity. Urine stains may be detected through the use of long-wavelength UV light, which will cause a yellow fluorescence on burlap bags and a pale, blue-white fluorescence on kraft paper.

The presence of rats can also be determined by gnaw marks. The incisor teeth of rats are strong enough to gnaw through metal pipes, unhardened concrete, sacks, wood, and corrugated materials to reach food. Teeth marks can be observed if gnawings are recent. A bumping noise at night, accompanied by shrill squeaks, fight noises, or gnawing sounds are clues that rodents may be present.

Control

Control of rodents, especially rats, is difficult because of their ability to adapt to the environment. The most effective method of rodent control is proper sanitation. Without an entrance to shelter and the presence of debris, which can nourish rodents, these pests cannot survive and will migrate to other locations. Without effective sanitation practices, poisons and traps will provide only a temporary reduction in a rodent population.

Prevention of Entry

Rat-proofing an establishment can be accomplished most effectively by the elimination of all possible entrances. Poorly fitting doors and improper masonry around external pipes can be flashed or covered with metal or filled with concrete to block entry of rodents. Vents, drains, and windows should be covered with screens. Because decay in building foundations will permit rats to burrow into buildings, masonry should be repaired, and fan openings and other potential entrances should be blocked.

Rodent control is enhanced by depriving them of a location to reside (harborage). Shapton and Shapton (1991) have suggested that outside equipment must be raised 23 to 30 cm clear of the surface to prevent rodent harborage. Shrubbery should be at least 10 m away from food facilities. Katsuyama and Strachan (1980) recommend that a grass-free strip 0.6 to 0.9 m in size be covered with a layer of gravel or stones 2.5 to 3.8 cm deep around food processing buildings. This feature helps to control weeds and rodents, and is convenient for the sanitation inspection rodent bait stations or traps placed against the building. Shapton and Shapton (1991) recommended that employees not eat on the food plant grounds because dropped food attracts rodents, birds, and insects.

Elimination of Rodent Shelters

Crowded storage rooms with poor housekeeping provide sheltered areas for rodents to build nests and reproduce. Rodents thrive in areas where garbage and other refuse are placed. These sheltered areas are less attractive to rodents if garbage is stored 0.5 m above the floor or ground. If waste containers are stored on concrete blocks, hiding places beneath them are eliminated. Waste containers should be constructed of heavy-duty plastic or galvanized metal with tight-fitting lids.

Housekeeping can be improved, with concomitant protection against rodent infestation, by storing foodstuffs on racks at least 15 cm above the floor or away from the walls. A white strip painted around the edge of the floor of storage areas reminds workers to stack products away from the walls and aids in the identification of rodent infestation through the presence of tracks, droppings, and hair.

Elimination of Rodent Food Sources

Proper storage of food and supplies combined with effective cleaning can aid in the elimination of food sources for rodents. Prompt cleaning of spills, regular sweeping of floors, and frequent removal of waste materials from the premises also reduce available food for rodents. Food ingredients and supplies should be stored in properly constructed containers that are tightly sealed.

Eradication

The more effective methods of eradicating rodents are poisoning, gassing, trapping, and the use of ultrasonic devices.

Poisoning

Poisoning can be an effective method of eradication; however, precautions are necessary because poison baits can be hazardous if consumed by humans. Examples of rodenticides are the anticoagulants, such as 3-(α Acetonylfurfuryl)-4-hydroxycoumarin (fumarin), 3-(α Acetonylbenzyl)-4-hydroxycoumarin (warfarin), 2-Pivaloyl-1,3-indandione (pival), brodifacoum, bromodiolone, and chlorophacinone. Because these multidose poisons must be consumed several times before death occurs, humans who accidentally consume poisoned bait once are not in danger.

The multiple-dose anticoagulants (chronic poisons), although safer than most other poisons, should be prepared and applied according to directions. The ideal locations for application are along rodent runways and near feeding sites. To protect against accidental consumption by humans, the baits should be placed in bait boxes or beneath shelters. Fresh bait should be put out daily for at least 2 weeks to ensure that the poison is effective.

Anticoagulant rodenticides are commercially available in several forms. They are sold as ready-to-use baits that can be placed in plastic or corrugated containers near rodent runways; in pellet form, mixed with grain for use in rodent burrows and dead spaces between walls; in small plastic packages for placement in rodent hiding places; in bait blocks; and as salts that are mixed with water. The sanitarian or pest control operator should record the location of all bait containers for easy inspection and replacement. If a bait is not consumed after two or more inspections, it should be relocated.

Anticoagulants have been extensively used to eradicate rats. One unfortunate result is that rats have become increasingly resistant to them. Consequently, new control strategies are being studied that utilize alternative cycles of anticoagulant and acute (fast-acting) rodenticides.

If immediate death of rodents is required, single-dose (acute) poisons, such as red squill and zinc phosphide, are available. These poisons can be mixed with fresh bait material, such as meat, cornmeal, and peanut butter. These baits should be prepared and administered according to directions provided by the manufacturer. Unfortunately, some of the single-dose poisons are effective against only Norway rats.

Baits should be deposited in several locations because rodents frequently travel only a limited distance from their shelter. If sufficient food and shelter are available, rats tend to stay within a radius of 50 m. Mice tend to journey about 10 m under similar conditions. If baits are dispersed too sparsely or are not strategically located, rodents may not locate the poison. Where signs of rodent activity are recent and numerous, baits should be dispersed liberally and replaced frequently. Rodents that are killed by single-dose poisons may die in their nests because they frequently carry food to this location; their death becomes apparent by the odor of decomposition. Dead rodents should be removed and burned or buried. Most mice are killed by compounds used for the eradication of rats.

Although use of bait is one of the most effective methods of eradication, rats that have suffered a toxic response by ingesting a poison, such as discomfort and pain but not death, may avoid the bait. They also become cautious if dead or dying rats are in the vicinity of a bait. Therefore, the most acceptable bait is the type with which the rat is most familiar.

Bait shyness and avoidance may be countered by the use of a prebait, a nonpoisoned bait introduced for approximately 1 week. Then the prebait is replaced with the same bait containing a rodenticide. Prebaiting is especially important if single-dose poisons are used but is not recommended when anticoagulants are incorporated. Because mice have weaker avoidance instincts than rats, prebaiting for mice is not necessary.

Tracking Powder

These compounds kill rats or, in the case of nontoxic powders, identify their presence and number. These powders may contain an anticoagulant or a single-dose poison. This poison kills rodents when they groom themselves after running through the powder. Such powders are effective if the food supply is abundant. It is best to use self-contained bait boxes placed inside the buildings where the food products are processed, prepared, or stored to restrict the spread of these poisoned baits. Tracking powders are less effective against rats than mice, but sodium fluorosilicate is an effective rodenticide (Hill, 1990).

Gassing

This technique should be used only if other eradication methods are not effective. If this approach is necessary, rodent burrows should be gassed with a compound such as methyl bromide only by a professional exterminator or a thoroughly trained employee. Rodent burrows should not be gassed if they are less than 6 m from a building because burrows can extend beneath a closely located building.

Trapping

This is a slow but generally safe method of rodent eradication. Traps should be placed at right angles to rodent runways, with the baited or trigger end toward the wall. Any food that appeals to rodents can be used as bait. Traps should be checked daily, with trapped rodents removed and bait replaced as needed. Trapping should be considered a supplementary procedure to other methods of eradication, and an abundance of traps should be used. When setting traps, the sanitarian should be aware of the rat's innate shyness and adaptability. Rats can avoid traps as effectively as they can bait. An effective mouse trap is the glueboard, which physically prevents a mouse from escaping by sticking to its feet. After use, the disposable tray and mouse are discarded by the pest control operator, and a new tray is placed in the most strategic location.

Ultrasonic Devices

This eradication method uses sound waves that are supposed to repel the entry of rodents into areas where the device is installed. Although this method can reduce the presence of rodents, with prolonged hunger, rodents ignore the sound barriers. Furthermore, ultrasound does not provide randomly and continually varying frequencies, which may be more effective.

BIRDS

Birds such as *Columba livia* (pigeons), *Passer domesticus* (sparrows), and *Sturnus vulgaris* (starlings) may present problems for the food facility. Their droppings are unsightly and can carry microorganisms detrimental to humans. Birds are potential carriers of mites, mycosis, ornithosis, pseudotuberculosis, toxoplasmosis, salmonellosis, and organisms that cause encephalitis, psittacosis, and other diseases. Insect infestations may also occur from those brought into the plant by birds.

A bird population can be reduced through proper management and sanitation. If sanitary practices are followed to remove food from the site, birds will not be attracted. Entry into buildings can be reduced through the installation of screens on doors, windows, and ventilation openings.

Trapping is generally considered an acceptable method of bird control. Wires that administer a mild electric shock and pastes that repel birds are also effective in preventing them from roosting near food establishments. However, electric wires are expensive and require frequent inspection and maintenance. Flashing lights and noisemaking devices have a limited effect on birds, which soon become accustomed to this equipment. Other techniques that may be effective if conducted repeatedly are removal of bird nests and spraying of birds with water as a form of harassment. The most effective procedure for eradication is employment of an exterminator who specializes in bird control. A professional exterminator provides expertise and equipment required for the safe use of chemicals to combat birds.

Bird density can be reduced through the use of commercially available chemical poisons, although these compounds should not be used inside a food establishment. Strychnine has been used in the past; however, its incorporation is restricted by some local regulations. Strychnine alkaloid is used at a concentration of 0.6% to coat baits such as cereal grains. Dead birds should be removed so that dogs and cats will not eat them and suffer from secondary poisoning. Another compound that controls bird density is 4-aminopyridine. In addition to killing birds, it causes the affected birds to make distress sounds and to behave abnormally, thus frightening away those that remain. Azacosterol is a temporary sterilant approved only for the control of pigeons. A biological control method such as this offers potential with less risk than other compounds but provides only a long-term solution, especially in a long-lived species such as pigeons. Minimal intermediate value from this compound is provided to the sanitarian who must rid a bird population immediately.

Birds can be controlled through trapping. Live decoys are required for maximal efficiency. Starlings have been effectively trapped by decoys and by an Australian crow trap. Tunnel traps and sparrow traps can also be effective. Pigeons can be trapped with a device containing bars that swing inward into a trap baited with grain. A major limitation of trapping is the cost of labor and materials.

USE OF PESTICIDES

Insecticides should not be sprayed in food areas during hours of operation. They should be applied only after the shift, over the weekend, or at other times when the food establishment is closed. Precautions should be taken to ensure against spattering or drift of the insecticide out of the treatment area to adjacent surfaces or onto food. Insecticidal dusts, which generally contain in dry form the same toxic compounds that are present in sprays, are also available. They require more skill in application than do sprays and should be administered only by professional pest control operators.

Prior to the use of insecticides approved for edible food products or supply storage areas, all exposed food and supply items should be covered or removed from the area to be treated. The equipment used in spraying inevitably will become contaminated and must be thoroughly cleansed before reuse. This is best accomplished by scrubbing with a cleaning compound and hot water, then rinsing. Products containing residual-type insecticides should not be used on any surfaces that come into contact with food. A fumigation procedure is not recommended unless it appears to be the only effective method, and even then only when it is carried out by a professional fumigator. Under no circumstances should regular plant personnel or supervisors attempt this type of work unless they are thoroughly trained. Even when professional fumigators are used, the plant managers should ensure themselves that all precautions have been taken in accordance with accepted safety practices.

The following precautions, suggested by the National Restaurant Association Education Foundation (1992), should be considered when applying pesticides:

1. Pesticide containers should be properly identified and labeled.
2. Exterminators employed should have insurance on their work to protect the establishment, employees, and customers.
3. Instructions should be followed when using pesticides. These chemicals should be used for only the designated purposes. An insecticide effective against one type of insect may not destroy other pests.
4. The weakest poison that will destroy the pests should be used with the recommended concentration.
5. Oil-based and water-based sprays should be used in appropriate locations. Oil-based sprays should be applied where water can cause an electrical short circuit, shrink fabric, or cause mildew. Water-based sprays should be applied in locations where oil may cause fire, damage to rubber or asphalt, or an objectionable odor.
6. Prolonged exposure to sprays should be avoided. Protective clothing should be worn during application, and hands should be washed after the application of pesticides.
7. Food, equipment, and utensils should not be contaminated with pesticides.
8. If accidental poisoning occurs, a physician should be called. If a physician is unavailable, a fire department, rescue squad, or poison control center should be contacted. If immediate assistance cannot be obtained, treatment should include induction of vomiting by inserting a finger down the throat, with a follow-up of 2 tablespoons of Epsom salts or milk of magnesia in water, followed by one or more glasses of milk and/or water. If the poison does not present immediate danger, no action should be taken until a physician arrives. Poisoning from heavy metals should be treated with the administration of a half-teaspoon of bicarbonate of soda in a glass of water, 1 tablespoon of salt in a glass of warm water (until vomit is clear), 2 tablespoons of Epsom salts in a glass of water, and two or more glasses of wa-

ter. If strychnine poisoning occurs, administer 1 tablespoon of salt in a glass of water within 10 minutes to induce vomiting, followed by 1 teaspoon of activated charcoal in half a glass of water. The victim should then be laid down and kept warm.

Chemical pesticides are not considered to be a substitute for effective sanitation. Rigid sanitary practices are more effective and more economical than are pesticides. Even with effective pesticides, pests will return when unsanitary conditions prevail.

To minimize possible contamination, a food facility should store on the premises only pesticides essential to control pests that present a problem to the establishment. Pesticide supplies should be checked periodically to verify inventories and to inspect product condition. The following storage precautions should be observed:

1. Pesticides should be stored in a dry area and at a temperature that does not exceed 35°C.
2. The area where pesticides are stored should be located away from food handling and food storage areas, and should be locked. These compounds should be stored separately from other hazardous materials, such as cleaning compounds, petroleum products, and other chemicals.
3. Pesticides should not be transferred from their labeled package to any other storage container. Storage of pesticides in empty food containers can cause pesticide poisoning.
4. Empty pesticide containers should be placed in plastic receptacles marked for disposal of hazardous wastes. Even empty containers are a potential hazard because residual toxic materials may be present. Paper and cardboard may be

incinerated, but empty aerosol cans should not be destroyed by burning. Local regulatory requirements related to restricted pesticides and general use and disposal should be followed.

INTEGRATED PEST MANAGEMENT

Because of limitations of chemical pesticides, integrated pest control programs based on predicted ecological and economic consequences have been developed. Most single insect control methods have not been successful, and insect resistance to pesticides has become extensive. Pesticide concentration is a common concern, especially after moisture removal steps in processing (Petersen et al., 1996). Thus, a variety of methods have been selected and integrated into a control program for the target pest. This program is called *integrated pest management* (IPM). Its major objective is to control pests economically through environmentally sound techniques, many of which use biological control. The goals of IPM are to use pesticides wisely and to seek alternatives to commonly used pesticides.

IPM implies that pests are "managed" and not necessarily eliminated. However, the ultimate objective of pest management in food processing is to prevent or eliminate pests. Several food processing and preparation firms have discovered the benefits of IPM as a means for pest control, due to the progress accomplished in the development and implementation of these methods since the early 1970s (Brunner, 1994). Economic, social/ psychological, and environmental advantages may be attained through IPM. Outlook for the acceptance of IPM methods is encouraging and should continue to improve over time with continued exposure. The apparent benefits are realized through lower costs, increased pest control, and reduced pesticide usage by up to 60% (Paschall et al., 1992). Pest control practices are classified as inspection, housekeeping, and physical, mechanical, and chemical methods. The integrated use of these practices in a complementary manner is essential for economical, effective, and safe pest management (Mills and Pedersen, 1990). A brief discussion of control practices follows.

Inspection

Inspection is a preventive, monitoring control measure that is time-consuming but important and cost-effective. Increased practice of IPM to replace chemical control practices has made inspection a more critical function. This function can identify existing problems and detect potential problems, and can monitor an ongoing sanitation problem. Both formal and informal inspections should be conducted periodically (e.g., monthly).

Formal inspections should be conducted with a predetermined frequency. These inspections should be thorough and should evaluate the overall progress and effectiveness of pest management. If well-qualified inspectors can be obtained from outside the plant (e.g., corporate staff inspector, consultant, or contracting inspection service representative), this resource should be used.

Informal inspections should be conducted by plant personnel assigned to specific work areas on an ongoing basis. Supervisory personnel should encourage and expect awareness of sanitation problems that may reduce pest control effectiveness among plant personnel as they conduct their normal tasks.

Inspections should include raw materials, manufactured or prepared products, site, facilities, and equipment. Inspectors should be equipped with a flashlight, equipment-opening tools, and sample containers. An inspection form should be devised as a guide and for recoding results. These forms provide written identification of potential problems and identification of problem areas.

Two types of pheromones, sex and aggregation pheromones, are being evaluated as a part of IPM programs for stored-product insects, and a few are being used principally for monitoring infestation (Mills and Pedersen, 1990). Pheromones are being considered because they are chemicals emitted by insects that are perceived and reacted to by others of the same or related species. Several stored-product pest pheromones have been identified and synthesized commercially. Aggregation pheromones are potential baits for monitoring devices utilizing sticky or pitfall traps.

Housekeeping

Mills and Pedersen (1990) suggested that standards of cleanliness and cleaning schedules must be established with direct accountability for cleaning activity. These authors suggested that, in many areas, cleaning must be continuous, as even small amounts of undisturbed product residues can attract infestation and provide adequate pest harborage.

Physical and Mechanical Methods

Because many pesticides once commonly used are no longer allowed in the control of pests, physical and mechanical methods have become more important. Examples are rodent traps, glue boards, and electric fly traps. Generally, these methods are noncontaminating and can fill some of the gaps in an IPM program left by reduced or restricted pesticide use. One of the effective methods is temperature manipulation, which is sometimes combined with forced-air movement. Because the optimal temperature for most insect species is 24°C to 34°C, variation above or below this range can reduce pest proliferation.

Insects depend as much on suitable moisture levels as on acceptable temperatures;

thus, moisture content is critical in determining whether proliferation occurs. Lower moisture content (especially below 12%) of foods discourages insect growth.

Several forms of radiation, such as radio frequencies, microwaves, infrared and ultraviolet light, gamma rays, X rays, and accelerated electrons can effectively disinfect food products, but not all of these methods are effective and practical. Gamma rays, X rays, and accelerated electrons have commercial applications for insect disinfection.

Chemical Methods

Pesticides and other chemicals, such as repellents, pheromones and sticky materials for traps, barriers, or repellency are incorporated when needed. Whoever applies pesticides must be trained to know the safe, approved, and effective use of each chemical. Application of restricted-use pesticides requires state certification of the application.

The EPA classifies pesticides as being either for general use or restricted use. Those classified as restricted use are more likely to adversely affect the environment or to injure the applicator. Thus, these pesticides can be purchased and used by only certified applicators or by persons directly under a certified applicator's supervision. Through an EPA-approved program, states train and certify applicators.

The pesticide storage area should be large enough to store normal supplies of pesticide materials adequately and neatly. This should be in a separate building, if possible, or stored in isolated areas from food. The area should be equipped with power ventilation exhausting to the outside and should never be cross-ventilated with food processing or food con-

tainer storage areas. This storage area should be totally enclosed by walls, and the door should be locked to prevent unauthorized entry. The storage environment should be dry, with the temperature controlled sufficiently to protect the pesticides. Pesticide containers should be stored with the label plainly visible and a current inventory maintained. Pesticide handling and application equipment should include rubber gloves, protective outer garments, and respirators such as dust masks or self-contained breathing apparatus (SCUBA) equipment.

IPM principles will be applied to future pest control programs because of the success of this program and increased environmental concerns associated with the indiscriminant use of chemical insecticides. The control of insects in commodities by the IPM technique influences the overall infestation levels in plants processing these materials in foods.

SUMMARY

Pests of major significance to the food industry include the German cockroach, American cockroach, Oriental cockroach, housefly, fruit fly, Norway rat, house mouse, pigeon, sparrow, and starling. Control of pests can be most effective through prevention of entry into food establishments and the elimination of shelter areas and food sources for subsistence and reproduction. If pests become established, pesticides, traps, and other control techniques are essential. These eradication devices should be considered a supplement to, rather than a replacement for, effective sanitation practices. Because pesticides are toxic, these compounds should be selected and handled carefully. Precautions during use, storage, and disposal are essential. Although a trained employee can handle pesticides, a professional exterminator should be employed for complex and hazardous applications.

STUDY QUESTIONS

1. What adverse effects do cockroaches have on a food facility?
2. How are cockroaches best controlled?
3. Why are flies so unsanitary?
4. How are flies destroyed most effectively?
5. What is the difference between a residual and a nonresidual insecticide?
6. How does an insect light trap destroy flies?
7. What are insect pheromones?
8. How are rats and mice controlled most effectively?
9. How are birds controlled most effectively?
10. What is integrated pest management?

REFERENCES

Association of Official Analytical Chemists, Inc. 1990. *Official methods of analysis*. 15th ed. Arlington, VA.

Brunner, J.F. 1994. IPM in fruit tree crops. *Food Rev Int* 10:135.

Hill, D.S. 1990. *Pests of stored products and their control*. Boca Raton, FL: CRC Press.

Katsuyama, A.M., and J.P. Strachan. 1980. *Principles of food processing sanitation*. Washington DC: The Food Processors Institute.

Mills, R., and J. Pedersen. 1990. *A flour mill sanitation manual,* 55. St. Paul, MN: Eagan Press.

National Restaurant Association Education Foundation. 1992. *Applied foodservice sanitation*. 4th ed. New York: John Wiley & Sons, in cooperation with the Education Foundation of the Restaurant Association.

Paschall, M.J., et al. 1992. Washington bugs out, integrated pest management saves crops and the environment. *J Am Diet Assoc* 92: 93.

Petersen, B., et al. 1996. Pesticide degradation: Exceptions to the rule. *Food Tech* 50: 221.

Shapton, D.A., and N.F. Shapton, eds. 1991. Buildings. In *Principles and practices for the safe processing of foods*, 37. Oxford: Butterworth-Heinemann.

SUGGESTED READING

Desmarchefier, J.M. 1994. Carbonyl sulfide as a fumigant for control of insects and mites. In *Proc 6th International Working Conference on Stored Product Protection,* 77–82. Canberra, UK: Australian AB International.

CHAPTER 12

Sanitary Design and Construction of Food Facilities

New and renovated food processing and food service facilities should be planned to enhance a hygienic operation and effective cleaning. Because most equipment and facilities are designed to feature functionality, hygienic design and construction principles should be emphasized to ensure a sanitary operation. A hygienically designed facility can enhance the wholesomeness of all foods and improve the effectiveness and efficiency of a sanitation program.

SITE SELECTION

Site selection plays an important role in the development of a hygienic operation. Food facilities should not be constructed near chemical plants that emit noxious odors or near salvage or water disposal operations. Food products that are relatively high in fat will readily pick up bad odors and flavors, and pathogenic microorganisms can be picked up by the wind and blown on the manufactured products unless special filters are added to the intake air systems. Drainage is important, as sites located close to standing water with poor drainage are more likely to have *Listeria monocytogenes* in the facility and on manufactured products. Large bodies of water will attract scavenger birds that carry *Salmonella*. Standing water provides a breed- ing environment conducive to insects and provides water to sustain the lives of rodents and other pests. A food manufacturing facil- ity should not be located near existing pest harborages for further protection against pathogenic microorganisms.

Troller (1993) suggested that the location of a food plant near small streams and drain- age ditches should be avoided, as should lo- cations near refuse dumps, landfills, and equipment storage yards. Land reclaimed from swampy ground or refuse disposal areas should not receive serious consideration.

The selected site should permit future ex- pansion. Overcrowded facilities are ineffi- cient and pose a sanitation-related liability. Water availability and adequate waste dis- posal facilities should be considered. Trees and foliage that provide food and/or harbor- age for birds should not be planted close to the buildings, and existing growth should be removed. Parking lots should be paved to pre- vent dust and should be well drained to facili- tate prompt removal of rainwater. A perim- eter chain-link type fence that surrounds the property should be considered.

SITE PREPARATION

Graham (1991a) recommended that toxic materials be removed, if present at the site, to prevent potential contamination. The site should be graded to prevent standing water, which provides breeding sites for insects (especially mosquitoes). Storm sewers should be provided. Many municipalities demand landscaping for aesthetic reasons; however, shrubbery should be at least 10 m from buildings to eliminate protection for pests such as birds, rodents, and insects. Grass should not be present within 1 m of building walls so that a pea gravel strip 7.5- to 10-cm deep can be laid over polyethylene or the equivalent to discourage rodent entry.

BUILDING CONSTRUCTION CONSIDERATIONS

Walls

The foundation and walls of a food processing or food service facility should be impervious to moisture, easily cleaned, and constructed to prevent rodent entry. Graham (1991b) recommended that a slab floor contain footers constructed with a rodent flange 60 cm below grade, extending 30 cm out at right angles to the foundation to prevent rats from burrowing under the floor slab and gnawing their way into the building. If a basement or cellar is planned, the floor should be tied directly to the solid wall foundation to create a solid box as a pest barrier.

The most appropriate walls are poured concrete, troweled smooth to a maximum of nine holes per square meter, none of which exceeds 3 mm. Poured concrete is more expensive and requires on-site construction of forms and finishing, but this material does not have seams that require the caulking that is needed for precast or tilt-up construction.

An alternative material is notched beams, notched precast wall panels, and double-tee precast roof panels. This technique involves precasting the wall panels and the roof sup-

port beams, complete with notches large enough to accommodate the precast double tees of the roof panels. By fitting inside the notch, dust-collecting flat surfaces on top of the beams or wall panels are eliminated. Caulking the spaces around the double tees creates a hygienic structure. Caution about precast, tilt-up, and concrete block construction is important. Use of a parting agent to enhance the removal of the panel or block from the form necessitates that the agent be tested to ensure compatibility with any wall covering (i.e., paint and epoxy). Incompatibility results in paint peeling.

If concrete block wall construction is incorporated, it must be a high-density type. Less porous material reduces moisture absorption and reduces microbial growth. An effective sealer can close pores to improve hygienic design. Graham (1991b) recommended that when concrete block is laid, the first course should have the center core filled with mortar to provide an effective seal against insects entering through the joint created at the junction with the foundation. Walls should be covered at the floor, to a minimum radius of 2.5 cm. The top course of block should be capped off to prevent access by rodents and insects.

Corrugated metal siding is not recommended because it is not reliable in stopping the entry of insects and rodents, and because this material is damaged easily. If corrugated metal is incorporated, the outside corrugation must be blocked and caulked at the top and at the foundation to discourage pest entry. Wall penetration for access by utilities should be sealed the same day that this operation is performed to reduce pest invasion.

Wet processing areas should have glazed ceramic tile to enhance the ability to clean inside walls. This material is resistant to food, blood, acid, alkali, cleaning compound, and sanitizer. Tile walls are expensive to install but inexpensive and easy to maintain. Epoxy paints over a compatible sealer provide additional protection.

Loading Dock

Loading docks and platforms should be constructed at least 1 m above the ground. The underside of the dock opening should be lined with a smooth, impervious material, such as plastic or galvanized metal, to prevent rodents from climbing into the building. Rodent access should be denied through a dock or platform overhang of 30 cm that will not permit a roosting location for birds. Pest entry is discouraged by the use of truck door seals and air curtains.

Roof Construction

A logical roof type for precast concrete wall panels is a precast double tee. This design is attractive and hygienic. Pitch and gravel roofs should not be installed over food processing or preparation areas, as they are difficult to clean. Low-moisture materials, such as grain, starch, and flour, can be carried out through vents and will attract birds and insects and encourage the growth of weeds, bacteria, molds, and yeasts. Graham (1991c) has recommended smooth membrane-type roofs because they can be swept, hosed, and kept clean more effectively than other roofs. Roof openings for air handling or other uses should be screened, flashed, or sealed to prevent the entry of contaminants such as insects, water, and dust. Roof opening caps and mounted air-handling units should be insulated with sandwich panel insulation, as open insulation is difficult to clean and can become infested with insects.

Windows

Effective environmental control and adequate lighting negates the need for windows, which can present a sanitation hazard, due to breakage and contamination from pests, dust, and other sources. Windows increase maintenance through required repair, cleaning, and caulking. If windows are installed, less difficulties exist if they cannot be opened and if they are constructed of unbreakable polycarbonate material (Graham, 1991c). Furthermore, the sill on the outside should be sloped at a 60° angle to prevent roosting and debris accumulation. The next best design for windows is to place them flush with the outside wall and to use the same slope for the inside sill. Some municipalities require windows to conform with local fire codes.

Doors

Doors present an opportunity for the entry of pests and airborne contaminants. A double-door entry design will reduce airborne contamination and pest entry. The exterior of the doors should be equipped with air curtains to enhance effective sanitation. Air curtains should have enough air velocity (minimum of 500 m/min) to prevent the entry of insects and air contaminants and should extend completely across the opening with a down-and-out sweep. Air curtains should be wired directly into the door opening switch to permit air movement simultaneous with the door opening and continue through its closing.

Ceilings

False ceilings are not recommended because the area above can become infested with insects and other contamination. If a dropped ceiling is installed, it should be constructed as if it is another floor sealed off from the processing area below and should contain utility runs, air handling ducts, and fans. Construction usually includes catwalks so that the maintenance crew can service the equipment or lines passing through the area.

This area should be kept pressurized to avoid dust infiltration.

Ceiling construction should be a smooth concrete slab of exposed double tees with caulked joints. If exposed structural steel is used over processing areas, it should be enclosed in concrete, granite, or the equivalent to avoid overhead areas that collect dust and debris or provide rodent runways or insect harborage. Metal panels should not be installed because their high heat transfer rate can cause moisture condensation. Furthermore, the metal expansion and contraction complicates the maintenance of seals at the joints, resulting in harborages for insects. Fiberglas batting should not be installed, as rodents live and thrive in it. Preferred insulation is Styrofoam, foam glass, and other insert materials. The hazards of asbestos prohibit its use.

Floors

Floors may range from plain, sealed concrete in warehouses to acid brick in high-impact, high-temperature, high-chemical-exposure areas. Monolithic floors are gaining in popularity because they are seamless, easier to apply, and less expensive than brick or tile. These floors are both epoxy- and polyurethane-based and are either rolled or troweled on by hand.

Floors in food facilities should be impervious to water; free of cracks and crevices, and resistant to chemicals. Although tile floors provide an acceptable surface, with heavy wear, grouting loss can occur, which results in the penetration of water. Plastic or asphalt membranes may be laid between the underlying concrete surface and the tile or brick.

PROCESSING AND DESIGN CONSIDERATIONS

Appropriate facility design incorporates a product flow that permits finished items from making contact with raw materials or unprocessed products. The ideal flow provides for raw materials and adjuncts to enter the process near the receiving dock, flowing sequentially into the preparation area, process area, packaging area, and to storage. Graham (1991d) has supported this design flow because it permits proper air pressure conditions to the overall plant efficiency. Some personnel doors support this concept because they are designed so that workers must pass from a "clean" to "less clean" area. Return to the cleaner area may require a uniform change and a sanitizing step, followed by entrance through an air lock or pressurized vestibule.

Processing equipment should have 1 m of clear space around it to facilitate maintenance and cleaning. A minimum of 0.5 m of clearance over each piece of equipment should be provided to permit effective cleaning. Floor-mounted equipment should be either sealed directly to the floor or mounted at least 15 cm from the floor. The processing layout should permit the location of equipment for accessibility to maintenance, sanitation, and inspection. Areas that are difficult to reach and clean are less likely to be cleaned frequently and thoroughly.

Airborne contamination is attributable to the cause of some pathogenic contamination. Unfiltered air and negative air pressure in areas where the product is exposed contribute to microbial contamination in the plant environment. Thus, air flow design is as important to hygiene as is the design and construction of floors, walls and ceilings. The zone with the highest pressure should be the area where the product is last exposed to the open air and packaged. The air flow from this zone is outward to the processing/preparation area and on to the storage zone. Dust collection is more effective if conducted under a positive pressure.

If an air-handling system is currently designed, the opening of an outside door provides an air stream exiting the building; whereas, in a negative air pressure situation, an opened door causes an incoming breeze containing outside contamination. The continual influx of unfiltered air complicates the overall cleaning of a plant, equipment, overhead pipelines, and other structural features.

PEST CONTROL DESIGN

The topography near a food facility should be sloped to permit water flow away from the building without the formation of puddles. Puddles provide available water for pests and attract them close to the facility. A rodent lip installed 60 cm down on the foundation and extending out 30 cm prevents rats from burrowing under the slab and entering the plant by chewing through felt expansion joints or through drains inside the building.

Cavities within walls should be avoided because they become nests for rodents and insects. All parts of the structure should allow easy cleaning of ledges, scale pits, and elevator pits. The installation of electric lines, cables, conduit, and electrical motors should eliminate harborage sites. Motor housings provide ideal nesting sites for mice. Ventilation stacks should be equipped with adequate screening to prevent pest entry as mice can enter through a hole approximately 6 mm in diameter, and the Norway rats (the largest rat) can go through a 12-cm hole.

Locker rooms and eating areas are vulnerable to pest entry because of traffic, food particles, and moisture. These facilities should be designed and constructed with interiors that can be cleaned, covered wall/floor junctions, and smooth, water-impermeable walls and washable floors. Drinking fountains, vending machines, and other fixtures should be mounted far enough away from the walls

for access to routine cleaning or mounted on casters for moving during cleaning. Locker tops should contain a 60° slope to avoid debris accumulation. These facilities should not open directly into a processing room or any area with exposed food. The toilet facilities should have a negative air pressure, and the internal air should be exhausted directly to the outside.

CONSTRUCTION MATERIALS

Stainless steel is the preferred material for food contact surfaces. This inert material resists corrosion, abrasion, and thermal shock; is cleaned easily; and is resistant to sanitizers. The high chromium content (12% or more of the steel) provides corrosion resistance, although other components, such as molybdenum, nitrogen, copper, titanium, niobium, sulfur, and carbon, contribute to structural and mechanical properties or corrosion resistance. The most commonly used stainless steel is type 304 of the 300 series. Type 316 contains approximately 10% nickel instead of the usual 8% and is used more frequently for corrosive products such as fruit juices and drinks. A variation of 316 is 316b, which offers more resistance to high-salt-content products.

SUMMARY

Sanitary design and construction of food facilities is essential to maintain a hygienic operation. A sanitary design begins with a site free of environmental contamination such as polluted air, pests, and pathogenic microorganisms. Site preparation is necessary to attain proper drainage and the reduction of contamination from the environment. All portions of a food facility must contain smooth, impervious surfaces that discourage pest entry. An appropriate process design incorporates a flow that prevents finished items from making contact with raw materials and unprocessed products.

STUDY QUESTIONS

1. Why is site selection important when building a food facility?
2. What site selection considerations should be adopted when building a food facility?
3. What site preparation should be conducted before building a food facility?
4. What are the desired characteristics for the walls of a food facility?
5. Why is corrugated metal siding not recommended for food facilities?
6. What roof construction is preferred for food facilities?
7. Why are windows not recommended for a food facility?
8. Why should air curtains be installed?
9. Why are false ceilings not recommended in food facilities?
10. What is the best flow design for food products?
11. What is the importance of positive air pressure in a food plant?
12. How can the welfare facilities of food facilities be designed to reduce pest entry?
13. Why is stainless steel superior to other materials for food facilities?
14. What are monolithic floors and why are they popular?

REFERENCES

Graham, D.J. 1991a. Sanitary design—a mind set. *Dairy Food Environ Sanit 11,* no. 7: 338–389.

Graham, D.J. 1991b. Sanitary design—a mind set (Part II). *Dairy Food Environ Sanit 11,* no. 8: 454–455.

Graham, D.J. 1991c. Sanitary design—a mind set (Part III). *Dairy Food Environ Sanit 11,* no. 9: 533–534.

Graham, D.J. 1991d. Sanitary design—a mind set (Part IV). *Dairy Food Environ Sanit 11,* no. 10: 600–601.

Troller, J.A. 1993. *Sanitation in food processing.* 2nd ed. San Diego, CA: Academic Press.

CHAPTER 13

Low-Moisture Food Manufacturing and Storage Sanitation

As with other food manufacturing units, it is essential that an effective and practical sanitation program be implemented in low-moisture food manufacturing plants. This practice is necessary to ensure that the operation complies with the U.S. Food and Drug Administration (FDA), state, and local requirements. Furthermore, rigid sanitation in low-moisture food manufacturing operations is needed to ensure that consumers are provided with safe and wholesome foodstuffs.

The Office of the Inspector General of the Department of Health and Human Services has indicated that low-risk food operations, such as bakeries, bottlers, and food warehouses, are becoming riskier because of ineffective inspection. Although firms engaged in interstate commerce are regulated by the FDA and subject to inspection by state and local authorities, where inspections are made, the surveillance is often cursory, with primary emphasis on birds, rodents, and insects. Firms operating under unsanitary conditions put the population's health at risk. Low-moisture food operations should organize and implement an effective sanitation program to provide their customers with wholesome products.

Effective sanitation in low-moisture food manufacturing is essential to maintain an acceptable operation. A tidy operation can be more efficient, assist in the promotion of branded products and company image, and, as suggested by Mills and Pedersen (1990), determine whether an operation remains profitable or even stays in business. As these authors have stated, failure to exercise proper sanitation can lead to customer dissatisfaction, decreased sales, and damage to a firm's reputation.

SANITARY CONSTRUCTION CONSIDERATIONS

Site Selection

Walsh and Walker (1990) suggested that sites should exhibit the following hygienic characteristics:

- Nearly level to a slight slope with adequate drainage
- Free of springs or water accumulation
- Accessible to municipal services (sewage, police, and fire)
- Remote from incinerators, sewage treatment plants, and other sources of noxious odors, pests or bacteria- producing enterprises
- Located within an air quality district tolerant of emissions from thermal processing
- Located away from areas prone to flooding, earthquakes, or other natural disasters

Exterior Design

The exterior should incorporate walls that are smooth, tight, impervious to water, and free of ledges and overhangs that could harbor birds. They should also contain sanitary seals against rodents and insects. Driveways should be paved and free of vegetation, trash, and water accumulation areas. Regular sweeping should be conducted to keep dust from blowing into storage areas. Other exterior design considerations may be found in Chapter 12.

Interior Design

Walls and Framing

Exposed structural members may be satisfactory in nonproduct areas, as long as they can be kept clean and dust free. Reinforced concrete construction is preferred for product areas, and interior columns should be kept to a minimum. Personnel doors should be fitted using self-closing devices (hydraulic or spring hinges) and screened. Gaps at door bases should not exceed 0.6 cm, and 20-mesh (minimum) screening should be incorporated.

Walls should be free of cracks and crevices, and impervious to water and other liquids to permit easy and effective cleaning. Wall finishes should consist of appropriate food-approved materials, as dictated by the function of each area. Glazed tile for surface finishes on processing area walls should be considered, with Fiberglas-reinforced plastic panels painted with epoxy or coated with other materials meeting the company and regulatory standards. Alternatives to painting in food areas should be considered because, although paint is inexpensive, it tends to crack, flake, and chip with age, and requires frequent maintenance replacement.

Insulation should be installed carefully in bakery facilities because it constitutes a potential dust and insect harborage. Even though inert, it should be applied to the outside of the building.

Ceilings

The use of suspended ceilings is satisfactory in nonfood areas if the space above the ceiling can be inspected and kept free of pests, dust, and other debris. Ceiling panels must be sealed into the grid but be easily removable. This feature is difficult to accomplish with most designs.

Suspended ceilings are not recommended in food production or handling areas. They can become a shelter for pests and may become moldy if wet, thus providing a source of contamination. In flour-handling areas of bakeries, dust may accumulate above the ceiling very rapidly, leading to insect, microbial, fire, and even explosion hazards.

Walsh and Walker (1990) recommend that overhead structural elements, such as bar joists and support members, be avoided whenever possible. Precast concrete roof panels provide a clean, unobstructed ceiling. Precast panels can be fabricated with a smooth interior surface, coated to resist dust accumulation, and easily cleaned. Overhead equipment supports; gas piping; water, steam, and air lines; and electrical conduits should be designed to avoid passing over exposed food areas, cluttering the ceiling, and dripping dust or moisture onto people, equipment, and product. A mechanical mezzanine to house utility equipment above can result in an easily cleaned ceiling, free from horizontal pipe runs and ductwork (Figure 13–1).

Floors

Floors in wet-washed areas must be impervious to water, free of cracks and crevices, and resistant to chemicals and acids. Yeast food solutions are particularly corrosive. Floor joints must be sealed, and wall junc-

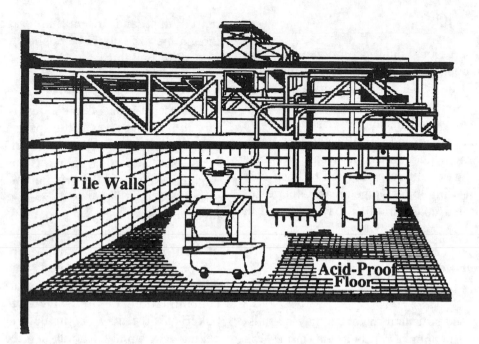

Figure 13–1 Specially designed mechanical mezzanine separates ductwork and utility support equipment from the bakery mixing room. This arrangement reduces the need for overhead cleaning, improves access for equipment maintenance, and enhances product safety.

tions must be covered and sealed. Expansive concrete should be used whenever possible to minimize the number of joints. The U.S. Department of Agriculture (USDA Food Safety and Inspection Service, 1986) recommendation that floors be sloped to drains with a pitch of 21 mm/m (approximately 2% grade) is an appropriate guideline to follow for proper wet cleaning and drainage. Process equipment should be connected to drain lines, and drip pans should be used to control floor spillage.

A perimeter setback of 0.5 m should be provided around all storage areas. White painted floor striping of setback spaces should be provided. Stored food must be segregated from nonfood items. Examples of products that must be segregated to avoid cross-contamination are bulk or palletized agricultural commodities and biologically ac-

tive materials (i.e., pesticides, petroleum products, paints, cleaning compounds, and aromatic hydrocarbons).

The ideal floor material depends on the operation and type of traffic. For packaging and oven areas of bakeries, reinforced concrete, coated or hardened to prevent dust, may be adequate. However, areas such as those for liquid fermentation and dough handling that are often wet-cleaned and exposed to hot water, steam, acids, sugar, and other ingredients or sanitizing chemicals should have a surface composition tailored to the use and abuse the floor is expected to receive.

Chemical-resistant floors are most appropriate for wet areas. Monolithic materials, such as epoxies or polyester, and tile or brick are recommended and are often less expensive. Toppings bonded directly to a substrate, such as concrete, should be used. They function as a resistant, watertight barrier protecting the concrete. However, they will crack if the same happens to the substrate, allowing liquids to enter. Only materials with proven success should be considered.

"Dairy" tile or pavers should be considered for areas with heavy traffic and those that come in contact with product or cleaning solutions. This material, when properly installed with acid-resistant bonds, is very durable and has minimal sanitary problems. It is cleaned easily and can be manufactured with a nonslip finish. It is an expensive option but can be the most economical.

Floors in specialized areas, such as coolers and freezers, must be constructed with appropriate materials designed for their intended uses and be properly insulated and ventilated. An uninsulated freezer floor will eventually permit the ground beneath to freeze deep and hard enough to cause cracking or buckling of the freezer floors, with resultant jamming of the doors.

Ventilation and Dust Control

Dust control is very important. Although organisms from unprocessed low-moisture materials are usually harmless, they have been found to contain *Salmonella*, pathogenic mold spores, and other undesirable organisms. The manufacturing process, by heating the product above the pasteurization temperature, usually kills vegetative organisms, but spores may survive in the interior, especially in relatively soft, high-moisture baked foods. Furthermore, finished food can become contaminated from raw material dust within the plant, especially in coolers and packaging equipment.

To maintain acceptability, facilities must be designed so that finished foods are not contaminated. This requires a superb sanitation design and follow-through procedures,

proper equipment arrangement, and proper ventilation and dust control. The proper selection of temperature/humidity controls will minimize the opportunity for bacterial growth.

Equipment Considerations

Equipment features that will enhance productivity include separating heating and cooling equipment from the process areas by using a mechanical mezzanine, high-efficiency motors and electrical equipment (see Figure 13–1), the latest technology in controls and automation to the maximum degree that is cost effective, and flexible modular design for responding to changing markets and business demands. Furthermore, all equipment should meet the latest requirements from regulatory or advisory agencies.

Sanitary Considerations

Because an operation such as bread making is a fermentation process, it is necessary for facilities such as bakeries to be easily maintained in a sanitary condition. Naturally occurring organisms must be prevented from fermenting the dough in competition with the desired yeast inoculum. An ineffective sanitary facilities design can result in the growth of wild microbial strains such as *Bacillus subtilis* or *mesentericus* "rope" formers, which can degrade product acceptability. Once established in the facility, these organisms are very difficult to remove totally and to control.

Sanitation features that are integrated into plant design were given increased emphasis by the FDA's promulgation of Good Manufacturing Practices (GMPs). For low-moisture food products, current GMPs (CGMPs) as they relate to the design and construction should:

1. Provide adequate space for equipment installation and storage of materials.
2. Provide separation of operations that might contaminate food.
3. Provide adequate lighting.
4. Provide adequate ventilation.
5. Provide protection against pests.

The best way to achieve these objectives is to keep the plant interior spaces simple and uncluttered. This characteristic facilitates sanitation, cleaning, and inspection. Plant sanitation criteria and engineering specifications should be thoroughly integrated into the layout by the design team, including the plant's technical and engineering staff, as well as that of a contracted design and engineering firm. A representative from the production staff should be consulted.

Other Considerations

Additional design considerations are mentioned in Chapter 12. Walsh and Walker (1990) suggested a partial list of items that should be considered in an effective hygienic design:

- Locate the plant management office and the laboratory centrally for proper supervision and quality control.
- Locate ingredients storage near the mixing and use areas.
- Locate secondary equipment, such as boilers and refrigeration machines, to minimize pipe and utility runs.
- Arrange manufacturing equipment for convenient cleaning in place (CIP).
- Use proven equipment or allow time for testing any equipment or process that does not have a known track record.
- Apply state-of-the-art controls and automation to the greatest extent that is cost effective. Be prepared to update the system as capabilities climb and prices fall.

- Check plant design for compliance with federal, state, and local regulations.

RECEIPT AND STORAGE OF RAW MATERIALS

Sampling for Acceptability

Because it is impractical to sample all of the raw materials being received, a sampling protocol should be devised to determine whether products should be accepted or rejected. A statistically valid sample is necessary to determine acceptance or rejection with reasonable confidence. More information about statistical sampling and statistical quality control is provided in Chapter 6.

Transport Vehicle Inspection

Inspection of low-moisture raw materials should begin with examination of the transport vehicle before, during, and after unloading. The overall condition of the vehicle should be appraised, and it should be checked for dead areas where product and dust can collect and harbor insects, whether the containers are full or empty. Areas adjacent to doors or hatches should be observed for insects. This inspection is accomplished by examination for crawling or flying insects and their tracks. It is important to check for nesting materials, odor, and fecal material. Pellets and odors may indicate rodents, and feathers or droppings may reveal contamination by birds.

Product Evaluation

To reduce contamination from raw materials being received, product evaluation is essential. Although moisture content may be determined objectively through the analysis for percentage of moisture, a subjective evaluation should also be conducted. A sour or musty odor can result from mold growth, which indicates a high moisture content in products such as cereal grains. Such a discovery indicates that additional inspection should be conducted, with sampling to identify the specific characteristics of the problem. Cereal grains with a moisture content above 15.5% should not be put in long-term storage because of potential insect development and mold growth. Evaluation of products being received should also include checking for pesticide odors that may be associated with the presence of insects. The inspection process should also determine whether the pesticide has made the product unacceptable.

Samples taken when materials are received should be evaluated to determine the amount of individual kernels that are damaged by insects. Further examination should be conducted to determine amounts of dust and other foreign material, webbing, evidence of molds and odors, live and dead insects, rodent droppings, and rodent-damaged kernels. These defects can be determined through visual inspection. Internal infestation in the form of immature insects inside of the kernels can be determined with X-ray equipment or by cracking-flotation methods. Samples should also be examined for rodent filth, such as droppings and hair.

The inspection of inbound goods is an appropriate prevention measure to reduce pest damage because incoming items can contaminate the end product. Because pests or their contamination can enter buildings as "hitchhikers," incoming ingredients, packaging materials, pallets, and machinery should be inspected. A food processor has the right to reject any materials coming into the plant or to hold any questionable shipment for further evaluation. The decisions regarding acceptance or rejection of shipments should be made by knowledgeable and dependable personnel.

Product Storage

Many of the low-moisture food processing plants store material such as grain for processing. According to Troller (1993), grain is the commodity stored in the greatest volume in the United States. Unfortunately, when grain is stored, it initially contains mold spores and insect eggs in enough quantity to infest and damage the product if specific environmental conditions occur. Physical damage to the kernel itself can allow entry of infesting or infecting agents. Biological damage from insects through penetration of the kernel permits fungal entry through inoculation of the inner tissues.

Mills and Pedersen (1990) have stated that grain to be placed in storage for more than 1 month should receive special treatment. In addition to being inspected for verification that infestation and infection has not occurred, it is necessary to maintain a maximum of 13.5% moisture content. These authors suggest that cleaning the grain before storage using aspiration or other methods can remove dockage, external insects, weed seeds, and foreign materials, and can improve its storability. Furthermore, as grain is being stored, chemical grain protectants can be applied to provide residual protection against insects.

Grain may be fumigated through the use of a modified atmosphere, such as carbon dioxide and nitrogen. Fumigation by inert gases is receiving more attention because of increased restrictions on the use of chemicals. Although inert atmospheres do not represent a residual hazard, the environment in a storage bin with an inert gas can be as deadly to humans as if it contained a lethal concentration of a chemical fumigant. Insect feeding and reproduction can be reduced in temperate regions if storage bins are equipped with aeration systems.

Control of dust in handling and storage of low-moisture foods can improve housekeeping and pest control. The containment of dust production reduces deposits on floors, walls, ledges, overhead objects, and equipment, with a resultant decrease in cleaning time. Dust control is enhanced through suction (reduced pressure) on grain handling equipment such as conveyors, receiving hoppers, bucket elevators, and bins, as well as at points in the handling system where product is transferred from one piece of equipment to another (e.g., from spout to conveyor belt, conveyor to bin, and bin to conveyor).

Mills and Pedersen (1990) suggested that the application of highly refined oils to grain as it goes to storage is an effective way to reduce dust when handling grain. Oil, which may be added to levels of up to 200 ppm, should be applied to the grain as closely as possible to the point of discharge from the transport vehicle to reduce dust formation and to provide a grain-protectant treatment.

Although the sanitation of root crops, such as potatoes, during storage is not as critical as for other foods, storage conditions must be controlled to prevent *Fusarium* tuber rot and bacterial soft rots. Well-ventilated storage rooms with concrete floors have enabled the potato storage industry to exert adequate control over its product. According to Troller (1993), 80% to 90% of the potatoes harvested each year are stored in such facilities.

The bulk storage of oils and shortenings normally occurs in large carbon steel or stainless steel tanks. Thus, appropriate sanitation can be attained by proper cleaning of these containers—washing with a strong alkaline solution or alkali and detergents before use. Hygiene conditions can be enhanced further through the nitrogen blanketing of process and deodorized oils. However, precautions are essential during the bottling and emptying of unprocessed, processed, and deodorized oils to prevent excessive splashing and agitation, which can potentially promote oxidative

deterioration. Cleaned bulk tanks (especially carbon steel tanks) should be recoated with oil to seal them for rust prevention.

Housekeeping during Product Storage

As with high- and intermediate-moisture foods, unsanitary conditions can be prevented through proper maintenance and housekeeping. Bulk storage areas (especially interior areas) should be maintained so that they are free of cracks or ledges that collect dust and other debris, which provide a conducive environment for insect growth, whether full or empty. Empty bins or other storage containers should be inspected for residues of product stored previously and for overall condition. Any residual material that can harbor or support insect growth should be removed before products are stored.

Tunnels, gallery floors, and associated areas of storage bins or similar facilities should be maintained in a sanitary condition. Periodic inspections are an essential part of effective sanitation for stored products. As in other areas, inspectors should examine the dust on the floors and walls for insect tracks, as well as for resting or flying moths. Mills and Pedersen (1990) suggested that inspection include examination for damp lower areas that collect dust and provide conditions conducive to the proliferation of molds, mites, and fungus-feeding insects. Further inspection should include checking for any unusual odors that could indicate mold, insects, or chemicals. It is especially important to inspect handling equipment, such as elevators and conveyors, that may harbor residual product. Unused equipment may retain residual material that will encourage insect growth and subsequent migration to storage areas or contamination of new product.

Storage areas require regular inspection to observe for live insects on product surfaces, floors, and walls. Thermocouple cables should be used for grain in extended storage so that temperature can be monitored. Increased temperature during storage should be investigated. Samples should be taken with probes or as the product is transferred to another location to determine whether the temperature rise could cause developing populations of insects or molds. Mold growth can normally be controlled through drying or blending with other dry products. If insects are present, treatment or fumigation should be conducted. Heating from insect infestation can also cause moisture to spread, with resultant mold development. Inspection should be accompanied by a complete record of inspections, cleaning and fumigation, or other corrections administered.

Sanitation requirements for product storage are similar to those for bulk storage. An orderly storage arrangement is essential to ease inspection and cleaning, and to reduce the potential for sanitation problems. According to Marriott et al. (1991), records of regular inspections and housekeeping are essential. Inspectors and other employees should be aware of the presence of pests and of eradication methods.

Storage practices suggested by Mills and Pedersen (1990) should be followed to ensure an effective sanitation program. They recommend that bags and cartons be stacked on pallets and spaced away from the walls and from each other for inspection and that the surrounding area be cleaned. Stock should be rotated to reduce insect infestation and rodent entrance. Inspections should include visual observation, using a light for looking in dark corners, under pallets, and between stacks. Insects may be detected while flying; crawling on walls, ceilings, and floors; and while hovering over bags and cartons. Product spoilage should be sifted to detect insects. (Information on insect and rodent control is provided in Chapter 11.)

Cleaning and inspection frequency depends on temperature and moisture conditions. Under ambient temperature (25° to 30°C) conditions, the life cycle of many insects that infest low-moisture grains and foods is approximately 30 to 35 days. Insect reproduction normally ceases when the storage temperature is below 10°C. When storage temperature increases, the cleaning and inspection interval should be decreased. Raw material or product temperature has more influence on insect growth than does ambient temperature. Areas where high moisture (humid) conditions exist will require more frequent inspections and cleaning. The potential for high-moisture conditions should be reduced through proper ventilation. Moist materials that remain static at room temperature or above will increase immediate development of molds, yeasts, and/or bacteria. Suction can be used to remove moist air.

Ledges and other locations that can accumulate static material should be eliminated. External supports, braces, and other construction features and/or equipment should be designed to prevent material accumulation. Dust can adhere to moist surfaces and provide an excellent habitat for molds.

Heat treatment (superheating) can combat pests in dry storage and production areas where unprocessed materials are stored. However, this practice is energy-intensive because of the amount of heat required to kill insects, especially during cold weather. Portable heating units may be used to superheat an individual piece of equipment that may be infested with insects. When designing new or renovated facilities, consideration should be given to the potential for heat treatment. Maintenance of a cold environment is less practical because of refrigeration costs and possible equipment or facility damage from freezing. It may be impractical to maintain a moderately low temperature that will retard insect inactivity.

Inspection of Raw Materials and Product Storage Area

Inspections should be conducted and reported. An inspection report format should be developed with a numerical scoring system. Scoring and rating values should be defined, with a description for each value. Foulk (1992) has suggested that warehouses be rated on three levels.

- *Acceptable* if most of the requirements are met.
- *Provisionally acceptable* when corrective measures can and will be taken to attain compliance with the established standards and upgrade to acceptable is possible.
- *Unacceptable* when any of several deviations from standards occur that will result in an unsanitary operation.

Inspection in the processing and storage areas should emphasize the identification of potential product contaminants and prompt and proper corrective action to prevent contamination. The lower minimum water activity (A_w) of low-moisture foods reduces the chance of microbial spoilage; thus, more emphasis should be placed on other forms of contamination, which are discussed in the following paragraphs.

Overhead areas should be examined for flaking paint, obstructions to cleaning, dust accumulation, and condensation. Ground-level, basement, and above-ground-level inspection should focus on broken window panes and absence of or damage to screens. Open windows or other entry avenues for pests are potential sources of contamination and should be reported and/or corrected on a continual basis. Evidence of pests, such as insect trails in dust, rodent droppings, and bird droppings or feathers should be identified through periodic inspection and inspection by employees on a continual basis. Evidence of pests should be reported so that appropriate action can be taken to identify the problem source and correct it. All employees should be alert for evidence of pest activity.

Inspection of equipment exteriors may be accomplished on a continual basis by operations personnel. Overhead equipment should be inspected regularly. Equipment interiors should also be examined periodically for sanitation-related problems during maintenance inspection. Some equipment contains dead spots where product can accumulate. Therefore, inspection of equipment should be performed routinely when the equipment is not in operation. Equipment, especially conveyors, should be constructed, if possible, so that interiors are accessible through clean-out openings or by easy disassembly. Also, this design also facilitates equipment cleaning during routine housekeeping. If feasible, equipment not in use should be removed from the facility. Equipment that is used infrequently should be left "open" so that any product filtering into it will pass through or will be observed easily. The openness feature will also enable easy and effective cleaning.

The condition of the facility itself should exclude contaminating factors, such as insects, rodents, and birds. Any defects discovered should be reported and corrected immediately.

CLEANING OF LOW-MOISTURE FOOD MANUFACTURING PLANTS

Cleaning in the manufacturing area of low-moisture food plants should be accomplished daily. Some of the cleaning should be done while the plant is in operation to ensure that the facility remains tidy. Yet, some of the cleaning tasks and most of the equipment cleaning (especially equipment interiors) should be done while the manufacturing portion is not in operation. Some of the required cleaning can be combined with routine maintenance operations. Easily stored, conveniently located equipment encourages employees to accomplish the cleaning necessary to control infestation.

Most of the cleaning equipment is easy to use. Hand brooms, push brooms, and dust and wet mops provide the basic equipment used for cleaning. Brushes, brooms, and dust pans remove the heaviest debris accumulations and function well on semismooth-surfaced floors. Dust mops provide a more rapid mans of cleaning on smooth floor surfaces with low levels of dust accumulation. In many production areas, vacuuming provides the most acceptable means of equipment cleaning. Vacuum cleaning provides one of the most thorough methods of cleaning because it removes light and moderate accumulations of debris from both smooth and irregular surfaces. The dust is contained and does not require a secondary means of collection. Smaller operations can more effectively utilize portable vacuum equipment, whereas larger facilities can benefit from an installed vacuum system. Although installed equipment is more expensive, centralized debris collection and disposal is more convenient, with more convenient access to difficult-to-reach areas. In large storage areas with nonporous floors, a mechanical scrubber or sweeper should be used to more efficiently and effectively maintain a clean environment.

A compressed air line is widely used to remove debris from equipment and other difficult-to-reach areas. Although Mills and Pedersen (1990) do not consider this method for cleaning facilities such as flour mills, they recognize that it provides easy cleaning of inaccessible areas in plant facilities and equipment. Furthermore, using compressed air is safer than depending on employees to work from a ladder with a brush. However, compressed air disperses dust from a specific location to a less confined area and may spread an infestation if it exists. Compressed air should be incorporated at low volume and low pressure to minimize dust dispersal. Employees who use compressed air should wear safety equipment, such as dust respirators and safety goggles.

Specialized tools are required for certain equipment cleaning. Cylindrical brushes are used for spouts. They can be either dragged through spouts by rope or cords or operated on flexible motorized shafts.

The maintenance of a tidy operation depends on proper organization and installation of equipment and on cleaning of individual pieces of equipment and of the surrounding area. Ingredients and supplies should be properly stacked in a designated storage area. Receptacles should be conveniently located for the disposal of bags, film, paper, and waste products from manufacturing, packaging, and shipping.

SUMMARY

Rigid sanitation practices are essential in low-moisture food manufacturing and storage facilities to maintain product acceptability and to comply with regulatory requirements. A sanitary operation should be complemented with appropriate facility site selection and hygienic design of the building and equipment. Unprocessed materials should be sampled during the receiving operation to verify that they are not infested with insects, molds, rodents, or other unacceptable contaminants. During storage, unprocessed and manufactured products should be protected from contamination through effective housekeeping practices. Storage areas require routine inspection to observe for microbial and pest infestation. Inspection and cleaning frequency of storage areas depends on temperature and humidity. Cleaning in the manufacturing area should be done daily. Cleaning equipment consists of basic cleaning tools for low-moisture product areas, including vacuum equipment, powered floor sweepers and scrubbers, and compressed air for certain applications.

STUDY QUESTIONS

1. What percentage slope should exist in wet-washed areas of low-moisture food plants?
2. What chemical-resistant floors are recommended in wet-washed areas?
3. What is the maximum percentage moisture for cereal grains placed in long-term storage to be protected against insects and molds?
4. How can dust be reduced in low-moisture food plants?
5. How can compressed air be used to clean in low-moisture food plants?

REFERENCES

Foulk, J.D. 1992. Qualification inspection procedure for leased food warehouses. *Dairy Food Environ Sanit* 12: 346.

Marriott, N.G., et al. 1991. *Quality assurance manual for the food industry*. Virginia Cooperative Extension, Virginia Polytechnic Institute and State University, Blacksburg. Publication No. 458-013.

Mills, R., and J. Pedersen. 1990. *Flour mill sanitation manual*. St. Paul, MN: Eagan Press.

Troller, J.A. 1993. *Sanitation in food processing*. 2nd ed. New York: Academic Press.

USDA Food Safety and Inspection Service. 1986. Inspected meat and poultry packing plants–A guide to construction and layout. *Agriculture handbook*, 570. Washington, DC: US Government Printing Office.

U.S. Food and Drug Administration. 1987. U.S. code of federal regulations Part 110. Current good manufacturing practices in manufacturing, processing, packaging, or holding human food. *Food and drugs*, Title 21 Subchapter B—Buildings and facilities. Washington, DC: US Government Printing Office.

Walsh, D.E., and C.E. Walker. 1990. Bakery construction design. *Cereal Foods World* 35, no. 5: 446.

CHAPTER 14

Dairy Processing Plant Sanitation

The dairy industry has the reputation for being a leader in the food industry in hygienic design and practices, as well as in the implementation of sanitation standards. Also instrumental in the leadership role has been recognition by the industry of the primary need for good sanitation practices to ensure improved stability and high quality of dairy products that require refrigeration.

The physical and chemical properties of dairy products, especially fluid items, have made possible the automated cleaning of the processing facilities. Some of the following components identified by Seiberling (1987) have been developed that contribute to automation:

- Permanent piping of nearly all "welded" construction has been installed to reduce the amount of manual cleaning of tubing and fittings.
- Control systems based on relay logic, dedicated solid-state controllers, small computers, and programmable logic controllers wired or programmed to control complex cleaning sequences have been developed.
- Automatically controlled cleaning-in-place (CIP) systems have provided a method to ensure uniformly thorough cleaning of tanks, valves, and pipes on a daily basis.
- Air-operated, CIP-cleaned sanitary valves have eliminated the manual cleaning of plug-type valves and provided for remote and/or automatic control of CIP solution flow.
- Silo-type storage tanks and dome-top processors have been designed to be cleaned effectively by CIP equipment.
- Processing equipment has been designed for CIP cleaning (homogenizers, plate heat exchangers, certain fillers, and the self-desludging centrifugal machine).

These components function most effectively when properly integrated into a complete cleaning system designed and installed for automated control of all cleaning and sanitizing operations.

The source of the milk supply is of major concern. Even the most effective pasteurization cannot upgrade quality or eliminate the problems created by undesirable bacteria in the raw supply. Although pasteurization is an effective weapon against pathogenic and spoilage microorganisms, it is only a safeguard measure and should never be used to cover up an unsanitary raw supply or improper sanitation.

Polyphosphates and synthetic surface-active agents have been responsible for changes in the cleaning operations to keep pace with new materials and cleaning and sanitizing equipment. These advances have enabled the formulation of specific cleaning compounds that adapt to water conditions, types of metals, and soil characteristics. They also have the buffering ability to minimize corrosion. They have opened up a new avenue of close union and intimate association between cleaning compounds and sanitizing agents to enhance the value of both phases of sanitization.

ROLE OF PATHOGENS

Despite of the industry's reputation for hygienic design and practices, pathogens have continued to invade dairy products. During 1985, a large outbreak of salmonellosis occurred in pasteurized milk. Other recent foodborne illness outbreaks from the ingestion of dairy products have included staphylococcal food poisoning caused by ice cream, the implication of campylobacteriosis that has occurred sporadically without a finite determination of the mode of transmission, and listeriosis from contaminated cheese. The latter outbreak was responsible for several deaths. As a result, the dairy industry has had to make a large number of product recalls at great expense. These events have brought the full force of the regulatory agencies upon the industry and motivated several dairy processors to invest heavily in the improvement in sanitation of their production facilities. These experiences have underscored the importance and urgency of effective sanitation programs. Because pathogens are discussed in Chapter 2, only *Listeria monocytogenes* and *Escherichia*

coli O157:H7, the pathogens of greatest concern in dairy products, will be discussed here.

Listeria monocytogenes

The discovery of *L. monocytogenes* in fermented and unfermented dairy products has prompted food manufacturers to renew their concern about plant hygiene and product safety. *Listeria monocytogenes* is widely distributed in nature and often carried in the intestinal tract of cattle. Approximately 5% of normal, healthy humans are fecal excretors of this microorganism. Approximately 5% to 10% of raw bovine milk is contaminated with *L. monocytogenes*. This microbe has been isolated from improperly fermented silage, leafy plants, and the soil—the latter being a reservoir of *Listeria* organisms.

Listeria recalls of ice cream and cheese products have precipitated major processing and sanitation operation changes in dairy processing plants. Many processors are voluntarily adopting Grade A standards required for the production of pasteurized milk. The importance of an effective sanitation program to combat *L. monocytogenes* has contributed to major increases in training, supervision, total employee count, and salaries of sanitation workers in dairy processing plants.

The epidemiologic implication of pasteurized milk in the Massachusetts listeriosis outbreak in 1983 and in the outbreak in Los Angeles in 1985, attributable to a Mexican-style soft cheese, led to the establishment of the U.S. Food and Drug Administration (FDA) standard methodology for detection of this pathogen. These events also contributed to a decision to conduct a large survey for pathogenic microorganisms in the dairy industry. This survey revealed that, in nearly all instances, post-processing contamination was responsible for contamination of *L. monocytogenes*.

Specific guidelines have been developed for controlling *L. monocytogenes* in dairy processing facilities in the United States. These guidelines stress the need to (a) decrease the possibility that raw products will contain *Listeria* organisms, (b) minimize environmental contamination in food processing facilities; and (c) use processing methods and sanitation techniques that will reduce the probability that this pathogen will occur in food.

Properly constructed and maintained facilities and equipment are fundamental to an effective cleaning and sanitation program for the control of *L. monocytogenes*. Construction characteristics that will be described in this chapter and in Chapters 15, 16, and 17 should be considered when planning a program for the control of this pathogen.

Listeria monocytogenes is sensitive to sanitizing agents commonly employed in the food industry. Chlorine-based, iodine-based, acid anionic, and/or quaternary ammonium-type sanitizers are effective against this pathogen when used at concentrations of 100 ppm, 25 to 45 ppm, 200 ppm, and 200 ppm, respectively. Although these concentrations may require adjustment to compensate for in-plant use (as may oxidation-reduction factors relating to water quality and hardness), recommended concentrations should not be markedly exceeded, as use of extremely concentrated sanitizing solutions heightens the danger to employees, increases the risk of chemical contamination of food, and, in some instances, causes corrosion of equipment.

Quaternary ammonium-based sanitizers are not recommended for use on food contact surfaces and should not be used in cheese factories, as lactic acid starter culture bacteria are inactivated rapidly by small residues of these sanitizers. In contrast, acid anionic and iodine-type sanitizers are best suited for equipment surfaces, with the former readily neutralizing excess alkalinity from cleaning compounds and preventing the formation of alkaline mineral deposits. The use of steam should be discouraged (due to energy costs) and, if used, should be confined to closed systems because of potential hazards associated with aerosol formation. Sanitizing with hot water is not recommended because of the high energy costs of heating water and because high temperatures cannot be maintained easily.

Effectiveness of a *Listeria* control program can be measured by conventional and routine, preoperative microbial monitoring, such as aerobic plate count and coliform count (see Chapters 2 and 6). However, industry experience has suggested that the most accurate measurement relies on specific testing for *Listeria* organisms in the plant environment. Environmental sampling should be organized to guide preoperative sanitation practices and direct management toward a *Listeria*-controlled operation.

Escherichia coli O157:H7

Outbreaks of this pathogen associated with raw milk have challenged investigators to further research this microorganism in dairy products. This pathogen can grow in cottage cheese and cheddar cheese but is inactivated by the pasteurization of milk.

Buchanan and Doyle (1997) suggested that alternative technologies to thermal processing can control *E. coli* O157:H7 while maintaining the acceptability of dairy products. A viable alternative technology for dairy, meat, and poultry products is ionizing radiation. This pathogen is relatively radiation-sensitive, and radiation pasteurization doses of 1.5 to 3.0 KGy appear to be destructive at the levels that are most likely to occur in ground beef (Clavero et al., 1994).

SANITARY CONSTRUCTION CONSIDERATIONS

The considerations most important to dairy plant sanitation are drainage and waste disposal. Storm and sanitary sewers must be adequate and readily available. In rural areas and municipalities with limited treatment facilities, dairy processors frequently must provide their own waste disposal facilities. An adequate supply of potable water and acceptable drainage and waste disposal are essential. Other considerations are mentioned in Chapter 12.

Floor Plan and Type of Building

The layout and construction of a dairy plant are subject to the approval of one or more regulatory agencies. All equipment and utensils should be purchased subject to the approval of the various regulatory authorities.

Ventilation is important, especially in areas where excess heat produced during processing must be removed. The ventilation should be tailored to the different types of rooms and should have the flexibility to meet the needs of any future alterations in production. It is frequently necessary to filter incoming air, especially if the plant is located in a heavy industrial area. Also, the control of humidity, condensation, dust, and spores should be considered.

Construction Guidelines

Unless construction is carefully planned, the structure and equipment can contribute to contamination. This problem can be helped by reducing overhead equipment to a minimum, which reduces contamination due to maintenance of this equipment. Overhead equipment is also difficult to clean. A separate service floor that will accommodate a major portion of the ducts, pipe works, compressors, and other equipment should be provided. This arrangement results in a clear ceiling, which is easy to clean and keep sanitary.

Some other design and construction characteristics that are conducive to effective sanitation are:

- All metal construction should be treated to withstand corrosion.
- Pipe insulation should be of a material that is resistant to damage and corrosion, and will endure frequent cleaning.
- Chronic condensation points should be protected by the installation of a drainage collection system.
- All openings should be equipped with air or mesh screens and tight-fitting windows.

Structural finishes should be of materials that require minimal maintenance. Walls, floors, and ceilings should be impervious to moisture. Floor materials should be resistant to milk, milk acids, grease, cleaning compounds, steam, and impact damage. Epoxy, tile, and brick are good choices. Paint should not be used if suitable alternatives exist. If paint is applied, it should be of a grade that is acceptable for food plants. Floor drains should be designed to control insect infestation and odors. A slope of approximately 2.1

cm/m is recommended to reduce accumulation of water and waste on the floor, which could hamper the sanitation and lead to growth of *L. monocytogenes*.

Both floor drains and ventilation systems may contribute to contamination from airborne microorganisms instead of acting as a sanitation barrier. However, a properly designed ventilation system with air filtration can improve air quality. Inexpensive filters can remove dust and other contaminants that would normally be drawn into these spaces or rooms.

Equipment should be designed and oriented for easy cleaning and reduction of contamination. In the past, equipment layout has been important to operational efficiency, with the effects on the sanitation operation of only secondary importance. The most critical considerations related to equipment sanitation include a location to permit sanitary operations between equipment and walls or partitions, an exterior with an easy-to-clean surface, and a design to permit effective sanitation between the equipment and floor. All equipment should be accessible, easy to clean, and designed for draining and sanitizing.

SOIL CHARACTERISTICS IN DAIRY PLANTS

In the dairy industry, soil consists primarily of constituents of minerals, lipids, carbohydrates, proteins, and water. Other soil constituents may be dust, lubricants, microorganisms, cleaning compounds, and sanitizers.

White or grayish films that form on dairy equipment are usually milkstone and water stone. These films usually accumulate slowly on unheated surfaces because of poor cleaning or use of hard water, or both. Calcium and magnesium salts precipitate when sodium carbonates are added to hard water. During cleaning, some of this precipitate may adhere to equipment, leaving a film of water stone. When proteins denatured by heat adhere to surfaces and other components absorb them, milkstone may form quickly on heated surfaces. Because they become less soluble at high temperatures, calcium phosphates from milk are present in large quantities. The nature of soil on heated and unheated surfaces usually differs in composition. Thus, each type of soil requires a different cleaning procedure. Milkstone is usually a porous deposit that will harbor microbial contaminants and eventually defy sanitizing methods. It can be removed through an acid cleaner to dissolve the alkaline minerals and remove the film. Heavy soil deposits require a stronger cleaning compound that do lighter soils. Also, freshly deposited soil on an unheated surface is more readily dissolved than the same soil that has dried or has baked on a heated surface.

Soil deposition can be reduced and subsequent removal eased by application of the following principles:

- Generally, product heating surfaces should be cooled before and immediately after emptying of heated processing vats.
- Foams and other products should be rinsed after the production shift and before they dry.
- Where possible and practical, the soil deposits should be kept moist until the cleaning operation starts.
- Rinsing should be accomplished with warm (not hot) water.

Soil deposition is increased in ultra-high-temperature heaters if milk contains high acidity. This problem is complicated by low-velocity movement and poor agitation during the operation. Preheating and holding at a high temperature reduce film deposition.

The nature of the surface determines the ease or difficulty of soil removal. Pits in corroded surfaces, cracks of rubber parts, and crevices in insufficiently polished surfaces protect soil and microorganisms from the effects of cleaning compounds and sanitizers. The type of soil to be removed determines the cleaning method and concentration and composition of the cleaning compound.

SANITATION PRINCIPLES

Cleaned and sanitized equipment and buildings are essential to the production, processing, and distribution of wholesome dairy products. The major part of the total cleaning cost is labor. Therefore, it is important to use appropriate cleaning compounds and equipment so that the sanitation program can be effectively administered in a shorter period of time and with less labor.

The sanitarian should know the time required to clean each piece of equipment with the mechanization and cleaning compounds available. Cleaning tasks should be assigned to specific employees, who should be made responsible for the equipment and area under their care. These assignments should be made through official notification or by posting the cleaning schedule or assignments on a bulletin board.

Role of Water

The major constituent of almost all cleaners, including those used by dairy plants, is water. Because most plant water is not ideal, the cleaning compounds selected should be tailored to the water supply or should be treated to increase the effectiveness of the cleaning compound. It is especially important to reduce suspended matter in water to avoid

deposits on clean equipment surfaces. Water hardness complicates the cleaning operation. Suspended matter and soluble manganese and iron can be removed only by treatment, whereas small amounts of water hardness can be counteracted by sequestering agents in the cleaning compounds used in the sanitation operation. If the water is hard or very hard, it is usually more economical to pretreat the water to remove or minimize water hardness.

Role of Cleaning Compounds

Like all other cleaning compounds, those used in cleaning dairy plants generally are complex mixtures of chemicals combined to achieve a specific desired purpose. The following cleaning functions are related to the role of cleaning compounds in dairy sanitation operations:

1. Prerinsing is conducted to remove as much soil as possible and to increase the effectiveness of the cleaning compound.
2. The cleaning compound is applied to the soil to facilitate subsequent removal through effective wetting and penetrating properties.
3. Solid and liquid soils from the surface to be cleaned are displaced by fat saponification, protein peptizing, and mineral dissolution.
4. Soil deposits are dispersed in the cleaning medium by dispersion, deflocculation, or emulsification.
5. Effective rinsing is conducted to prevent redeposition of the dispersed soil onto the cleaned surface.

The value of a cleaning compound is most accurately determined by measuring the area that can be cleaned efficiently with minimal costs. High-cost cleaning compounds are frequently the most economical because of labor, energy, and cleaning compound savings.

More discussion of cleaning compounds is provided in Chapter 7.

Application of Cleaning Compounds

Identification of the optimal external energy factors and application methods is necessary to facilitate cleaning. If cleaning is done by hand, strong acids and alkalies should be avoided because they irritate human skin. Instead, emphasis should be placed on external energy, such as heat and force. Superb results depend on the use of circulation cleaning and on whether it is done in or out of place. Table 14–1 provides a guide to the most appropriate cleaning compound, cleaning procedure, and cleaning equipment for the major cleaning applications.

Role of Sanitizers

After cleaning, sanitizers should be applied to the cleaned surface to help destroy microorganisms. Of the many methods for sanitizing (see Chapter 8), those most frequently used in dairy plants are steam, hot water, and chemical sanitizers.

Steam Sanitizing

Steam sanitizing is accomplished by maintaining steam in contact with the product contact surfaces for a designated time. The effective procedures have been found to be 15 minutes of exposure when the condensate

Table 14–1 Optimal Cleaning Guides for Dairy Processing Equipment

Cleaning Applications	Cleaning Compound	Cleaning Medium	Cleaning Equipment
Plant floors	Most types of self-foaming, or foam boosters added to most moderate to heavy-duty cleaners	Foam (high-pressure, low-volume should be used with heavy fat or protein deposits)	Portable or centralized foam cleaning equipment with foam guns for air injection into the cleaning solution
Plant walls and ceilings	Same as above	Foam	Same as above
Processing equipment and conveyors*	Moderate to heavy-duty alkalies that may be chlorinated or nonalkaline	High-pressure, low-volume spray	Portable or centralized high-pressure, low-volume equipment; sprays should be rotary hydraulic
Closed equipment	Low-foam, moderate to heavy-duty chlorinated alkalies with periodic use of acid cleaners as follow-up brighteners and neutralizers	CIP	Pumps, fan or ball sprays, and CIP tanks

*Packaging equipment can be effectively cleaned with gel cleaning equipment.
Source: Adapted from Stenson and Forwalter, 1978.

leaving the assembled equipment is at 80°C. This method of sanitizing has limited utility because it is difficult to maintain a constant required temperature and because the energy costs are excessive. Steam application can also be more dangerous than other sanitizing methods and is not usually recommended.

Hot-Water Sanitizing

Hot water is pumped through the assembled equipment to bring the product surfaces in contact with water at a given temperature for a specified time. Water temperature maintained at 80°C at the equipment outlet for 5 minutes serves as the sanitizer. This technique is expensive because of the required energy costs.

Chemical Sanitizing

This method is accomplished by pumping an acceptable sanitizer such as the halogens (usually chlorine or iodine compounds) through the assembly for at least 1 minute. This technique requires contact of the sanitizer with all of the possible product surfaces. Because contact of the sanitizer with the surface is essential, the application method in dairy operations is important.

For large-volume, mechanized operations, the sanitizer can be applied through sanitary pipelines by *circulation,* or pumping of a sanitizing solution through the system. The appropriate amount of sanitizing solution is prepared in a container and pumped. A slight back-pressure should be built up in the system to ensure contact with the upper inner surface of the pipeline.

Small operations that cannot justify mechanization can sanitize by the *submersion* of equipment, utensils, and parts in the sanitizer solution. This process normally involves submersion for approximately 2 minutes, then draining and air drying on a clean surface.

Closed containers, such as tanks and vats, can be easily and effectively sanitized by *fogging*. The strength of the sanitizing solution

should be twice that of the ordinary use solution and it should be given at least 5 minutes of exposure.

If the sanitizer is applied by *spraying*, all surfaces should be contacted and completely wetted. As with fogging, the sanitizing solution strength should be twice that of the ordinary use solution.

If mechanized sanitizing equipment is unavailable, large open containers, such as cheese vats, can be sanitized by *brush application*. All areas should be touched with the brush. This method has high labor costs.

Sanitized surfaces should not be rinsed with water; otherwise, equipment and utensils can be recontaminated with aerobic microorganisms that reduce product stability. Furthermore, other recontamination of the sanitized surfaces should be avoided.

Cleaning Steps

Dairy operations require eight cleaning steps:

1. *Cover electrical equipment.* Covering material should be polyethylene, or equivalent.
2. *Remove large debris.* This task should be accomplished during the production shift and/or prior to prerinsing.
3. *Disassemble equipment as required.*
4. *Prerinse.* Prerinsing can effectively remove up to 90% of the soluble materials. This operation also loosens tightly bound soils and facilitates penetration of the cleaning compound in the next cleaning step.
5. *Apply cleaning compound.* This step can be simplified through proper selection and use of processing equipment and cleaning equipment, proper location of equipment, and reduction of soil accumulation. Further reduction of soil buildup is possible through use of the

minimum required temperature for heating products a minimum amount of time; cooling product heating surfaces, when practical, before and after emptying of processing vats; and keeping soil films moist by immediate rinsing of foam and other products with 40° to 45°C water and leaving it in the processing vats until cleaning.

6. *Postrinse*. This step solubilizes and carries away soil. Rinsing also removes residual soil and cleaning compounds, and prevents redeposition of the soil on the cleaned surface.

7. *Inspect*. This step is essential to verify that the area and equipment are clean and to correct any deficiencies.

8. *Sanitize*. A sanitizer is added to destroy any residual microorganisms. By destruction of microorganisms, the area and equipment contribute to less contamination of the processed products.

Other Cleaning Applications

When mechanized cleaning is not practical, hand cleaning should be done, following these guidelines:

- Cleaning application should involve a prerinse of water at 37° to 38°C.
- The cleaning compound used should have a pH of less than 10 to minimize skin irritation. The temperature of the cleaning solution should be maintained at 45°C. Solution-fed brushes can be used effectively with hand cleaning operations. Filler parts and other parts that are difficult to clean should be cleaned with cleaning-out-of-place (COP) equipment to move the surface lubricant and other deposits more effectively.
- The postrinse operation should use water tempered to 37° to 38°C, with subsequent air drying.

- The sanitizing operation should include a chlorine sanitizer applied by a spray or dip.

Table 14–2 classifies and summarizes special considerations for various types of dairy plant hand cleaning equipment.

CLEANING EQUIPMENT

Cleaning of dairy facilities involves physical removal of soil from all product contact surfaces after each period of use, with subsequent application of a sanitizer. Although surfaces that contact nonproducts are less critical, they must be cleaned. The techniques for cleaning dairy plants vary depending on the plant size. The major portion of a large-volume plant is cleaned by some CIP system. This cleaning technique is the recognized standard for cleaning pipelines, milking machines, bulk storage tanks, and most equipment used throughout the processing operation. Because the normal period of use for dairy processing plant equipment is less than 24 hours, this equipment and the area are cleaned daily. Longer and continued use of piping and storage systems can reduce the cleaning frequency to once every 3 days.

CIP and Recirculating Equipment

Effectiveness of the CIP approach depends on the process variables, time, temperature, concentration, and force. Rinse and wash time should be minimized to conserve water and cleaning compounds but should be long enough to remove soil and to clean effectively and efficiently. Time is affected by temperature, concentration, and force. An energy-efficient CIP system can reduce cleaning costs by over 35% with approximately 40% less energy consumption.

A salmonellosis outbreak in pasteurized milk during the 1980s that was allegedly

Table 14–2 Special Considerations for Hand Cleaning Dairy Plant Equipment

Equipment	Recommended Cleaning Procedures
Weigh tanks (can receiving and/or in-plant can transfer)	Rinse immediately after milk has been removed; disconnect and disassemble all valves and other fittings; wash weigh tank, rinse tank, and fittings; sanitize prior to next use.
Tank trucks, storage tanks, processing tanks	Remove outlet valve, drain, rinse several times with small volumes of tempered (38°C) water, remove other fittings and agitator; brush or pressure-clean vats, tanks, and fittings; rinse and reassemble after sanitizing fittings just before reuse. Thoroughly clean manhole covers, valve outlets, slight glass recesses, and any air lines. High-pressure sprays are preferable to keep the cleanup personnel out of the tanks or vats and to minimize damage to surface and contamination of cleaned surfaces.
Batch pasteurizers and heated produce surfaces	Lower temperature to below 49°C immediately after emptying product; immediately rinse, with brushing to loosen burned-on products. If the vat cannot be rinsed, fill with warm (32°–38°C) water until cleaning. Clean the same as for other processing vats.
Coil vats	Although not in general use, they are difficult to clean because of inaccessibility of some surfaces of the coil. After prerinsing, fill with hot water until bottom of coil is covered. Add cleaning compounds and rotate coil while all exposed coil surfaces are brushed.
Homogenizers	Prerinse while the unit is assembled; dismantle and clean each piece; place clean parts on a parts cart to dry. Sanitize and reassemble prior to use.
Sanitary pumps	After use, remove head of pump and flush thoroughly with tempered (38°C) water; remove impellers and place them in the bucket containing a cleaning solution of 49°–50°C. Wash intake and discharge parts and chamber. Brush impellers and place them in a basket on a parts table to dry.
Centrifugal machines	Non-CIP types must be cleaned by hand. Rinse with 38°C water until discharge is clear. Dismantle, remove bowel and discs, and rinse each part before placing in the wash vat. A separate wash vat is desirable for separator and clarifier parts to avoid damaging the discs and other close-tolerance parts. Each disc should be washed separately, rinsed, and drained thoroughly. If a separator is used intermittently during the day, it should be rinsed after each use, with at least 100 L of tempered water. Use of a mild alkaline wetting agent can improve rinsing efficiency.

caused by a CIP cross-connection between raw and pasteurized products has been responsible for the installation in many dairies of a third, completely separate CIP system for the receiving area of the plant.

Temperature of the cleaning solution for CIP equipment should be as low as possible and still permit effective cleaning with minimal use of the cleaning compound. Rinse temperature should be low enough to avoid deposits from hard water.

Force or physical action determines how effectively the cleaning compound is introduced to the areas to be cleaned and how it is controlled by the system design. Adequate force (or physical action) can be ensured by the selection and utilization of appropriate high-pressure pumps to provide sufficient turbulence in and through pipelines and storage tanks, achieving maximum efficiency.

CIP operations in dairy plants are normally divided into two major categories: spray cleaning and line cleaning. Other closed circuits, such as high-temperature short-time (HTST) units, are frequently used.

Although many types of spray devices are utilized in the dairy processing industry, permanently installed fixed-spray units are more durable than are portable units and rotating or oscillating units. Other advantages include no moving parts, stainless steel construction, and less performance difficulty, due to minor variations in supply pressure.

The line cleaning principle can involve product piping CIP circuits with readily available points from which a circuit can be fed and to which it can be returned. Return lines from storage tanks to a return pump should have an approximate 2% pitch continuously toward the return pump inlet. Control of pressure and flow should be provided for each spray device.

Shell and tube heat exchangers that are equipped with return-bend connections of CIP design can be incorporated into CIP piping circuits or may be cleaned independently as a separate operation. Triple tube-type tubular heat exchangers can be installed so that they will be self-draining. Plate-type heat exchangers are more widely used than are tubular units because of ease of inspection, flexibility of design, and ease of adaptation to new applications.

In CIP, the cleaning compound must be applied forcefully enough to provide intimate association with the soiled surfaces, and it must be continuously replenished. Various forms of CIP equipment systems are available. (The basic forms are discussed in Chapter 9.) Some CIP systems have been modified to permit use of final rinses as the cleaning solution for makeup water of the following cleaning cycle and to segregate and recover initial rinses to minimize waste discharges.

Installations since the mid-1970s have incorporated CIP systems that combine the advantages of the flexibility and reliability of single-use systems with water and solution recovery techniques that aid in reducing the amount of water required for a cleaning cycle. The intent of these systems is to recover the spent cleaning solution and the postrinse water from one cleaning cycle for temporary storage and reuse of the detergent-rinse water mixture as a prerinse for the subsequent cleaning cycle. This approach reduces the total water requirement of spray-cleaning systems by 25% to 30%, as compared with alternative approaches. Through this technique, steam consumption is reduced by 12% to 15% and cleaning compound consumption by 10% to 12%, because a prerinse of the spent solution adds heat to the vessel as it removes the soil. If a CIP recirculating unit is used to clean equipment with a large quantity of insoluble soil, a powered strainer, centrifuge, or settling basin may be incorporated in the return system to prevent this material from recirculating and impairing the spray action. Proper operation of the entire CIP system should be verified from data collected on recording charts, which can be stored for future reference.

COP Equipment

The following steps are recommended when COP equipment is used in dairy plants:

1. A prerinse with tempered water at 37° to 38°C to remove gross soil
2. A wash phase through circulation of a chlorinated alkali cleaning solution for approximately 10 to 12 minutes at 30° to

65°C for loosening and eradicating soil not removed during the prerinse phase

3. A postrinse with water tempered to 37° to 38°C to remove any residual soil or cleaning compound

Cleaning of Storage Equipment

Appropriately designed storage tanks with properly installed spray devices are essential for effective spray cleaning. The fixed-based spray that is permanently installed has become more prominent in the industry than the rotating and oscillating spray devices. It requires less maintenance, is constructed of stainless steel without moving parts, and endures. Performance of this unit is not affected by minor variations in supply pressure, and spray is continuously applied to all of the surfaces. Cylindrical and rectangular tanks can be properly cleaned when sprayed with 4 to 10 L/min/m² of internal surface, with patterns designed to spray the upper one-third of the storage container. Because the equipment contains heating or cooling coils with complex agitators, a special spray pattern is normally required, as is a subsequent increase in pressure and volume to cover all of the surfaces.

The vertical silo-type tank requires flow rates of 27 to 36 L per linear meter of tank circumference. Because of the difficulty in reaching the spray devices for occasional inspection and cleaning, nonclogging disc sprays are normally used in this type of storage vessel. Although most spray cleaning is conducted with standard sprays, special devices such as disc sprays, ball sprays, and ring sprays are available for use with vacuum chambers, dryers, evaporators, and complex vessels with special processing features.

Cleaning of large tanks that use spray devices differs from line cleaning applications because prerinsing and postrinsing are generally accomplished through use of a burst technique in which water is discharged in three or more bursts of 15 to 30 seconds each, with complete draining of the tank between successive bursts. This procedure is more effective in removing sedimented soil and foam than is continuous rinsing, and it can be accomplished with less water consumption.

The soil deposited in storage tanks and processing vessels is more variable than that associated with piping circuits; thus, cleaning techniques for this equipment are more diverse. For lightly soiled surfaces, such as those of storage tanks for milk or low-fat milk byproducts, effective cleaning can be accomplished through a three-burst prerinse of tempered water. Recirculation of a chlorinated alkaline detergent of 5 to 7 minutes at 55°C, application of a two-burst postrinse at tap water temperature, and recirculation of an acidified final rinse for 1 to 2 minutes at tap water temperature also contribute to effective cleaning. Recirculation time and temperature may be increased slightly for more viscous products with a higher content of fat and total solids.

Soil components from cold surfaces differ from those of burned-on deposits, which contain higher protein and mineral contents. Burned-on soil requires increased cleaning compound concentration and solution temperatures of up to 82°C, with an application time of up to 60 minutes. Excessive amounts of burned-on deposits can also be cleaned effectively with application and circulation of a hot alkaline detergent and a hot acid detergent solution.

Table 14–3 lists the typical concentration of cleaning compounds and sanitizers for various cleaning applications. Although variations can exist, the suggested concentrations should be considered.

Cleaning programs depend on the properties of the product passing through the system during production. In addition to cleaning applications previously discussed, the follow-

Table 14–3 Typical Concentrations for Various Cleaning Applications

Cleaning Applications	Chlorinated Cleaning Compounds (ppm)	Acid/Acid Anionic Chlorine Sanitizers (ppm)
Milk storage and transportation tanks	1,500–2,000	100
Cream, condensed milk, and ice cream storage tanks	2,500–3,000	100–130
Processing vessels for moderate heat treatment	4,000–5,000*	100–200
Heavy "burn-on"	0.75–1.0% (causticity)	Acid wash at pH 2.0–2.5

*An acid rinse after cleaning should also be considered.

ing approach is recommended for the following processing systems:

Milk, skim milk, and low-fat products processing equipment. Because of the mineral content of these products, the equipment can be cleaned effectively by recirculation of an acid detergent for 20 to 30 minutes, with follow-up by direct addition of a strong alkaline cleaner, which is then recirculated for approximately 45 minutes. An intermediate rinse of cold water may be alternated between the acid and alkaline cleaners.

Cream and ice cream processing equipment. These products, which contain a higher percentage of fat and a lower percentage of minerals, can be cleaned more effectively if an alkaline cleaner is first recirculated for approximately 30 minutes. The concentration of the alkaline solution may range from 0.5% to 1.5% causticity. The acid is generally added to produce a pH of 2.0 to 2.5. A practical rule of thumb is to use a cleaning solution temperature during the recirculating period that is adjusted to approximately 5°C higher than the maximum processing temperature used during the production shift.

Cheesemaking Area and Equipment

The two main types of spoilage of hard and semihard varieties of cheese are surface growth of microorganisms (usually molds) and gas production of microorganisms growing in the body of cheese. *Penicillium* accounts for up to 80% of spoilage cases, and other common spoilage species are *Alternaria, Aspergillus, Candida, Monilia,* and *Mucor.* Mold spoilage reduction may be accomplished through sterile filtration of air, ultraviolet disinfection of handling surfaces, and antimycotic coating of packaging material. *Enterobacteriaceae, Bacillus, Clostridium,* and *Candida* are some common microorganisms responsible for gas production. According to Varnam and Sutherland (1994), soft cheeses can be affected by gram-negative bacteria, such as *Pseudomonas fluorescens, P. putida,* and *Enterobacter agglomerans;* by diarrheagenic strains of *E. coli,* which come from wash water or added ingredients; and by gram-positive bacteria, such as *L. monocytogenes.*

Milk should be stored in tanks constructed with materials and designs that are easy to clean. However, silo tanks that are large and cannot be cleaned using normal cleansing methods should be equipped with CIP methods and cleaned every time that they are emptied. They should be rinsed with water to remove gross soils and washed with detergent solutions, rinsed, and sterilized. Acid solutions should be incorporated when tank materials permit their use. Chemical sterilization is the preferred method, and steam sterilization should be avoided.

As with other dairy processing plants, piping should be carefully laid out to prevent cross-contamination between pasteurized and unpasteurized milk. Separate CIP equipment should be provided for both products. Cleaning and sterilization can be achieved by circulating materials such as sodium hydroxide and nitric acid (Varnam and Sutherland, 1994).

Brine tanks should be lined with a noncorrosive material, such as tiles or plastics. Brines should be maintained at the correct strength to reduce the growth of halophilic microorganisms. The walls, floors, and ceilings of ripening rooms and cheese storage areas should be washed with fungicide solutions.

Increased outbreaks of *L. monocytogenes*, *S. aureus*, and *Yersinia enterocolitica* cause concern because these organisms can attach to surfaces and cross-contaminate food products or expose workers to contamination if surfaces are not cleaned and sanitized properly. Because disinfectants affect microorganisms differently and at different concentrations, tests should be conducted to determine the appropriate disinfectants and concentrations at each step of the cheese manufacturing process.

SUMMARY

Plant layout and construction affect microbial contamination and overall wholesomeness of the product. It is especially important to ensure that clean air and water are available and that surfaces in contact with dairy foods do not react with the products.

Soils that are found in dairy plants include minerals, proteins, lipids, carbohydrates, water, dust, lubricants, cleaning compounds, sanitizers, and microorganisms. Effective sanitation practices can reduce soil deposition and effectively remove soil and microorganisms through the optimal combination of chemical and mechanical energy and sanitizers. This condition is accomplished through the appropriate selection of clean water, cleaning compounds, cleaning and sanitizing equipment, and sanitizers for each cleaning application. A current trend has been toward modification of CIP systems to permit final rinses to be utilized as makeup water for the cleaning solution of the following cleaning cycle and to segregate and recover initial product-water rinses to minimize waste discharges. Every processing facility should verify the effectiveness of its cleaning and sanitation program through daily microbial analyses of both product and various equipment and areas.

STUDY QUESTIONS

1. What construction characteristics are needed for effective sanitation in dairy plants?
2. What temperature is necessary to hot-water sanitize dairy processing equipment?
3. How is chemical sanitizing of dairy processing equipment accomplished?
4. What are the two major categories of CIP operations?
5. What kind of brushes are best for cleaning dairy processing equipment?

REFERENCES

Buchanan, R.L., and M.P. Doyle. 1997. Foodborne disease: Significance of *Eschericia coli* O157:H7 and other interohemorrhagic *E. coli*. *Food Technol* 51, no. 10: 69.

Clavero, M.R.S., et al. 1994. Inactivation of *Eschericia coli* O157:H7, *Salmonellae*, and *Campylobacter jejuni* in raw ground beef by gamma irradiation. *Appl Environ Microbiol* 60: 2069–2075.

Seiberling, D.A. 1987. Process/CIP engineering for product safety. In *Food protection technology*, 181. Chelsea, MI: Lewis Publishers.

Stenson, W.S., and Forwalter. 1978. Selection guide to cleaning and sanitizing compounds. *Food Processing* 39: 34.

Varnam, A.H., and J.P. Sutherland. 1994. *Milk and milk products: Technology, chemistry and microbiology*. New York: Chapman & Hall.

SUGGESTED READINGS

Jowitt, R. 1990. *Hygienic design and operation of food plant*. Westport, CT: AVI Publishing Co.

Marriott, N.G. 1990. *Meat sanitation guide II*. American Association of Meat Processors and Virginia Polytechnic Institute and State University, Blacksburg.

CHAPTER 15

Meat and Poultry Plant Sanitation

Meat and poultry are perishable foodstuffs, and red meat has a relatively unstable color. If sanitary practices are poor, there will be increased microbial damage resulting in the degradation of color and flavor. An effective sanitation program is essential to reduce discoloration and spoilage with a resultant increase in shelf life.

Sanitation in the meat and poultry industry requires good housekeeping, beginning with the live animal or bird and continuing through serving the prepared product. The sanitation program should be thoroughly planned, actively enforced, and effectively supervised. An effective program should be headed by an individual accountable to top management for the sanitary condition of the plant and wholesomeness of the product. The most successful program is effected through cleaning inspection by trained personnel who are directly responsible for the sanitary condition of the plant and equipment.

ROLE OF SANITATION

Meat and poultry nourish microorganisms that cause discoloration, spoilage, and foodborne illness. Methods of processing and distribution are responsible for the increased exposure of these products to microbial contamination. For example, many of today's

merchandising techniques depend on appearance to sell the product. Improved sanitation is responsible for reduced contamination and increased product stability.

There are many obvious reasons for maintaining high standards of cleanliness in meat and poultry facilities. The following are a few that are important:

- These products are vulnerable to attack by microorganisms present under unsanitary conditions.
- Microorganisms cause product discoloration and flavor degradation.
- Self-service merchandising of aerobically packaged fresh meat and poultry places a premium on intensive sanitation to increase shelf life.
- Improved sanitary conditions reduce waste because less discolored and spoiled product has to be discarded.
- Immaculate sanitary conditions can improve the image of a firm, whose reputation depends on product condition. A sanitary product is more wholesome and superior in appearance to tainted merchandise.
- Increased emphasis on food nutrition and sanitation by regulatory agencies and consumers suggests a need for an effective sanitation program.

- Employees deserve clean, safe working conditions. Sanitary and uncluttered surroundings improve morale, productivity, and product turnover.
- The established trend toward increased centralized processing and packaging dictates a need for increased emphasis on sanitation. Increased processing and handling necessitate a more intensive sanitation program.
- Sanitation is good business.

Effect on Product Discoloration

Biochemical discoloration is related to the amounts of oxygen and carbon dioxide present. Figure 15–1 illustrates how the partial pressure of oxygen affects the myoglobin chemical state, which ultimately influences muscle color. High carbon dioxide partial pressure can cause a gray or brownish discoloration by association of carbon dioxide with myoglobin at the free binding site, and the rate of metmyoglobin formation increases with decreasing oxygen pressure.

A major cause of discoloration is related to microorganisms. Microbes consume available oxygen at the product surface, which reduces available oxygen needed to maintain the muscle pigment myoglobin in the oxymyoglobin state. Oxidation can cause an abnormal brown, gray, or green discoloration of meat by oxidation of the ferrous iron of the heme compound to the ferric state and direct attack by oxygen on the porphyrin ring. The color of fresh meats becomes unacceptable when metmyoglobin reaches approximately 70% of the surface pigment. Formation of metmyoglobin is accelerated by decreased oxygen pressure as a result of oxygen consumption through growth of aerobic microorganisms. The critical partial pressure for oxygen has been found to be 4 mm. Below this level, rapid oxidation to metmyoglobin occurs.

Research has suggested that the primary role of bacteria in meat discoloration is the reduction of the oxygen tension in the surface tissue. This conclusion has been based on the following observations:

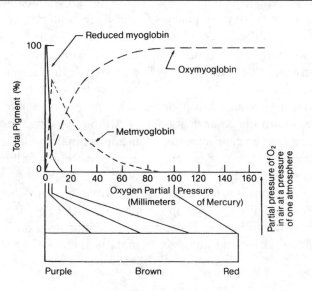

Figure 15–1 Relationship of partial oxygen pressure to myoglobin chemical state. *Source:* Kropf, 1980.

1. Rate of oxygen uptake on the muscle tissue surface is related to microbial activity and color change.
2. Oxidation to metmyoglobin occurs at intermediate levels of oxygen demand of the surface tissue. With high respiration rates, reduction to myoglobin occurs, correlating with similar changes under controlled oxygen atmospheres.
3. Pigment oxidation and reduction are controlled by adjustment of oxygen level in the storage atmosphere with a light load of microorganisms.
4. Agents inhibiting high oxygen uptake rates in exposed tissues preserve color under atmospheric conditions but are ineffective under low oxygen pressures.

These observations result in the conclusion that the reduction of oxygen in muscle tissue by microbial growth or by physical effects can produce an increase in reduced myoglobin through oxidation by metabolic hydrogen peroxide produced by muscle tissue or by bacteria. With oxygen tension reduced to a low enough level, hydrogen peroxide formation is nil, and no oxidation will occur. This condition indicates that the dissociation of the oxy compound increases as oxygen tension decreases. Fresh meat pigments are more vulnerable to discoloration at oxygen tensions below that of air at atmospheric pressure.

Clearly, the growth of bacteria from poor sanitation contributes to muscle color degradation through reduced oxygen concentration and ultimate discoloration. Various genera and species of microorganisms differ in their effect on pigment alteration; however, improved cleanliness can delay the development of high numbers of microbes. Those who handle meat should strive to minimize the initial microbial load.

Meat and Poultry Contamination

During the slaughter, processing, distribution, and food service cycle, food items are handled frequently—often as many as 18 to 20 times. Because almost anything contacting meat and poultry can serve as a contamination source, the risk of this condition occurring rises each time these products are handled.

When alive, a healthy animal possesses defense mechanisms that counteract the entrance and growth of bacteria in the muscle tissue. After slaughter, the natural defenses break down, and there is a race between humans and microbes to determine the ultimate consumer. If the handling is careless and ineffective, the microbes win. Those involved with sanitation must create a less favorable environment for the microorganisms. (Chapter 3 discusses contamination sources during slaughter and processing.)

Approximately 1 billion microorganisms are contained in a gram of soil attached to the hide of a live animal. A gram of manure contains approximately 220 million microbes. Sticking knives contaminated with bacteria introduce contamination through the wound. An animal's heart may beat for 2 to 9 minutes after sticking, thereby permitting thorough distribution of microbes. Unwashed animals have approximately 155 million microorganisms/cm^2 of skin where the jugular vein is cut.

Although the temperature of a scalding vat is approximately 60°C, the microbial load is approximately 1 million bacteria per liter of water. The dehairing operation for hogs is responsible for microorganisms being beaten into the surface skin.

Contamination during evisceration of animals is increased because the stomach and intestinal contents are loaded with microorganisms. A major contamination source for meats in the slaughterhouse is rumen fluid, which averages 1.3 billion microorganisms per milliliter.

Carcass surface counts of microorganisms average 300 to 3,000/cm². Beef and pork trimmings contain 10,000 to 500,000 bacteria per gram, depending on contamination and sanitation practices. Cutting boards on fabricating tables normally contain approximately 77,500 bacteria per square centimeter. Slicers, conveyors, and packaging equipment may increase the contamination of processed meats by 1,000 to 50,000 bacteria per gram, depending on sanitation practices.

Pathogen Control

In the past, meat and poultry products have accounted for 23% of foodborne illness outbreaks and 27% of the cases of outbreak-associated foodborne disease for which a food vehicle was implicated. During the same period, meat and poultry were associated with 10% and 5%, respectively, of the reported foodborne outbreak death. The primary etiologic agents for meat-related outbreaks are *Staphylococcus aureus* (20%), *Salmonella* species (18%), and *Clostridium perfringens* (10%). The etiologic agents for poultry-related outbreaks are *Salmonella* (26%), *S. aureus* (13%), and *C. perfringens* (11%).

During the scalding, evisceration, rinsing, and chilling phases of poultry processing, the carcasses are quite vulnerable to contamination from species of *Salmonella* and *Campylobacter*, *Aeromonas hydrophila*, *Listeria monocytogenes*, and other microorganisms of public health concern. *Campylobacter* has presented a serious problem for the poultry industry because it is commonly present on raw poultry. Poultry has been implicated in campylobacteriosis that has occurred sporadically without a finite determination of the mode of transmission. The design of poultry processing equipment, especially the plucking equipment, is such that adequate cleaning is difficult. The major risk in evisceration is the spilling of the gut content onto the carcass. Furthermore, the knife and hands of the meat inspector are often heavily contaminated. *Campylobacter jejuni* will spread during the slaughtering process. Regardless of the type of slaughtering, heavily infected poultry flocks may result in a contamination rate of 100% for the finished product. Immersion chilling poses a contamination threat because of entrapment of microorganisms in skin channels and with the swelling of collagenous material in the neck flap area. These highly contaminated carcass parts should be trimmed to lower the microbial load. Current research results indicate that rinsing poultry carcasses removes a small amount of *Salmonella* organisms that may be present. Species of *Salmonella* and *Campylobacter* affix themselves to the skin and flesh of poultry so tightly that they become part of food intended for human consumption.

Shapton and Shapton (1991) emphasized the need for cleaning of roofs over food manufacturing areas. Process equipment and exhaust stacks may be vented through the roof. If feasible, roof-mounted process equipment should be enclosed with a floor to separate it from the processing area. Particles, especially hygroscopic powders, can deposit on the roof, especially if it is flat. When left unattended, this area may attract birds, rodents, or insects, which are known carriers of *Salmonella* organisms and of *L. monocytogenes*. Pools of water will encourage these pests. A minimum slope of 1% is recommended to ensure drainage.

The incidence of *L. monocytogenes* is approximately 15% to 50% for poultry carcasses, 20% of dry sausage and fresh sausage, and 10% or more of ground beef samples evaluated. Growth can also occur in some cooked meat products after packaging. A significant portion of fresh meats used as raw materials for processed products can be contaminated with this psychrotrophic pathogen and point to the importance of preventing postprocessing recontamination of ready-to-eat products. Table 15–1 illustrates the incidence of *L. monocytogenes* in post-heat-processing environments in 41 meat plants.

Listeria monocytogenes is often found around wet areas and cleaning aids, such as floors, drains, wash areas, ceiling condensate, mops and sponges, brine chillers, and at peeler stations. Thus, control of *Listeria* organisms in processing plants is essential to reduce the potential of postprocessing contamination. One cannot control the growth of this pathogen through refrigeration at 4° to 5°C (a common storage temperature) because this microbe can survive at a 0°C storage environment. Doyle (1987) has suggested that novel, nontraditional approaches, such as the use of antimicrobial agents, reduced temperature (<2°C) storage, reformulation of products (reduced minimum water activity [A_w], pH, etc.), or postprocessing pasteurization of products may need to be incorporated for the control of such psychrotrophic pathogens in foods.

Pathogens such as *L. monocytogenes* can be better controlled through the reduction of cross-contamination. Employees who work in the raw and finished product areas, such as smokehouses and water- and steam-cooking areas, should change outer clothing and sanitize their hands or change gloves when moving from a raw to finished product area. Utensils and thermometers that are used for raw and finished products should be sanitized each time they are used. Frequent cleaning with floor scrubbers is essential. If ceiling condensate is present, it should be removed by a vacuum unit or a sanitized sponge mop. Cleaned floors that do not dry before production startup should be vacuumed or squeegeed.

The following guidelines should be considered when planning for the control of *L. monocytogenes* in meat, poultry, and other food plants.

Table 15–1 Incidence of *Listeria monocytogenes* in Meat Plant Postheat Processing Environments

Location	Percentage Positive for *L. monocytogenes*
Floors	37
Drains	37
Cleaning aids	24
Wash areas	24
Sausage peelers	22
Food contact surfaces	20
Condensate	7
Walls and ceilings	5
Compressed air	4

Source: Wilson, 1987.

Layout and Plant Design

In addition to principles mentioned in Chapter 12, the following should be considered.

1. Plant layout should prevent pests and vermin and should control the movement of *L. monocytogenes* between raw and cooked product areas. Examples are employee traffic patterns, support and supervisory staff movement, and food-handling activities.
2. Air and refrigeration movement equipment should be designed for easy cleaning and sanitizing. Ready-to-eat areas should have a positive air pressure design.

3. All equipment and other surfaces should be easily cleaned and sanitized with smooth, nonporous surfaces.
4. Floors should be surfaced with materials that are easily cleaned and will not encourage water accumulation.

Equipment Design

1. Only regulatory approved materials should be incorporated.
2. Equipment should be properly maintained and stored.

Process Control

1. If the process does not contain an *L. monocytogenes* kill step, the operation should be designed to reduce contamination.
2. The kill step (if applicable) should be a critical control point in the Hazard Analysis Critical Control Points (HACCP) program.

Operation Practices

1. Employees should be educated about good manufacturing practices (GMPs), HACCP, and the responsibilities of each.
2. Equipment should be provided to maintain sanitary conditions such as (a) foot baths, (b) hand dips, (c) hair nets, and (d) gloves.
3. Contamination sources, especially in ready-to-eat areas, should be eliminated.
4. Management should be educated to support GMPs and HACCP.

Sanitation Practices

1. An adequate number of employees, time, and supervision should be provided for cleaning and sanitizing.

2. Written cleaning and sanitizing procedures should be developed and posted for each area in the plant.
3. Environmental sampling programs to verify the effectiveness of cleaning and sanitizing should be established.

Verification of L. monocytogenes Control

1. A microbial assay of weekly samples from plant areas, equipment, and the air supply should be conducted. It is especially important to sample points between the kill step and packaging.
2. Samples can be composited to reduce the analysis cost. If a composite sample is positive, a follow-up analysis of individual samples is necessary to determine which equipment is the contamination source.

The following suggestions for *Listeria* control in meat plants are adapted from those provided by Douglas (1987):

1. Mechanically or manually scrub floors and drains daily. Drains should contain a "quat plug" or be rinsed with disinfectants daily.
2. Scrub walls weekly.
3. Clean the exterior of all equipment, light fixtures, sills and ledges, piping, vents, and other areas in the processing and packaging areas that are not in the daily cleaning program.
4. Clean cooling and heating units and ducts weekly.
5. Caulk all cracks in walls, ceilings, and window and door sills.
6. Scrub and clean raw material areas at least as often as the processing and packaging areas are cleaned.
7. Keep hallways and passageways that are common to raw and finished product clean and dry.

8. Minimize traffic in and out of processing and packaging areas and establish plant traffic patterns to reduce cross-contamination from feet, containers, pallet jacks, pallets, and fork trucks.

9. Change outer clothing and sanitize hands or change gloves when moving from a "raw" to a finished product area.

10. Change into clean work clothes each day at the plant. Provide some pattern of color coding to designate the various plant areas.

11. Have visitors change into clean clothes provided at the plant.

12. Provide a monitoring program of the plant environment to measure the effectiveness of the *Listeria* control procedures.

13. Enclose processing and packaging rooms so that filtered air comes in and these areas are under positive pressure.

14. Clean and sanitize all equipment and containers before being bringing them into processing and packaging areas.

Temperature Control

Meat and poultry spoil when held at a high temperature. Temperature affects the rate of chemical and biochemical reactions, and, especially, the lag phase of the growth pattern of microorganisms. The rates for both microbial and nonmicrobial spoilage increase to approximately 45°C. Above 60°C, microbial spoilage usually does not occur. (Microbial growth kinetics are discussed in Chapter 2).

Microorganisms grow most rapidly between 2° and 60°C. This temperature range is considered the critical zone, or the danger zone. Meat and poultry must be stored out of this temperature zone and should be taken through this range as quickly as possible when a temperature change is necessary (as when cooking and chilling). Storage temperature below the critical zone does not effectively destroy bacteria but does reduce the rate of growth and multiplication of microorganisms. Below the critical zone, bacteria are less active, and some death can occur through stress.

Processing and storage at a colder temperature will reduce spoilage and growth of any microorganisms present on equipment, supplies, or other materials in the storage and processing areas. Under unsanitary conditions with improper temperature control, certain species of *Pseudomonas* will double in number every 20 minutes. At this rate, in 12 hours, 1 bacterium can become 280 trillion. Meat and poultry are generally expected to avoid spoilage twice as long at 0°C than at 4.4°C, and at least four times as long at 0°C than at 10°C.

Air curtains should be installed, especially when truck doors must be left open, to prevent refrigeration loss where the plant is under positive pressure. Entry of insects and dust is reduced with the use of air curtains. The air velocity should be at least 488 m/min, measured at a distance of 910 mm above the floor. For personnel entrances, the air stream should be continuous across the entire width of the opening, with a thickness of at least 254 mm and a minimum velocity of 503 m/min, measured 910 mm above the floor (Shapton and Shapton, 1991).

SANITATION PRINCIPLES

An efficient cleaning arrangement can reduce labor costs up to 50%. Proper construction and equipment selection are critical for the most effective cleaning operation. It is important that the floors, walls, and ceilings be constructed of impervious material that can be easily cleaned. Floors should be sloped with a minimum of 10.5 mm/m. (Discussion of cleaning equipment appears in Chapter 9.)

Hot-Water Wash

Because soil from meat and poultry is primarily fat and protein deposits, a hot-water wash is not an effective cleaning method. Hot water can loosen and melt fat deposits but tends to polymerize fats, denature proteins, and complicate removal of protein deposits by binding them more tightly to the surface to be cleaned. The main advantage of a hot-water wash system is minimal investment of cleaning equipment. Limitations of this approach include increased labor requirements and water condensation on equipment, walls, and ceilings. It is difficult to remove heavy soil with this system.

High-Pressure, Low-Volume Cleaning

High-pressure, low-volume spray cleaning is a viable method in the meat and poultry industry because of the effectiveness with which it removes tenacious soils. With this equipment, the operator can more effectively clean difficult-to-reach areas with less labor, and the cleaning compound is more effective at a lower temperature.

This hydraulic cleaning system can be provided by portable units that can be easily moved throughout the plant. This portable equipment can be utilized for cleaning parts of equipment and building surfaces and is especially effective for conveyors and processing equipment when soaking operations are impractical and hand brushing is difficult and labor-intensive.

The metering device and controls of a centralized high-pressure cleaning unit are illustrated in Chapter 9. A dispensing nozzle for this equipment is pictured in Figure 15–2.

Foam Cleaning

Foam is particularly beneficial in cleaning large surface areas of meat and poultry plants

Figure 15–2 High-pressure hose with a female, stainless steel, quick-connect, heavy duty, dead-man, shutoff-type spray gun extension wand. Courtesy of Diversey Lever, 255 E. 5th Street, Suite 1200, Cincinnati, Ohio, 800-233-1000.

and is frequently used to clean transportation equipment exteriors, ceilings, walls, piping, belts, and storage containers. Portable foam equipment is similar in size and cost to portable high-pressure units. Centralized foam cleaning applies cleaning compounds by the same desirable features as a centralized system.

Gel Cleaning

This equipment is similar to high-pressure units, except that the cleaning compound is applied as a gel rather than as a high-pressure spray. Gel is especially effective for cleaning packaging equipment because it clings to the surfaces for subsequent soil removal. Equipment cost is similar to that of portable high-pressure units.

Combination Centralized High-Pressure, Low-Volume, and Foam Cleaning

This system is the same as centralized high pressure except that foam can also be applied through the equipment. This method offers the most flexibility because foam can be used on large-surface areas, and high pressure can be applied to belts, conveyors, and hard-to-reach areas in a meat or poultry plant. Equipment costs for this system range from $15,000 to over $150,000, depending on size.

Cleaning-in-Place (CIP)

With this closed system, a recirculating cleaning solution is applied by installed nozzles, which automatically clean, rinse, and sanitize equipment. Benefits of CIP systems are discussed in Chapter 9. The use of CIP systems in the meat and poultry industry is limited. This equipment is expensive and lacks effectiveness in heavily soiled areas. CIP cleaning has some application in vacuum thawing chambers, pumping and brine circu-

lation lines, preblend/batch silos, and edible and inedible fat-rendering systems. Figure 15–3 illustrates a CIP application principle for washing shackles, rollers, and chains in poultry plants. The motor and drive components are mounted on a base plate. As the shackles pass between two rotating brushes, they are cleaned. The brushes can be lifted above the rail when not in use.

CLEANING COMPOUNDS FOR MEAT AND POULTRY PLANTS

Acid Cleaners

Information about strongly and mildly acid cleaners is provided in Chapter 7.

Strongly Alkaline Cleaners

Examples of strongly alkaline compounds are sodium hydroxide (caustic soda) and silicates having high $N_2O:SiO_2$ ratios. The addition of silicates tends to reduce the corrosiveness and to improve the penetration and rinsing properties of sodium hydroxide. These cleaners are used to remove heavy soils, such as those found in smokehouses.

Heavy-Duty Alkaline Cleaners

The active ingredients of these cleaners may be sodium metasilicate, sodium hexametaphosphate, sodium pyrophosphate, and trisodium phosphate. The addition of sulfites tends to reduce the corrosion attack on tin and tinned metals. These cleaners are frequently used with CIP, high-pressure, and other mechanized systems found in meat and poultry plants.

Mild Alkaline Cleaners

Mild cleaners are frequently in solution and are used for hand cleaning lightly soiled areas in meat and poultry plants.

Figure 15–3 Shackle washer for cleaning shackles, rollers, and the chain in poultry processing plants. Courtesy of Chemdyne Corp.

Neutral Cleaners

Information about these and other cleaning compounds is discussed in Chapter 7.

SANITIZERS FOR MEAT AND POULTRY PLANTS

To obtain maximum benefits from use of a sanitizer, it must be applied to surfaces that are free of visible soil. Soils of special concern are fats, meat juices, blood, grease, oil, and mineral buildup. These deposits provide areas for microbial growth, both below and within the soil, and can hold food and water necessary for microbial proliferation. Chemical sanitizers cannot successfully penetrate soil deposits to destroy microorganisms.

Steam

Steam is an effective sanitizer for most applications. Many operators mistake water vapor for steam and fail to provide adequate exposure to create a sanitizing effect. Steam should not be used in refrigerated areas because of condensation and energy waste, and it is unsatisfactory for continuous sanitizing of conveyors.

Chemical Sanitizers

Chlorine is one of the halogens used for disinfecting, sterilizing, and sanitizing equipment, utensils, and water. The inorganic chlorine compounds most frequently used in sanitizing meat and poultry operations are the following:

- *Sodium and calcium hypochlorite*: These are more costly than elemental chlorine, but are more easily applied. Hypochlorous acid is an active germicidal agent, and the activity of hypochlorites is pH dependent. Alkalinity decreases as the germicidal activity increases.

- *Liquid chlorine*: This sanitizer is for use in the chlorination of processing and cooling waters to prevent bacterial slimes.
- *Chlorine dioxide*: This is an effective bactericide in the presence of organic matter because it does not react with nitrogenous compounds. The residual effect is also more persistent than that of chlorine. However, this sanitizer should be generated on site.
- *Active iodine* solutions, like active chlorine solutions, can be sanitizers. Iodophors are very stable products with much longer shelf lives than hypochlorites and are active at a low concentration. These sanitizing compounds are easily measured and dispensed, and they penetrate effectively. Their acid nature prevents film formation and spotting on equipment. Solution temperature should be below 48°C because free iodine will dissipate.
- The *quaternary ammonium* compounds are widely used on floors, walls, equipment, and furnishings of meat and poultry plants. The "quats" are effective on porous surfaces because of their penetration ability. A bacteriostatic film that inhibits bacterial growth is formed when quats are applied to surfaces. Those sanitizers and compounds containing both an acid and a quat sanitizer are most effective in controlling *L. monocytogenes* and mold growth. Quats may be temporarily used when a mold buildup is detected.
- The use of *acid sanitizers* combines the rinsing and sanitizing steps. Acid neutralizes the excess alkalinity from the cleaning residues, prevents formation of alkaline deposits, and sanitizes. Acid sanitizers effectively kill both gram-positive and gram-negative bacteria.

Other information about sanitizers may be found in Chapter 8.

SANITATION PRACTICES

General Instructions

All personnel should practice good personal hygiene, as discussed in Chapter 4. They should wear freshly laundered clothes and stay away from meat and other processing equipment if they are ill. All cleaning and sanitizing compounds should be kept in an area accessible only to the sanitation supervisor, manager, and superintendent, and they should be allocated only by the sanitation supervisor. Misuse of these compounds will raise cleaning costs, inhibit effective cleaning, and possibly result in personal injury and equipment damage. The water temperature should be locked in at 55°C.

Instructions provided with the portable or centralized high-pressure or foam cleaning system should be followed. Cleaning compounds should be applied according to instructions or recommendations provided by the vendor. (Chapter 7 provides a discussion related to safety precautions that should be followed when handling cleaning compounds.)

The sanitation supervisor should inspect all areas nightly while the cleanup crew is on duty. All soiled areas should be recleaned prior to the morning inspection by the regulatory agency.

Chlorine papers should be used to check the sanitizing solution if automatic make-up or instructions are not available. These test papers, normally packaged in vials of 100 strips each, include directions for use and are available through most cleaning compound suppliers. Other check systems for monitoring sanitation are also available and are discussed in Chapter 6. More information on these systems may be obtained from firms that sell cleaning compounds and monitoring systems.

Recommended Sanitary Work Habits

Sanitary workers should follow these general practices:

1. Store personal equipment (lunch, clothing, etc.) in a sanitary place and always keep storage lockers clean.
2. Wash and sanitize utensils frequently throughout the production shift and store them in a sanitary container that will not be in contact with floors, clothing, lockers, or pockets.
3. Do not allow the product to contact surfaces not sanitized for meat and poultry handling. If any particle contacts the floor or other unclean surface, it should be thoroughly washed.
4. Use only disposable towels to wipe hands or utensils.
5. Wear only clean clothing when entering production areas.
6. Cover the hair to prevent product contamination from falling hair.
7. Remove aprons, frocks, gloves, or other clothing items that may come in contact with the product before entering toilets.
8. Always wash and sanitize hands when leaving the toilet area.
9. Stay away from production areas when a communicable disease, infected wound, cold, sore throat, or skin disease exists.
10. Do not use tobacco in any production area.

HACCP

HACCP has been adopted by the U.S. Department of Agriculture, which regulates the hygienic production of meat and poultry products. (Additional discussion of HACCP, including implementation of this concept, is included in Chapters 1, 5, 16, and 20).

HACCP does not necessarily include major investments or expensive microbial or other techniques. The options chosen are selected on the basis of their efficacy and costs. An example would be control options for the pasteurization step in pork or turkey ham processing. Design/maintenance and process control are very successful and relatively inexpensive. In contrast, microbial examination is expensive and lacks effectiveness.

An example of using HACCP in a meat or poultry operation would be the development of a simple flow chart of the meat and poultry production line. The flow pattern is a long sequence of events, with steps that are difficult or impossible to control. Many relevant factors related to hazards of each step can be identified and critical control points determined.

Livestock and Poultry Production

It is possible to produce animals in a specific pathogen-free (SPF) environment to reduce contamination. Contamination can also be reduced through administration of bacterial cultures that exclude pathogens from the gut flora by competition.

Farm Environment

The farm environment (its pastures, streams, manure, etc.) contributes to the recycling of excreted pathogens to cause a cycle of infection, excretion, and reinfection. Sanitation practices must be established to improve hygiene in this portion of the flow chart.

Transportation

The stressful conditions of live animal transportation may cause pathogen carriers to spread these microorganisms. The challenge is to incorporate sanitary practices during transportation to reduce contamination in the processing plant.

Lairage

Stress during this phase of the flow chart can cause changes in the microbial flora composition of the intestinal tract, with the emergence and shedding of *Salmonella* organisms. Showering of animals can reduce stress and contamination.

Hide, Pelt, Hair, or Feather Removal

The protective coats of meat animals can and frequently do contain species of *Salmonella* and other detrimental microorganisms. New procedures and equipment modification are necessary to reduce contamination.

Evisceration

Intestinal spillage and viscera rupture can occur. In poultry slaughtering, a series of water or sanitizer sprays can be applied to reduce contamination. Red meat carcasses can also be decontaminated. The efficacy of spraying has not been totally resolved because this operation does not completely remove microorganisms and can spread contamination over the carcass.

Inspection

A meat inspector should use a sanitizer for the hands and knife because they can contaminate dressed carcasses.

Chilling

Control of chilling parameters (air temperature, air movement, relative humidity, and filtering air) can reduce microbial growth. Drying of the carcass surface is important in the suppression of microorganisms (e.g., *Campylobacter* species). Use of a solution that is not heavily contaminated and trimming of the neck flap area of poultry carcasses after chilling can reduce the microbial load.

Further Processing

Chilled carcasses and cuts should not be exposed to an unchilled environment. The equipment used in this operation should be hygienically designed and sanitized before use. Safe and wholesome adjuncts should be used in processed meats formulations.

Packaging

The appropriate packaging material will protect the product from contamination. Proper storage temperatures must be maintained.

Distribution

The method of distribution must be rapid and clean. An effective temperature and sanitary environment must be maintained. The transportation environment should be monitored for sanitation and temperature control.

SANITATION PROCEDURES

It is advantageous to have written cleaning instructions for specific operations. Documentation of procedures is especially useful when supervision changes are made and for training of new employees. As mechanization increases, cleaning methods become correspondingly more detailed and complicated. Detailed cleaning operations should be written and posted in the plant.

Prior to adopting a cleaning procedure, it is essential to become familiar with the operation of all production and cleaning equipment. In addition to providing the necessary information, this can lead to improvements in methods that are used or should be incorporated. The following are examples of cleaning procedures that could be used for distinct operations and areas in a plant. These examples are only guidelines. Every cleaning application should be adapted to the prevailing conditions. Although this step will not be mentioned, hoses and other equipment should be returned to their proper locations after cleanup.

Livestock and Poultry Trucks

FREQUENCY   After each load has been hauled.
PROCEDURE
1. Immediately after removing livestock or poultry from trucks, scrape and remove all manure that has accumulated from the premises.
2. Clean the truck beds, wheels, and frame by washing down the racks, floors, and frames with water to completely remove all manure, mud, and other debris, completely disinfecting with a quaternary ammonium sanitizer spray or by cleaning and sanitizing in one operation by spray-cleaning with an alkaline detergent sanitizer.

Livestock Pens

FREQUENCY   As soon as possible after each lot has been removed.
PROCEDURE
1. After the livestock are taken from each pen, clean the manure from the floors and walls, and remove it from the plant premises.
2. Every 4 months, scrape all dried manure and loose whitewash from the gates and partitions. Sweep cobwebs from the ceilings, and whitewash the interior of the pens. Mix a cresylic acid-type sanitizer with the whitewash slurry.

3. If a contagious disease is brought into the pens, quarantine the diseased animals and destroy them separately from the healthy livestock. Remove the manure completely from the surrounding pen area (using a hose if necessary), and disinfect the pens by spraying with a quaternary ammonium sanitizer.

A general cleaning procedure for slaughter and processing areas encompasses (1) gross physical removal of debris, (2) prerinsing and wetting, (3) cleaning compound application, (4) rinsing, (5) inspection, (6) sanitizing, and (7) prevention of recontamination. The first step is essential to reduce time and water requirements and can minimize the biological load on the sewage system. Physical removal of debris also reduces splashing of large particles during the second step. The significance of the other steps has been previously alluded to and will be discussed in other chapters. The role of these cleaning procedures will be further illustrated in the cleaning applications to follow.

Slaughter Area

FREQUENCY   Daily. Debris should be periodically removed during the production shift.
PROCEDURE
1. Pick up all large pieces of extraneous material and transfer the matter to receptacles.
2. Cover all electrical connections with plastic sheeting.
3. Briefly prerinse all soiled areas with 50° to 55°C water. Start working water from the ceiling and walls and the upper portion of all equipment, and continue to direct all extraneous matter down to the floor. Avoid direct contact of water with motors, outlets, and electrical cables.

4. Apply an alkaline cleaner through a centralized or portable foam system, using water that is 50° to 55°C. The system should be designed and operated to reach all framework, undersides, and other difficult-to-reach areas. Allow 5 to 20 minutes of exposure prior to the rinse. Although foam requires less labor, high-pressure equipment for application is more effective in penetrating hard-to-reach areas of equipment and may be more effective in the removal of *L. monocytogenes*.

5. Rinse ceilings, walls, and equipment within 20 minutes after application of the cleaning compound. Use the same rinse pattern as for prerinse and cleaning compound application, with 50°to 55°C water.

6. Inspect all equipment and surfaces and touch up as necessary.

7. Apply an organic sanitizer to all equipment with a centralized or portable sanitizing unit. If a chlorine sanitizer is used here or for applications that follow, the solution should be at least 50 parts per million (ppm) of chlorine.

8. Remove, clean, and replace drain covers.

9. Apply a white edible oil to surfaces subject to rust corrosion. Any further use of oil here or for applications that follow is discouraged because the protective film contributes to microbial growth.

10. Clean any specialized equipment in this area according to the manufacturing firm's recommended cleaning procedure (if available).

11. Avoid contamination during maintenance and equipment setup by requiring maintenance workers to carry a sanitizer and to sanitize where they have worked.

Poultry Mechanical Eviscerators

FREQUENCY   Daily. A continuous or intermittent sanitizer spray should be provided to reduce contamination.

PROCEDURE

1. Pick up all large pieces of extraneous material, and transfer the matter to receptacles.

2. Cover electrical connections with plastic sheeting.

3. Briefly prerinse this equipment with 50° to 55°C water.

4. Apply an alkaline cleaner through a centralized or portable foam system, using 50° to 55°C water. Allow 10 to 20 minutes of exposure time prior to rinse-down with 40° to 50°C water.

5. Inspect all areas and conduct any necessary touch-ups.

6. Apply 200-ppm chlorine (or other organic sanitizer) with a centralized or portable sanitizing unit.

7. Avoid contamination during maintenance, as described previously.

Poultry Pickers

FREQUENCY   Daily.

PROCEDURE

1. Pick up all large debris and transfer the matter to receptacles.

2. Cover electrical connections with plastic sheeting.

3. Briefly prerinse this equipment with 50° to 55°C water.

4. Apply a heavy-duty alkaline cleaner through a centralized or portable foam system on the shower cabinets. Shackles should go into the tank with the same cleaner.

5. After cleaning compound exposure for approximately 20 minutes, rinse down with 40° to 50°C water.

6. Remove residual feathers and other debris by hand.
7. Because of the rubber fingers, apply 25-ppm iodophor as a sanitizer through a centralized or portable sanitizing unit.

Receiving and Shipping Area

FREQUENCY Daily.
PROCEDURE
1. Cover all electrical connections, scales, and exposed product with plastic sheeting to prevent water and chemical damage.
2. Briefly rinse the walls and floors with 50° to 55°C high-pressure water. The wall-rinse motion must be from top to bottom and side to side, with extraneous matter worked to the floor. This prerinse is designed to remove heavy soil deposits and to wet the surfaces.
3. Apply an acid cleaning detergent through a slurry or foam gun. Recommended spray temperature is 55°C or less. High-pressure output (for this and other cleaning operations) is 25 to 70 kg/cm² and 7.5 to 12 L/min at the wand.
4. Within 20 minutes of the cleaning compound application, apply a high-pressure rinse with 50° to 55°C water.
5. Remove, clean, and replace drain covers in the proper position after rinse-down.

Processed Products, Offal, and Storage Cooler

FREQUENCY Weekly. Processed meats, offal, and hanging meat should be rotated so that half of a section at a time can be cleaned each week.
PROCEDURE
1. Clean each section, when empty, with a reliable floor cleaner. Apply slurry or foam via high pressure.

2. Rinse thoroughly with 55°C or lower temperature water at high pressure within 20 minutes of detergent application. Do not splash water on hanging meat in the section not being cleaned. Work all debris to the floor from overhead fixtures and walls.
3. Squeegee the floor where water has accumulated to prevent it from freezing.
4. Remove, clean, and replace drain covers.

Fabricating or Further Processing

FREQUENCY Daily.
PROCEDURE
1. Pick up all large pieces of lean, fat, bones, and other extraneous matter, and deposit them in a receptacle.
2. Cover all electrical connections with plastic.
3. Prerinse all soiled surfaces with 55°C water. Start at the bone conveyor top and work all extraneous matter down to the floor. Avoid hosing motors, outlets, and electrical cables.
4. Following wash-down and subsequent heavy soil removal, apply an alkaline cleaner through a centralized or portable high-pressure, low-volume system, using 50° to 55°C water. The system should be effectively used to reach all framework, table undersides, and other difficult-to-reach areas. Allow 5 to 20 minutes of soak time prior to rinse-down. Alternative equipment for cleaning compound application is a foam unit. This unit rapidly applies the cleaner but does not penetrate as well as does high-pressure, low-volume equipment and may be less effective in the removal of *L. monocytogenes*.
5. Rinse all equipment within 20 minutes after cleaning compound application.

Using the same rinse pattern as with prerinse and cleaning compound application, spray 50° to 55°C water on one side of a fabricating line at a time.

6. Thoroughly inspect all equipment surfaces and conduct any necessary touch-up.
7. Apply an organic sanitizer to all clean equipment with a centralized or portable sanitizing unit.
8. Remove, clean, and replace all drain covers.
9. Apply a white edible oil to surfaces subject to rust or corrosion.
10. Avoid contamination during maintenance, as described previously.

If a bone shelter or hopper exists, it should also be cleaned, as outlined in the preceding steps. This operation should be performed twice a week during winter months and daily during the summer.

Processed Products Area

FREQUENCY   Daily.
PROCEDURE
1. Dismantle all equipment and place the parts on a table or rack. Disconnect all stuffing pipes.
2. Pick up all large pieces of meat and other extraneous matter and deposit in a receptacle.
3. Cover all electrical connections with plastic.
4. Prerinse all soiled surfaces with 55°C water. Start at the top of all processing equipment, and direct all extraneous matter down to the floor. Avoid direct hosing of motors, outlets, and electrical cables.
5. Following wash-down and subsequent heavy soil removal, apply an alkaline cleaner through a centralized or portable

high-pressure, low-volume system, using 50° to 55°C water. The system should effectively reach all framework, tables, other equipment undersides, and other difficult-to-reach areas. Soak time prior to rinse-down should be 5 to 20 minutes. Although foam is less effective in penetration, it is a viable cleaning medium and is easily applied.

6. Rinse all equipment within 20 to 25 minutes after cleaning compound application. Using the same prerinse pattern as with the prerinse and detergent application, spray 50° to 55°C water on one side of each piece of processing equipment at a time.
7. Thoroughly inspect all equipment surfaces and touch up as necessary.
8. Apply an organic sanitizer to all clean equipment with a centralized or portable sanitizing unit.
9. Remove, clean, and replace drain covers.
10. Apply a white edible oil only to surfaces subject to rust or corrosion.
11. Avoid contamination during maintenance as described previously.

Fresh Product Processing Areas

FREQUENCY   Daily.
PROCEDURE
1. Dismantle all equipment, and place the parts on a table or rack. Disconnect all stuffing pipes.
2. Remove large debris from equipment and floor and deposit it in a receptacle.
3. Cover mixer and packaging equipment with plastic.
4. Briefly prerinse all soiled surfaces with 50° to 55°C water to remove heavy debris and to soak exposed surfaces. Guide hoses to force all debris toward the closest floor drain.

5. Apply an alkaline cleaner through centralized or portable high-pressure, low-volume cleaning equipment, using 50° to 55°C water. Foam, gel, or slurry may be incorporated to introduce the cleaning compound. Cleaning compound application must cover the entire area—equipment, floors, walls, and doors.
6. Rinse the area and equipment within 20 to 25 minutes after detergent application.
7. Inspect the area and all equipment. Touch up as needed.
8. Remove, clean, and replace drain covers.
9. Sanitize all clean equipment with an organic sanitizer using a centralized or portable sanitizing unit.
10. Apply a white edible oil only to surfaces subject to rust or corrosion.
11. Avoid contamination during maintenance as described previously.

Processed Products Packaging Area

FREQUENCY Daily.
PROCEDURE
1. Dismantle all equipment, placing the parts on a table or rack.
2. Remove large debris from equipment and floors and place in a receptacle.
3. Cover packaging equipment, motors, outlets, scales, controls, and other equipment with plastic film.
4. Prerinse all soiled surfaces with 55°C water to remove heavy debris and to soak exposed surfaces. Hoses should be guided to force all debris toward the closest floor drain.
5. Apply an alkaline cleaner through centralized or portable high-pressure, low-volume cleaning equipment using 50° to 55°C water. Foam, gel, or slurry

may be incorporated to introduce the cleaning compound. Cleaning compound application must cover the entire area—equipment, floors, walls, and doors.
6. Rinse the area and equipment within 20 to 25 minutes after application of the cleaning compound, using the same pattern of movement as used when applying the cleaner.
7. Inspect the area and all equipment. Touch up as needed.
8. Remove, clean, and replace drain covers.
9. Sanitize all clean equipment with an organic sanitizer using a centralized or portable sanitizing unit.
10. Apply a white edible oil only to surfaces subject to rust or corrosion.
11. Avoid contamination during maintenance as described previously.

Brine Curing and Packaging Area

FREQUENCY Daily.
PROCEDURE
1. Pick up all large debris and place in a receptacle.
2. Cover all electrical connections, scales, and exposed product with plastic sheeting.
3. Prerinse the area and all equipment with 55°C water.
4. Place an acid cleaner in the shrink tunnel (if used), and circulate for approximately 30 minutes during prerinsing.
5. Rinse shrink tunnel (if present) before detergent application.
6. Place all prerinse debris in a receptacle.
7. Apply an alkaline cleaner through a foam or slurry cleaning system, using 50° to 55°C water.
8. Rinse with 55°C water within 20 minutes after detergent application.

9. Inspect the area and equipment and touch up as needed.
10. Remove, clean, and replace drain covers.
11. Sanitize all clean equipment with an organic sanitizer applied through a centralized or portable system.
12. Apply a white edible oil only to those parts subject to rust or corrosion.
13. Avoid contamination during maintenance as described previously.

Dry Curing Areas (Curing, Equalization, and Aging)[1]

FREQUENCY After product input and at the end of designated cure or equalization period.

PROCEDURE
1. Sweep floors.
2. Remove pallets and other portable storage equipment, and rinse cure and other debris with 50°C water.
3. Hose down vacated areas with 50°C water.
4. Clean trolleys, trees, and other metal equipment used as outlined for wire pallets and metal containers or trolleys.
5. Sanitize cleaned areas according to manufacturer requirements with a quaternary ammonium compound for its residual effect.
6. Spray aging rooms once every 3 months with a synergized pyrethrin spray to reduce insect infestation. Follow mixing and application directions on the pesticide label.

---

[1]To reduce mold growth, filtered air or air conditioning with a filter is recommended for aging rooms.

Smokehouses

FREQUENCY After the end of each smoke period.

PROCEDURE
1. Pick up large debris and place in a receptacle.
2. Apply an alkaline cleaning compound recommended for cleaning smokehouses through a centralized or portable foam system. A high-pressure unit may be used where foam cannot penetrate. Figure 15–4 illustrates a unit used for cleaning a smokehouse.
3. Rinse the area within 20 to 30 minutes after cleaning compound application. Start at the ceiling and walls, and work all extraneous matter down to the floor drain.
4. Inspect all areas, and touch up where needed.
5. Apply an iodophor or quaternary ammonium sanitizer with a centralized or portable sanitizing unit around the entry area to reduce air contamination.

Smokehouse Blower

FREQUENCY After each use cycle.
PROCEDURE

Blades
1. Remove the blower housing access panel and drain plugs; soak with an alkaline solution.
2. Start the blower and flush with steam.
3. Stop the blower and flush again with water.
4. Repeat the operation until the equipment is clean.

Housing
1. Soak the inside of the plenum well, and wash the blower evolute wall with the alkaline cleaning solution.

Figure 15–4 A portable atomizer that covers all stainless steel surfaces in a smokehouse and reduces cleaning time by 60%–75%. Courtesy of Birko Chemical Corporation.

2. Flush the housing with steam, then with water. Repeat until the housing is clean.
3. Replace drain plugs and access panel.

Smokehouse Steam Coils

FREQUENCY Depends on amount of use.

PROCEDURE

Coils

1. Open the coil chamber access door and soak with an alkaline cleaning solution, brushing vigorously wherever possible.
2. Flush the coils with steam, then with water. Repeat until the metal is shining.

Chamber around Coils

1. Brush the cleaning solution on the inside of the chamber walls.
2. Use 55°C water to flush the chamber wall clean.
3. Close the coil chamber access door.

Smokehouse Ducts and Nozzles

FREQUENCY Depends on amount of use.

PROCEDURE

Outside Ducts

1. Remove the ductwork at the back of the house and remove carbon deposits. Dis-

assembly is not necessary if the ducts have access panels.

2. Spray the inside surface with an alkaline cleaning solution.
3. Flush the outside ducts clean with 90°C water or steam, followed by hot water. Repeat until the metal is exposed.

Inside Return Ducts

1. Mark the positions of the slide panels over the return ports so that the panels may be set back to their original openings for a balanced house.
2. Open the ports all the way and use as access doors for applying an alkaline cleaning solution to the ducts.
3. Use 90°C water for flushing the return ducts. Repeat until the metal shows.
4. Reset the slide panels to the originally marked positions.

Inside Jet Ducts

1. Open side access panels (or drop hinged panel, depending on type of house).
2. Soak the inside ducts and nozzles with a cleaning solution.
3. Use water at 90°C to flush these ducts clean. Repeat until the metal is exposed.
4. Close the access panel (or hinged panel).

Exhaust Stack

1. Disassemble the stack (or open access panels).
2. Soak the stack interior with an alkaline cleaning solution.
3. Flush the stack with 90°C water or steam, followed by hot water. Repeat until the metal shows, especially in the damper area.
4. Reassemble the stack (or close the access panels).

Smoke Generator

FREQUENCY   Depends on amount of use.
PROCEDURE

Filter

1. Soak the filter in an alkaline cleaning solution.
2. If mineralization has occurred, cut the frame apart, and clean the leaves individually. Reweld the frame after cleaning. Avoid warping.

Baffle and Cascade Chamber

1. Mechanically or hand brush the baffles (especially the edges) with a wire brush.
2. Scrape the edges of the cascade water outlet.

Wash Chamber

1. Disassemble the duct connecting the smoke generator to the house.
2. Remove soot and ash from the chamber below the filter.
3. Clean the duct and chamber surface until the metal shows.

Spiral Freezer

FREQUENCY   After use.
PROCEDURE See instructions for specific equipment to be cleaned.
PRECAUTIONS

1. To minimize friction, regularly wash the spiral with a foaming cleanser.
2. When the track is warm, wipe with a cloth dampened with a detergent solution. If the track is cold, a dry cloth may be used. Tie the cloth to the underside of the conveyor belt and let it be drawn through the spiral.
3. Defrosting the evaporator coil alone is insufficient for cleaning. Coils may ap-

pear clean, but grease, oils, salts, food adjuncts, and organic materials often remain hidden on internal surfaces. Therefore, it is necessary to clean and sanitize contaminated sites with warm water and a pH-balanced detergent. Cleansing solutions typically include an etching agent, a degreaser, inhibitors, metal protectors, stabilizers, and water. A mildly alkaline cleanser is often recommended for cleaning the evaporator coil.

4. If the freezer has been supplied with a recirculating CIP system, use a low-foaming detergent. Otherwise, a high-foaming detergent is best. A chemical supplier should be consulted to determine the best chemical cleanser.

Wash Areas

FREQUENCY Daily.
PROCEDURE See instructions for specific equipment to be cleaned.
PRECAUTIONS
1. Use a separate wash area for raw and cooked product equipment to reduce the spread of *Listeria* and spoilage microorganisms.
2. Provide this operation in an area where clean equipment does not cross fresh-product areas of the plant.

Packaged Meats Storage Area

FREQUENCY At least once per week and more often in a high-volume operation.
PROCEDURE
1. Pick up large debris and place in a receptacle.
2. Sweep and/or scrub with a mechanical sweeper or scrubber, if available. Use

cleaning compounds provided for mechanical scrubbers, according to directions provided by the vendor.

3. Use a portable or centralized foam or slurry cleaning system with 50° to 55°C water to clean areas heavily soiled by unpackaged products or other debris. Cleaning through rinsing-down should follow as previously described for production and processing areas.
4. Remove, clean, and replace drain covers, if present.

Low-Temperature Rendering (Edible)

FREQUENCY Daily.
PROCEDURE
1. Remove all large pieces of fat and tissue from the grinding equipment and store in a cool area for the next production shift.
2. Drain the system so that no lard, tallow, or melted fat remains.
3. The entire system should be flushed with 55° to 60°C water to remove heavy accumulations of deposits from the equipment and piping.
4. Disconnect the system where possible to allow the water and scrap to drain from each piece of equipment. Dismantle dead ends and T-joints in the piping to allow scrap accumulations to be removed from these sections.
5. Open the equipment and dismantle where possible to allow cleaning of all surfaces that come in contact with the product. Place parts, pipe sections, and other sections in a sink or truck to soak in an alkaline cleaning solution. Follow specific instructions from the manufacturer for dismantling and cleaning the processing equipment.

6. Remove large scraps of product from the interior of the equipment.
7. Spray-clean all exposed surfaces of the equipment throughout the system with an alkaline detergent sanitizer. Take special care to remove all possible product from the interiors of augers, pump screws, cutters, grinders, centrifuge chambers, and tanks. Spray-clean the cooling rollers where they are operating without refrigeration. Clean parts and pipe sections in a truck with a scrub brush and an alkaline cleaning solution.
8. Clean the centrifugal equipment and piping that cannot be dismantled to allow the interior surfaces to be spray-cleaned by circulating a solution of a heavy-duty alkaline cleaner through the equipment and piping. While circulating the cleaning solution, operate the centrifuges at reduced speeds to provide a scrubbing action in the system. Although CIP equipment is expensive, this system can be effectively utilized in this cleaning application, due to the potential savings of labor.
9. Circulate the cleaning solution for at least 30 minutes.
10. Drain the system and flush with 55° to 60°C water until the effluent is free of scraps.
11. Transfer all scraps flushed out of the equipment to the inedible department.

Wire Pallets and Metal Containers

FREQUENCY   Prior to use.
PROCEDURE
1. Use high-pressure water at 55°C or lower as a prerinse.
2. Preferably, apply an alkaline cleaner with a foam unit. If foam is unavailable, use a high-pressure, low-volume unit.

Whichever method is used, never spray more containers than can be rinsed before the cleaning compound dries.
3. Use a high-pressure spray of 55°C water as a rinse.
4. Inspect all rinsed containers and reclean as needed.

Trolley Wash

FREQUENCY   Depends on the physical appearance.
PROCEDURE
1. Skim off excess waste material from the cleaning solution.
2. Check the cleaning solution strength with a test kit. If it registers under the recommended strength, add the appropriate compound and retest.
3. Open the main steam valve. Maintain a solution temperature of 82° to 88°C.
4. Lower the trolleys into the tank.
5. After the trolleys have soaked for 25 to 30 minutes, remove them, and rinse thoroughly. Rinse in a rinse tank or over the drain area so that the cleaning solution will not be diluted.
6. Inspect the clean trolleys. Place the unsatisfactory ones on a rack for recleaning.
7. Place the clean trolleys in an oil bath while another rack is being cleaned.
8. Place the oiled trolleys over a drip pan or allow sufficient drip time while suspended over the oil tank.

Offices, Locker Rooms, and Rest Rooms

FREQUENCY Offices, daily; locker rooms and rest rooms, at least every other day.
PROCEDURE
1. Cover electrical connections with plastic sheeting.

2. Clean areas with a foam or high-pressure unit (or scrub brush and/or mop).
3. Within 20 minutes after cleaning compound application, rinse with 55°C water.
4. If the cleanser and rinse do not clean dirty areas or if drains are not present, hand scrub with scouring pads.

Garments

FREQUENCY   Daily.
PROCEDURE
1. Place dirty garments into the washer-extractor. Do not load the washer beyond its rated capacity.
2. Place the programmer dial at the start of the cycle and push the "On" and "Run" selector buttons. The drum programmer will automatically select the wash time and water temperature. Cleaning compounds are dispensed into the wash cycle immediately. Many detergents are available for this application. An example would be a mixture of 1 kg of a commercial laundry compound and 0.25 kg of a chlorine bleach per 65 kg of dry weight load. Bleach should not be used when washing gloves.
3. After the wash-extract cycle, remove the garments and place them in the dryer. Set aside garments not thoroughly cleaned for rewashing. Do not load the dryer beyond its rated capacity.
4. Set the temperature at 121°C for 30 minutes. Dry gloves for only 20 minutes.
5. Place dried garments in a clean wire crib or equivalent container. They need not be folded.

TROUBLESHOOTING TIPS

• *Discoloration of floors*: To restore the original color of darkened concrete floors, spread a bleach solution on them and allow it to stand for at least 30 minutes. A mechanical scrubber can then be used to finish cleaning the floor.
• *White film buildup on equipment*: This condition is caused when too much cleaning compound is used, when the equipment is not being properly rinsed, or when the water is hard.
• *Conveyor wheels freezing*: The cleaning water temperature is probably too high. Wheels lose lubricant at about 90°C. The cleaning temperature should not exceed 55°C.
• *Sewer lines plugged*: Sediment bowls are probably not being cleaned daily and/or floor sweepings are being flushed into sewer pipes.
• *Yellow protein buildup on equipment*: This condition may be caused by water temperature used in cleaning being too high. Brushing away all organic material will remove daily buildup. If heated soil is allowed to remain long on equipment, however, rubbing with steel wool will remove it. *To avoid trouble, do not spray*: liver slicers, cube steak machines, electronic scales, patty machines, any electrical outlet, motor, or equipment with open connections (cover all possible outlets with polyethylene bags), wrapping film or containers, or wrapping units.

SUMMARY

An efficient cleaning system can reduce labor costs in meat and poultry plants by up to 50%. The optimal cleaning system depends on the type of soil and type of equipment present. High-pressure, low-volume cleaning equipment is normally the most effective for removing heavy organic soil, especially when deposits are located in areas that are difficult to reach and penetrate. However, foam, slurry, and gel cleaning have become more prominent because cleaning is quicker and cleaners are easier to apply using these media. Because of high equipment costs and cleaning limitations, CIP systems are typically limited primarily to applications that involve large storage containers.

In meat and poultry plants, acid cleaning compounds are used most frequently to remove mineral deposits. Organic soils are more effectively removed by alkaline cleaning compounds. Chlorine compounds provide the most effective and least expensive sanitizer for destruction of residual microorganisms. However, iodine compounds give less corrosion and irritation, and quaternary ammonium sanitizers have more of a residual effect. Appropriate cleaning procedures depend on the area, equipment, and type of soil.

STUDY QUESTIONS

1. How do microorganisms affect meat color?
2. What is the function of air curtains?
3. What are the limited uses of CIP equipment in a meat or poultry plant?
4. Why is chlorine dioxide an effective sanitizer in meat and poultry plants?
5. Why does the meat and poultry sanitarian need to know something about HACCP?
6. How can the discoloration of darkened concrete floors be removed?
7. What causes a white film buildup on equipment in a meat and poultry plant?
8. What causes a yellow protein buildup on equipment in a meat and poultry plant?

REFERENCES

Douglas, G.S. 1987. *Listeria* and food processing. *Technics/Topics 11*: 2.

Doyle, M.P. 1987. Low-temperature bacterial pathogens. *Proc Meat Ind Res Conf,* 51. Washington, DC: American Meat Institute.

Kropf, D.H. 1980. Effects of retail display conditions on meat color. *Proc 33rd Annual Reciprocal Meat Conf,* 15.

Shapton, D.A., and N.F. Shapton, eds. 1991. Buildings. In *Principles and practices for the safe processing of foods,* 37. Oxford: Butterworth-Heinemann.

Wilson, G.D. 1987. Guidelines for production of ready-to-eat meat products. *Proc Meat Ind Res Conf,* 62. Washington, DC: American Meat Institute.

SUGGESTED READINGS

Anon. 1976. *Plant sanitation for the meat packing industry.* Office of Continuing Education, University of Guelph and Meat Packers Council of Canada.

Marriott, N.G. 1990. *Meat sanitation guide II.* American Association of Meat Processors and Virginia Polytechnic Institute and State University, Blacksburg.

Price, J.F., and B.S. Schweigert. 1987. *The science of meat and meat products.* 3rd ed. Westport, CT: Food & Nutrition Press.

CHAPTER 16

Seafood Plant Sanitation

Seafood processors should be familiar with microorganisms that cause spoilage and foodborne illness. Also, they need to know about characteristics of various types of soil, effective cleaning compounds and sanitizers, available cleaning equipment, and effective cleaning procedures. This knowledge is essential for maintaining a hygienic operation. Each processor should be equally familiar with existing federal, state, and local public health regulations.

Regulatory requirements are by no means the only reason that the seafood processor should practice strict sanitary procedures. Another important factor is the consumer's increased awareness of nutritional value, wholesomeness, and processing conditions of all foods, including seafood.

Sanitation programs in the seafood industry are essential to provide the processor with guidelines that will give the consumer a high-quality, wholesome food. Because these guidelines relate to the facility and work practices, proper planning of new, expanded, and renovated plants should be considered. Every production phase of the distribution chain, from harvest to the consumer, must ensure that only wholesome products are provided to the ultimate consumer. Effective sanitation contributes to the maintenance of desired seafood quality.

SANITARY CONSTRUCTION CONSIDERATIONS

A hygienically designed plant can enhance the wholesomeness of all foods and dramatically improve the effectiveness and efficiency of the sanitation program. Even a well-designed plan is not a safeguard against microbial infection or other contamination unless it is accomplished by sound maintenance and sanitation. In a hygienic operation, the employer or management team should ensure good housekeeping and should be constantly vigilant against ineffective sanitary practices for all physical facilities, unit operations, employees, and materials. Chapter 12 contains design and construction considerations to supplement those to be discussed here.

Site Requirements

A clean and attractive site is necessary. Clean premises should be maintained for a satisfactory public image, to promote the individual firm and the industry. First impressions of a site are important to regulatory personnel and to the public, who are favorably impressed by a clean, neat, and orderly plant. The condition of the plant premises frequently reflects the caliber of the plant hygienic practices. According to the U.S. Food

283

and Drug Administration (FDA), areas that are inadequately drained "may contribute to contamination of food products through seepage or foodborne filth and by providing an environment conducive to the proliferation of microorganisms and insects." Excessively dusty roads, yards, or parking lots constitute a contamination source in areas where food is exposed. Improperly stored refuse, litter, equipment, and uncut weeds or grass within the immediate vicinity of the plant buildings or structures may provide a breeding place or harborage for rodents, insects, and other pests.

The site should be equipped with the capability to dispose of the seafood plant wastes. Solids, liquids, vapors, and odors emanating from a plant present a poor image and can result in legal action by either regulatory groups and/or concerned citizens. Waste disposal facilities must be designed to meet federal, state, and local requirements.

The site must also supply an ample amount of potable water for plant operations. If water is drawn from wells, analysis for mineral content and microbial load should be conducted, and the water must meet the standards established by the appropriate regulatory agency. After water use, adequate provisions should be made for wastewater discharge.

Construction Requirements

It is especially important to use materials that will not absorb water, are easily cleaned, and are resistant to corrosion and other deterioration. All openings should be equipped with air or mesh screens to prevent entry of insects, rodents, birds, and other pests. Facilities should be large enough to provide an organized and tidy operation and to facilitate an effective sanitation program. A brief discussion of sanitary features of various construction phases will be covered to provide guidelines for establishing a hygienic facility designed for effective cleaning.

Floors

Floors should be constructed of hard and impervious material, such as waterproof concrete or tile. The material should be durable and have a surface that is even enough to prevent accumulation of debris but not smooth enough to cause hazards from slipping and falling. A rough finish or use of embedded abrasive particles can reduce accidents. A frequently used surface is a water-based acrylic epoxy resin that provides a durable, nonabsorbent, easy-to-clean surface that can double the life of the concrete floor. This finish should contain an abrasive material to provide a skid-resistant surface. Although the cost is nearly prohibitive, acid brick floors are known to be satisfactory and durable.

Floor Drains

A drainage outlet should be provided in the processing area for each 37 m² of floor space. As with other processing plants, floors in the processing areas should have a slope to a drainage outlet of 2%. It is imperative that this slope be uniform, with no dead spots to trap water and debris. All drains should contain traps. Drainage lines should have an inside diameter of at least 10 cm and should be constructed of cast iron, steel, or polyvinyl chloride tubing. State and local codes should be checked to verify that these materials are permitted. Drainage lines should be vented to the outside air to reduce odors and contamination. All vents should be screened to prevent entrance of pests into the plant. It is also recommended that contamination be further reduced by connection of drain lines from toilets directly into the sewage system instead of into other drainage lines.

Ceilings

Ceilings should be at least 3 m high in work areas. A material impervious to moisture should be selected. One acceptable material is Portland-cement plaster, with joints sealed by flexible sealing compound. A false ceiling can prevent debris from overhead pipes, machinery, and beams from falling onto exposed seafoods or on any other foods.

Walls and Windows

Walls should be smooth and flat. Construction should consist of a nonabsorbent material such as glazed tile, glazed brick, smooth-surface Portland-cement plaster, or other nonabsorbent, nontoxic material. Concrete walls are satisfactory if they contain a smooth finish. Although painting is discouraged, a nontoxic paint that is not lead-based can be applied. Window sills, if present, should be slanted at a 45° angle to reduce debris accumulation.

Entrances

Entrances should be constructed of rust-resistant materials with tightly soldered or welded seams. Double-entry screened doors should be provided for outside entrances, as well as air curtains (or equivalent) over outside doorways in the processing areas.

Processing Equipment

Processing equipment should have a durable, smooth finish that is easily cleaned. Surfaces should be free of pits, cracks, and scale. The equipment should be designed to prevent contamination of products from lubricants, dust, and other debris. In addition to hygienic design for cleaning ease, equipment should be installed and maintained to facilitate cleaning of equipment surfaces and surrounding areas.

Where metal construction is essential, stainless steel should be used to protect seafood or other edible products. Galvanized metal is discouraged because it is not sufficiently resistant to the corrosive action of seafood products, cleaning compounds, or salt water. However, galvanized construction can be economically used for handling of waste materials. If galvanized material is used, it should be smooth and have a high-quality dip.

Cutting boards should be fabricated of a hard, nonporous, moisture-resistant material. They should be easy to remove for cleaning and should be kept smooth. This material should be abrasion- and heat-resistant, shatterproof, and nontoxic. Cutting boards should not contain material that will contaminate seafoods.

Conveyor belts should be constructed of moisture-resistant material (such as nylon or stainless steel) that is easy to clean. Conveyors should be designed to eliminate debris-catching corners and inaccessible areas. This equipment, like other processing equipment, should be easily broken down for cleaning. Cleaning is facilitated through use of sealed or closed steel tubing, instead of angle or channel iron. Drive belts and pulleys should be protected with guard shields that are easily removed during cleaning. Motor mounts should be elevated enough to permit effective cleaning. Motors and oiled bearings should be located so that oil and grease will not come in contact with the product.

As with other food plants, stationary equipment should not be located within 0.3 m of walls and ceilings, so that access for cleaning is available. Equipment should be mounted at the same distance above the floor or have a watertight seal with the floor. All wastewater should be discharged through flumes or tanks, so that it is delivered with an uninterrupted connection to the drainage system without flowing over the floor.

CONTAMINATION SOURCES

The environment at a seafood plant location can contribute to contamination within the plant, as well as contamination to the products. The processing equipment, containers, and work surfaces are other contamination sources. An effective sanitation program is necessary to reduce contamination and to monitor program effectiveness.

Because seafood involves so many varieties of flesh foods, the amount of contamination varies among species. The initial contamination source can be the raw product, especially if the product is improperly harvested and subjected to unsanitary practices on a vessel or truck. Delayed refrigeration after harvest and other improper handling between harvesting and processing can result in produce decomposition and increase the microbial load.

Moore (1988) has suggested that seafood quality, including microbial load, will be satisfactory for processing the day after harvesting if:

• Chilling begins immediately after harvesting
• Chilling reduces product temperature to 10°C within 4 hours
• Chilling continues to approximately 1°C

Storing fish at 27°C or higher for 4 hours, with subsequent chilling to 1°C, will provide an acceptable product for only 12 hours.

Workers contribute to contamination, especially through unsanitary practices. Other sources of contamination are processing equipment, boxes, belts, tools, walls, floors, utensils, supplies, and pests. Contaminants of greatest concern are those that come in direct contact with ready-to-eat products. Therefore, effective cleaning and sanitizing of equipment are vital. Scombroid contamination is associated with some of the dark-fleshed, fast-swimming fish. This contamination could be properly called *histamine poisoning* and causes an allergic reaction. Nardi (1992) indicated that scombrotoxin is always associated with temperature abuse and resultant decomposition, so it is entirely avoidable. Undercooked shellfish can be contaminated with *Vibrio vulnificus* and can contain viral infections from hepatitis A.

Studies of fishery products serving as foodborne vehicles for listeriosis have been less focused than for some other foods in the past. However, Weagant et al. (1988) reported that 35 of 57 seafood samples tested contained *Listeria* organisms, and 15 of these were positive for *L. monocytogenes*. Samples positive for *L. monocytogenes* included raw and cooked shrimp, lobster tails, crab meat, squid, finfish, and surimi analogs.

SANITATION PRINCIPLES

A seafood sanitation program must encompass proper handling of the sanitation tasks as well as personnel allocation.

Personnel Allocations

In addition to the need for adequate cleaning methods and seafood facilities, a well-qualified sanitarian is required. Although the seafood plant manager is ultimately responsible for an effective sanitation program and the production of wholesome products, sanitation employees who are trained to maintain a clean plant must be provided. Employees should be adequately instructed in seafood product knowledge and in proper sanitary techniques, so that they are informed of the importance of the effect of proper sanitation on product wholesomeness. Any employee with a contagious illness should not work around processing areas, even during cleanup (see Chapter 4 for further discussion related to employee health requirements).

The typical seafood processing plant should have one or more employees responsible for daily inspection of all equipment and processing areas for hygienic conditions. Any sanitation deficiencies should be corrected before production operations are initiated.

Cleaning Schedule

A cleaning schedule with sequential cleaning steps is essential. The schedule should be adopted for each area of the plant and should be followed. Continuous-use equipment, such as conveyors, flumes, fileting machines, batter and breading machines, cookers, and tunnel freezers, should be cleaned at the end of each production shift. If there are no refrigerated areas, batter machines and other equipment in contact with milk or egg products should be cleaned at 4-hour intervals by draining the batter, flushing the batter reservoir with clean water, and subsequently applying a sanitizer. At the end of the production shift, this equipment should be disassembled, and all parts should be cleaned and sanitized. These parts, as well as portable equipment, should be stored off of the floor in a clean environment to protect against splash water, dust, and other contamination sources.

The following steps apply when cleaning seafood plants:

1. Cover electrical equipment with polyethylene or equivalent film.
2. Remove large debris and place it in receptacles.
3. Manually or mechanically remove soil deposits from the walls and floors by scraping, brushing, or by the action of a hose from mechanized cleaning equipment. Proceed from the top to the bottom of equipment and walls, toward the floor drains or exit.
4. Disassemble equipment as required.
5. Conduct a prerinse for wetting action and removal of large and water-soluble debris, with water at 40°C or lower. This temperature is important. A higher temperature can cause denaturation of seafood residues and other proteins, with subsequent baking onto the contact surface.
6. Apply a cleaning compound that is effective against organic soil (usually an alkaline cleaner) by portable or centralized high-pressure, low-volume, or foam equipment. The temperature of the cleaning solution should not exceed 55°C. Cleaning compounds such as sodium tripolyphosphate, tetrasodium pyrophosphate (a general-purpose cleaner), or a chlorinated alkaline detergent are usually considered satisfactory. More than one cleaner should be incorporated because of the nature of the soiled equipment material characteristics. (Chapter 7 discusses appropriate cleaning compounds for various cleaning applications. Chapter 9 provides a detailed discussion of the optimal cleaning equipment for various cleaning applications.)
7. After the cleaning compound has been applied and given approximately 15 minutes to aid in soil removal, rinse the equipment and area with water that is 55° to 60°C. Hotter water is more effective in removing fats, oils, and inorganic materials, but the cleaning compound aids in emulsification of these solids. Also, a higher water tem-

perature contributes to higher energy costs and more condensation on the equipment, walls, and ceilings.

8. Inspect equipment and the facility for effective cleaning, and correct deficiencies.

9. Ensure that the plant is microbially clean through application of a sanitizer. Although chlorine compounds are the most economic and most widely used, other methods (as discussed in Chapter 8) are available. Table 16–1 provides the recommended concentrations for various sanitizing operations. Sanitizers can be most effectively applied by use of a portable sprayer in small operations or with a centralized spraying or fogging system in large-volume operations. Chapter 9 also discusses the various equipment available for applying sanitizers.

10. Avoid contamination during maintenance and equipment setup by requiring maintenance workers to carry a sanitizer and to use it where they have worked.

Further information about cleaning compounds, sanitizers, and cleaning equipment is provided in Chapters 7, 8, and 9.

RECOVERY OF BY-PRODUCTS

Waste management, including the recycling of seafood waste products, has become increasingly important. In addition to the economic considerations, an effective recovery system can contribute to a more hygienic operation. Today, many food processors are recycling and/or reducing their liquid discharges.

Innovations in water conservation are:

- Wastewaters used for noncontaminating purposes in one area of a food processing operation are now being redirected to other areas that do not require potable water.
- Closed water system food processing operations in which all process waters are continuously filtered to remove solid materials have been established.
- Dry conveying equipment has been utilized to replace water transport of solids.

Table 16–1 Recommended Sanitizing Concentrations for Various Applications

Application	Available Chlorine (ppm)	Available Iodine (ppm)	Quaternary Ammonium Compounds (ppm)
Wash water	2–10	Not recommended	Not recommended
Hand dip	Not recommended	8–12	150
Clean, smooth surfaces (rest rooms and glassware)	50–100	10–35	Not recommended
Equipment and utensils	300	12–20	200
Rough surfaces (worn tables, concrete floors, and walls)	1,000–5,000	125–200	500–800

VOLUNTARY INSPECTION TO ENHANCE SANITATION

A voluntary seafood inspection program has been established for the seafood industry. The program is conducted by the U.S. Department of Commerce (USDC), National Marine Fisheries Service. The objective is to provide a systematic sanitation inspection program designed to assist the industry in developing and maintaining a high degree of sanitation continuity. The program has been credited with achieving increased production and consumption of wholesome seafoods.

Product examination during each production stage, including packaging, is performed under contract inspection to verify compliance with all requirements, standards, or specifications. This program is voluntary.

Many large firms operating under USDC inspection have designated a management representative to supervise the plant's cleaning operation in order to maintain an effective sanitation program. The representative is authorized to work directly with USDC inspection personnel in all areas related to sanitation. This cooperative arrangement has been responsible for the overall caliber of the seafood sanitation programs and for the awareness of effective hygiene in firms producing seafoods under USDC inspection.

The National Marine Fisheries Service (NMFS) can furnish a voluntary inspection service that is paid for by the firms that elect to participate. The NMFS has established a program of developing a Hazard Analysis Critical Control Point (HACCP) system of control for cooked, ready-to-eat, refrigerated seafood that was recommended by the National Academy of Sciences, Institute of Medicine. According to the National Marine Fisheries Service (1989), this organization is working with the seafood industry to implement model programs in several processing facilities.

Hazard Analysis Critical Control Point Models

The seafood industry has conducted in-plant testing of HACCP models developed for breaded and cooked shrimp. Testing has been conducted at nine plants for each commodity representative of production volume and, to the extent possible, geographic distribution of the plants across the United States (Anon., 1988). HACCP is now required in seafood plants.

Four raw fish workshops conducted by the National Marine Fisheries Service developed an HACCP model for each region that identified between 23 and 26 steps with 5 to 11 critical control points. The HACCP model for breaded shrimp production identified 30 process steps, with 9 identified as critical. Similar evaluations were made through analysis of cooked and raw shrimp processing. This surveillance model is designed to develop a seafood products inspection program to protect consumers, based on the HACCP concept.

SUMMARY

A hygienically designed plant can improve the wholesomeness of seafoods and the sanitation program. The location of the seafood plant can contribute to the sanitation of the facility. The design and construction materials used in the plant and equipment are also critical to an effective sanitation program.

Personnel allocation and an organized cleaning schedule with required cleaning steps are essential in maintaining a hygienic operation. This portion of the sanitation program should be matched with the most effective cleaning compounds, cleaning equipment, and sanitizers. The sanitation operation can frequently be enhanced by the recovery of by-products, adoption of recommendations provided by regulatory agencies, and participation in voluntary inspection programs.

STUDY QUESTIONS

1. How much floor slope should exist in seafood processing plants?

2. How much chlorine sanitizer should be applied to equipment and utensils in seafood plants?

3. How much quaternary ammonium sanitizer should be applied to equipment and utensils in seafood plants?

4. How much iodine sanitizer should be incorporated in a hand dip for seafood plants?

5. What is the maximum cleaning solution temperature for a seafood plant?

6. What is the maximum rinse temperature for a seafood plant?

REFERENCES

Anon. 1988. HACCP models developed for seafood species under NMFS program. *Food Chem News* 30, no.13: 21.

Moore, R.E. 1988. Harvesting and pre-process handling of food-grade menhaden: An interim report. In *Fatty fish utilization: Upgrading from feed to food*. ed. N. David, 163. UNC Sea Grant College. Publication no. 88-04.

Nardi, G.C. 1992. Seafood safety and consumer confidence.

Food Protection Inside Report. 8: 2A.

National Marine Fisheries Service (NMFS). 1989. *Plan operations—model seafood surveillance project*. Washington, DC: National Marine Service, Office of Trade and Industry Services.

Weagant, S., et al. 1988. The incidence of *Listeria* species in frozen seafood products. *J Seafood Prot* 51: 655.

SUGGESTED READINGS

Jowitt, R. 1980. *Hygienic design and operation of food plant*. Westport, CT: AVI Publishing.

Marriott, N.G. 1980. *Meat sanitation guide II*. American Association of Meat Processors and Virginia Polytechnic Institute and State University, Blacksburg.

CHAPTER 17

Fruit and Vegetable Processing Plant Sanitation

An effective sanitation program for fruit and vegetable processing facilities requires the same basic components needed in other food operations: appropriate cleaning compounds and sanitizers, effective cleaning procedures, and effective administration of the sanitation program. The ultimate goal is to provide a finished product that is sanitary and wholesome.

CONTAMINATION SOURCES

Effective preservation of fruits and vegetables depends on the prevention of contamination by spoilage-causing and food-poisoning microorganisms during production, processing, storage, and distribution. It is important to consider raw materials as a potential source for food spoilage microorganisms and as a contributor to bacterial pools within a processing plant.

Federal laws mandate that processed foods shipped interstate be free of pathogenic microorganisms. The normal sterilization process for commercially canned foods is sufficient to destroy pathogenic bacteria that may exist in the container at the time of sterilization. Also, washing and peeling operations contribute to the physical removal of organisms. Therefore, if the canning and freezing processes are properly conducted, the fin-

ished product should be wholesome. Chapter 3 provides more information on the contamination of raw materials.

Raw Materials

Raw materials are exposed to many unclean sources and can provide additional contamination in the receiving, raw material storage, and processing areas.

Soil Contamination

Heat-resistant bacteria are present in the ground and can cause "flat sour" and other spoilage of canned vegetables if washing is not thorough. Microbial population is affected by the degree of wind, humidity, sunlight, and temperature, as well as by domestic and wild animals, irrigation water, bird droppings, harvesting equipment, and workers. Most pathogens are introduced to fruits and vegetables via irrigation shortly before harvesting and before the sun dehydrates and destroys pathogens.

Air Contamination

Contaminated air contributes to less sanitary raw products. Besides normal microorganisms and pollutants found in the air, this medium serves as a transport of pathogens. Infiltration of unclean air into the processing plant can be improved by the use of air filters.

Pest Contamination

Certain pests can invade fruits and vegetables during the process of forming on the tree or vine. Contamination by pests can be expressed through the spread of viruses, spoilage bacteria, and pathogens, as well as by physical damage. Infesting microorganisms frequently remain inactive because of the protective skin layer of fruits and vegetables and because of the low availability of moisture (measured as minimum water activity [A_w]) on the surface. As these products reach maturity or shortly thereafter, profound changes in the medium can cause spoilage. The action of pests, such as the pollinating fig wasp (*Blastophaga psenes*), introduces microbes that persist and develop in quantity throughout the ripening period until the fruit is mature. Although a portion of the microorganisms introduced does not cause spoilage, these microbes attract other organisms, such as *Drosophila*, which carries spoilage yeasts and bacteria. When the protective covering of fruits and vegetables is broken by bruises, mechanical injury, or by attack of insects, microorganisms can enter readily.

The presence of coliforms on processing-grade fruit as it arrives at the processing plant is not truly indicative of the amount of these microorganisms in the manufactured juice or of positive evidence of unsanitary conditions in the processing plant. However, the presence of lactic acid bacteria constitutes an accurate index of processing sanitation for high-quality frozen citrus products. Lactic acid bacteria are a more accurate indicator of unsanitary conditions caused by inadequate cleaning because these microorganisms are the most likely to accumulate in the bacterial pools that can exist when proper sanitation practices are not followed.

Use of recirculated water is not recommended for washing fruits and vegetables because of the contamination caused through a rapid buildup of microorganisms in the wash water. The effectiveness of chlorination of the wash water is minimal because bacterial spores exhibit resistance to chlorine. The benefit of chlorinated water for recirculation is further reduced through absorption of free chlorine and subsequent neutralization by the accumulated organic content of the water.

SANITARY CONSTRUCTION CONSIDERATIONS

A well-designed processing plant does not eliminate microbial infiltration unless the design incorporates hygienic features, such as easy-to-clean areas and equipment with optimal cleaning features and instructions. If the processing plant is newly constructed, expanded, or renovated, functional layouts, mechanical and plumbing layouts, and equipment and construction specifications should be reviewed by all professional personnel associated with the processing organization—mechanical engineers, industrial engineers, food chemists, microbiologists, sanitarians, and operations personnel. This approach permits integration of operating procedures and process control (frequently called *quality control*).

Construction of new and expanded fruit and vegetable processing plants must reflect hygienic design because most of today's plants are volume-oriented. High-volume plants operate under the principle that greater capacity is attained by pushing more materials through a larger-capacity production pipeline. With increased mechanization, there has been less emphasis on manual cleaning and visual inspection, and more reliance on a cleaning-in-place (CIP) system. However, there is still limited use of CIP equipment in fruit and vegetable processing plants, except in the manufacture of juices. This concept also incorporates more emphasis on mechanized startup and shutdown of production equipment and cleaning and sanitizing equipment. This approach provides less opportunity for human error but also reduces the possibility of spotting a performance error in cleaning.

High-volume processing plants, by design, operate with longer production periods and much greater product volume flow than do lower-volume plants. There is much more microbial buildup in the plant because of the longer dwell time and larger volume output. To reduce the microbial buildup, safe levels should be set by a saturation device that senses the buildup, stops production, and triggers an automatic cleaning procedure. It is suggested that this device would be activated only under excessive buildup, such as 150% of normal conditions.

Sanitary design features are necessary to minimize downtime for cleaning and sterilizing. The need for maximum utilization of equipment and facilities and for minimum discharge of sewage has mandated that the minimum effective cleaning approach to a process cycle is a short cleaning time and less effluent discharge from cleaning.

More mechanization and automation has been developed for cleaning tasks to equipment previously done by hand. Prior to CIP cleaning, machines and storage equipment were disassembled every production day and hand-cleaned. After CIP cleaning was made available, control was initially conducted through a control panel with pushbuttons. Increased automation has incorporated use of an automatic panel with computer-controlled timers to provide automatic startup and cutoff of cleaning, rinsing, and sanitizing. (Addi-

tional features of the CIP cleaning system have been previously discussed in Chapter 9.)

One of the most important features of hygienic design is the absence of crevices (narrow and deep cracks or openings) and pockets (large cracks and openings) in the construction of buildings and equipment. Crevices frequently present greater cleaning obstacles than do pockets because penetration and access are more of a problem.

Principles of Hygienic Design

Minimum standards should be adopted when constructing or remodeling a fruit or vegetable processing plant. Effective hygienic design should incorporate the following principles:

- Equipment should be designed so that all surfaces in contact with the product can be readily disassembled for manual cleaning or CIP.
- Exterior surfaces should be constructed to prevent harboring of soil, pests, and microorganisms on the equipment, as well as on other parts of the production area, including walls, floors, ceilings, and hanging supports.
- Equipment should be designed to protect food from external contamination.
- All surfaces in contact with food should be inert to reaction with food and under conditions of use and must not migrate to or be absorbed by the food.
- All surfaces in contact with food should be smooth and nonporous to prevent accumulation of tiny particles of food, insect eggs, or microorganisms in microscopic surface crevices.
- Equipment should be designed internally, with a minimum number of crevices and pockets where soil particles may collect.

The interior and exterior of the plant should have the following sanitary features:

- Ledges and dirt traps should be avoided.
- Projecting bolts, screws, and rivets should be avoided to reduce the accumulation area for debris.
- Recessed corners and uneven surfaces and hollows should be avoided to reduce accumulation areas for debris.
- Sharp edges should be avoided to reduce soil accumulation.
- Unfilled edges should be avoided to reduce soil accumulation of debris and subsequent microbial contamination.
- Proofing against pest entry through double-door construction, heavy-duty strips, and self-closing mechanisms is essential.

Certain pitfalls should be avoided when a processing plant is being built, expanded, or renovated to minimize contamination from external sources. Requirements may change as technology advances. Thus, the layout should reflect maximum flexibility and should accommodate existing systems that are compatible with the proposed plant. The following points should be considered as a means of reducing contamination:

- Adequate storage space should be provided for raw materials and supplies. With inadequate storage space, contamination from the packaging material of supplies can occur. Sufficient space is also needed for thorough screening of raw material because foreign bodies may accommodate these products. Segregated materials that are contaminated should be salvaged and cleaned to prevent the spread of contaminants. Tainting can occur when raw materials share the same storage area as cleaning and maintenance materials.

- Separate storage space should be provided for finished products. Insufficient space may dictate use of the production area for this function. This practice can cause cross-contamination of the raw materials being processed.
- Congestion in areas of open food production should be eliminated. Insufficient space complicates cleaning and maintenance and increases contamination and risks of personnel injury and equipment damage.
- Short and direct routes for waste removal are necessary so that waste is not transported through open production areas. This design is especially critical because of the unsanitary condition of equipment used for waste collection.
- Location of the returned goods area is important. These foodstuffs are frequently infested and may be partially decomposed. It is essential to isolate these products from all raw material and production areas.
- Control of the environment in the perimeter should be exercised to reduce pests and to provide cleaner air through location of the waste collecting, waste treatment, and incineration areas as far as possible from the plant. This control also includes adequate surface drainage to prevent accumulation of water, outside surfaces that are easily cleaned, control of weed and grass growth, and control of stocks of surplus supplies and equipment.
- Employee personal hygiene is essential (discussed in detail in Chapter 4).

CLEANING CONSIDERATIONS

As with other food plants, management has the legal and moral responsibility to provide the consumer with a wholesome product. An effective sanitation program is needed to provide a clean environment for processing.

Housekeeping

Housekeeping relates to orderliness and tidiness. Careful arrangement of supplies, materials, and clothing contributes to a tidier operation, reduces contamination, and makes cleaning easier. Attention to neatness and orderliness contributes to the performance of responsibilities.

Although the responsibility for housekeeping should be assigned to the sanitarian, the maintenance of good housekeeping depends on the cooperation of all employees—production, maintenance, and sanitation. Cooperation is needed to ensure that trash containers, tools, supplies, and personal belongings of employees are kept in the proper place. Convenient location of trash receptacles is necessary to encourage that anything not likely to be used further be discarded immediately.

Insects, rodents, and birds increase contamination. A knowledge of their biological characteristics and habits is necessary for their control. Sanitary practices can eliminate nutrition and protection for pests and, thus, can provide an important means of control. Hygienic design—air and mesh screens and filling of holes, cracks, and crevices—will discourage pests from entering the plant. Periodic inspection for the presence of pests is another prevention technique. (Methods of detection and other discussion related to pests are included in Chapter 11.)

Waste Disposal

Wastes can be handled more effectively and salvaged more efficiently as by-products if solid and liquid wastes are separated. Solid wastes can be frequently separated by some method of pickup and/or transfer of solid ma-

terials before being flushed into drains or gutters. The liquid waste that is flushed away is usually handled as liquid waste and is treated as effluent, according to methods discussed in Chapter 10. Some food processing plants are using waste by-products. The citrus juice industry uses more than 99% of the raw material for juices, concentrates, or dried cattle feed. It has increased in salvage efficiency and has reduced the cost of solid and liquid waste disposal.

Water Supply

As with other cleaning applications, an abundant, high-quality water supply is necessary to produce a wholesome product and to effectively clean the plant. In addition to being used as a cleaning medium, water is important as a heat-transfer medium, and it is used in the processed products.

The sanitary condition of water should be monitored daily for two criteria: bacterial content and organic or inorganic impurities. Bacterial content serves as a guide for acceptability for use in contact with the food or any surface responsible for indirect contamination. The effectiveness of water in washing the product or equipment is dependent on organic and inorganic impurities.

CLEANING OF PROCESSING PLANTS

A hygienic product results from rigid sanitation and effective destruction of microbes during processing. Conventional fruit and vegetable canning operations may be characterized as pouring food into containers (i.e., metal, glass, or plastic), followed by sealing and heat treatment. This heat treatment is referred to as *terminal sterilization* and is designed to eliminate extremely large numbers of *Clostridium botulinum* spores and to reduce the chance of survival of the much more heat-resistant spores of spoilage organisms. This condition is called *commercial sterility*. The process of aseptic packaging is sometimes called *aseptic canning*. In the aseptic process, the food and containers are commercially sterilized separately. The food is cooled to an acceptable filling temperature with subsequent filling and sealing of the containers under aseptic conditions.

The microbial destruction (kill step) during terminal sterilization is given to sealed containers and, because of the excellent control that is technically possible over container integrity, conventional canning is safe technology. This technology is also suitable for the HACCP approach, which can provide the required positive assurance of safety.

Aseptic packaging is a relatively new technology; thus, development of test methods is important. Active areas of development and concern are package integrity and maintenance of sterility, package performance in distribution, package sterilization techniques, and package residual. An on-line continuous monitoring method is needed. Several methods are available for measurement of concentration levels of H_2O_2 solutions (Shapton and Shapton, 1991).

An efficient layout of cleaning equipment is essential to reduce cleaning labor. It is much easier to install cleaning equipment when the processing equipment is put in place. The type of soil found in fruit and vegetable processing plans is most easily cleaned by portable cleaning systems in small plants and by a combination of CIP and centralized foam cleaning in large plants.

Hot-Water Wash

Water provides transport of cleaning compounds and suspended soil. Sugars, other carbohydrates, and other compounds that are relatively soluble in water can be cleaned rather effectively with water. The main ad-vantage of a hot-water (60° to 80°C) wash for fruit and vegetable processing plants is minimal investment of cleaning equipment. Limitations of this cleaning method include labor requirements, energy costs, and water condensation on equipment and surroundings. This cleaning technique is not effective in the removal of heavy soil deposits.

High-Pressure, Low-Volume Cleaning

High-pressure spray cleaning has utility in the fruit and vegetable processing industry because of the effectiveness with which heavy soils can be removed. Difficult-to-reach areas can be cleaned more effectively with less labor, and there is increased effectiveness of the cleaning compounds below 60°C. Water temperature should not exceed 60°C because high-temperature sprays tend to bake the soil to the surface being cleaned and to increase microbial growth. More discussion on this cleaning method is provided in Chapter 9.

Foam Cleaning

Portable foam cleaning is widely used because of the ease and speed of foam application in cleaning ceilings, walls, piping, belts, and storage containers in fruit and vegetable processing plants. Equipment size and cost is similar to that of portable high-pressure units.

Centralized foam cleaning applies cleaning compounds by the same technique used in portable foam equipment. The equipment is installed at strategic locations throughout the plant. The cleaning compound is automatically mixed with water and air to form a foam, which is applied at various stations installed throughout the plant.

Gel Cleaning

Here, the cleaning compound is applied as a gel rather than as a high-pressure spray or foam.

Gel is an especially effective medium for cleaning canning and packaging equipment because it clings for subsequent soil removal.

Slurry Cleaning

This method is identical to foam cleaning, except that less air is mixed with the cleaning compound. A slurry is more fluid than foam and penetrates uneven surfaces in a canning plant more effectively, but it lacks the clinging ability of foam.

Combination Centralized High-Pressure and Foam Cleaning

This system is the same as a centralized high-pressure system, except that foam can also be applied through the equipment. This method is more flexible because foam can be used on large surface areas, and high pressure can be applied to belts, stainless steel conveyors, and hard-to-reach areas in a canning plant.

Cleaning in Place

With this closed system, a recirculating cleaning solution is applied by nozzles that automatically clean, rinse, and sanitize equipment. However, this equipment is expensive and ineffective on heavily soiled areas. Nevertheless, CIP cleaning has application in vacuum chambers, pumping and circulation lines, and large storage tanks. Higher-volume operations are better adapted to CIP cleaning because labor savings provide a quicker payout of the equipment. Additional information about cleaning equipment is provided in Chapter 9.

CLEANERS AND SANITIZERS

Soil remaining on equipment or at any location in the plant after cleaning is contaminated with microorganisms. Thorough physical cleaning of all equipment and rooms is necessary to prevent microorganisms from contacting chemical sanitizers. (Readers are referred to Chapter 7 for additional information on cleaning compounds.) Residual soil can also reduce the strength of chemical sanitizing solutions.

Combination cleaners (detergent-sanitizers) are used most frequently with smaller operations that perform manual cleaning at a temperature below 60°C. If the cleaning medium temperature exceeds 80° C, the solution will destroy spoilage microorganisms and most pathogenic bacteria without application of a chemical sanitizer. Chemical sanitizers most applicable to fruit and vegetable processing plants are classified as halogen compounds, quaternary ammonium compounds, and phenolic compounds.

Halogen Compounds

Chlorine and its compounds are the most effective sanitizers of the halogens for sanitizing food processing equipment and containers, and for disinfecting water supplies. Calcium hypochlorite and sodium hypochlorite are two of the most frequently used sanitizers in fruit and vegetable processing plants. Although elemental chlorine is less expensive on an available chlorine basis, calcium hypochlorite and sodium hypochlorite are easier to apply in low concentrations. Hypochlorite solutions are sensitive to changes in temperature, residual organic matter, and pH. These compounds are quick-acting and less expensive than other halogens but tend to be more corrosive and irritating to the skin. Additional information about chlorine and iodine sanitizers is provided in Chapter 8.

Quaternary Ammonium Compounds

Quaternary ammonium compounds ("quats") are effective against most bacteria and molds. These compounds are stable as a dry powder, a concentrated paste, or in solution at room temperature. They are heat stable, water soluble, colorless, odorless, noncorrosive to common metals, and nonirritating to the skin in normal concentrations. These compounds are more active if soil is present than are other sanitizers, and they express the greatest antimicrobial activity in the pH range of 6.0 and above. The quats have limited bacterial effectiveness when combined with cleaning compounds or when dissolved in hard water.

Phenolic Compounds

These compounds are used most frequently in the formulation of antifungal paints and antifungal protective coatings, instead of as sanitizers applied after cleaning. Phenolic compounds have limited utility in fruit and vegetable plants because of their low solubility in water.

CLEANING PROCEDURES

As with other food plants, a rigid set of procedures cannot be adopted for use in every fruit and vegetable processing plant. Procedures depend on plant construction, size, operations, age, and condition. Those discussed here are used only as guidelines and should be adapted to the actual cleaning application.

Facilitating Effective Cleaning

Employees can adopt continuous practices to facilitate and expedite effective cleaning. The following practices are recommended to aid in cleaning:

1. Reduce burn-on through careful, controlled heating of vessels.

2. Promptly rinse and wash equipment after use to reduce drying of soil.
3. Replace facility gaskets and seals to reduce leakage and splatter.
4. Handle food products and ingredients carefully to reduce spillage.
5. Work in an orderly manner to keep areas tidy throughout the operating period.
6. During a breakdown, rinse equipment and cool to 35°C to arrest microbial growth.
7. During brief shutdowns, keep washers, dewatering screens, blanchers, and similar equipment running and cooled to 35°C or below.

Preparation Steps for Effective Cleaning

To facilitate effective cleaning, it is necessary to prepare equipment and the area for cleaning:

1. Remove all large debris in the area to be cleaned.
2. Dismantle equipment to be cleaned as much as possible.
3. Cover all electrical connections with a plastic film.
4. Disconnect lines where possible or open cutouts to avoid washing debris onto other equipment that has been cleaned.
5. Remove large waste particles from equipment by use of an air hose, broom, shovel, or other appropriate tool.

Processing Areas

FREQUENCY Daily.
PROCEDURE
1. Prerinse all soiled surfaces with 55°C water to remove extraneous matter

from the ceilings and walls to the floor drains. Avoid direct hosing of motors, outlets, and electrical cables.

2. Apply a strongly acid cleaner through portable or centralized foam cleaning equipment. A centralized system is more appropriate for large plants. Portable equipment is more practical for smaller plants. For heavily soiled areas, cleaning compounds are more effective if applied by portable or centralized high-pressure cleaning equipment. If metal other than stainless steel is present, the acid cleaning compound should be replaced with a heavy-duty alkaline cleaning compound. Hand brushing may be necessary to remove tenacious soil deposits left from foam cleaning. The cleaning compound should reach all framework, table undersides, and other difficult-to-reach areas. Soak time for the cleaning compound should be 10 to 20 minutes.

3. Rinse surfaces within 20 minutes after application of the cleaner to remove cleaning compound residues. The same rinse pattern as with prerinse and cleaning compound application should be followed by the application of 50° to 55°C water.

4. Thoroughly inspect all surfaces and conduct any necessary touch-ups.

5. Apply a chlorine compound sanitizer to clean equipment with centralized or portable sanitizing equipment. The sanitizer should be sprayed as a solution that contains 100 ppm of chlorine. Water pipes used for recirculating wash water and for pumping peas, corn, and other vegetables, as well as brines and syrup, should be sanitized by the same method. Frequently drain, clean, and sanitize water storage tanks to reduce microbial buildup.

6. Thoroughly backwash and sanitize water filters and water softeners.

7. Eliminate scale (as needed) from the surfaces of pipeline blanchers, water pipes, and equipment to reduce the chance of thermophiles and other microorganisms being harbored.

8. Remove, clean, and replace drain covers.

9. Apply a white edible oil only to surfaces subject to rust or corrosion. Further use of oil is discouraged because the protective film harbors microorganisms.

10. Avoid contamination during maintenance by requiring maintenance workers to carry a sanitizer and to use it where they have worked.

Large processing plants can effectively utilize a CIP system for cleaning piping, large storage tanks, and cookers. The CIP system can be used as an alternative to steps 1, 2, 3, and 5 above.

Packaged Storage Areas

FREQUENCY   At least once per week where processed products are stored and more frequently in a high-volume operation. Daily in areas where raw products are stored.

PROCEDURE

1. Pick up large debris and place in receptacles.

2. Sweep and/or scrub with a mechanical sweeper or scrubber, if one is available. Use cleaning compounds provided for mechanical scrubbers, according to directions provided by the vendor.

3. Use a portable or centralized foam or slurry cleaning system with 50°C water to clean areas heavily soiled, unpack-

aged products, or other debris. Rinse as described for the processing areas.

4. Remove, clean, and replace drain covers.
5. Replace hoses and other equipment.
6. Wash and sanitize vegetable boxes after each trip. Replace wooden husker and cutter bins with metal containers, which should be cleaned and sanitized.

EVALUATION OF SANITATION EFFECTIVENESS

A sanitation program must be evaluated to determine the effectiveness of cleaning and sanitation. Performance data not only measure sanitation effectiveness, but also provide documentation of the program being conducted. Sanitation goals and checks are vital in the determination of sanitation effectiveness.

Sanitation Standards

To evaluate sanitation procedures, a yardstick measuring the current performance against past performance and desired goals should be used to determine progress. Sanitation standards, derived through visual inspections and microbial counts, can be established. This approach has limitations due to variations, especially in microbial counts. Visible contamination and microbial load are not always highly correlated. However, the sanitarian can compensate for variables and still effectively evaluate the program.

Inspections can be conducted by the sanitarian or by a sanitation committee consisting of the sanitarian, production superintendent, and maintenance supervisor. Evaluations should be made in writing. A form that uses a numerical rating system is considered the most appropriate. The report should be divided by areas with specific sanitary aspects itemized in each area, as shown in Exhibit 17–1. The completed re-

port should be provided to each supervisor associated with the inspected areas.

Laboratory Tests

The sanitarian must know the genera, characteristics, and sources of microorganisms found in the plant before laboratory tests have applicable value. With this knowledge, laboratory tests can be a monitoring device to evaluate the effectiveness of a sanitation program. The sanitarian should strive to reduce the total count of microorganisms found on clean equipment and among processed products but should also recognize that total plate count is not always highly correlated with spoilage potential or with the presence of microorganisms of public health concern. It is important to identify microorganisms, such as coliforms, as indicators of contamination or thermophiles and certain mesophiles as potential spoilage microbes. Large numbers of spore-formers can also be significant because these bacteria can reduce shelf life, and certain microorganisms can cause foodborne illness.

Spot checks for microbial load can verify opinions formed through visual inspections. Microbial sampling of products and equipment from various stages of manufacturing can identify trouble spots in the processing control cycle. Use of laboratory tests further utilizes the "think sanitation—practice sanitation" concept.

SUMMARY

An effective sanitation program for fruit and vegetable processing facilities requires hygienic design of facilities and equipment, training of sanitation personnel, use of appropriate cleaning compounds and sanitizers, adoption of effective cleaning procedures, and effective administration of the sanitation program—including evaluation of the pro-

Exhibit 17–1 Sanitation Evaluation Sheet for Food Processing Plants

Name of Plant:	*Location:*	*Date:*

Scoring System: 1 = Unsatisfactory, 2 = Poor, 3 = Fair, 4 = Satisfactory

Location	Score	Comments

1. Premises
   Property outside of building
   Waste disposal facilities
   Other
2. Receiving
   Dock
   Containers
   Conveyors
   Floors, walls, ceilings, and gutters (or drains)
   Other
3. Preparation
   Washers and flumes
   Conveyors
   Graders and snippers
   Blanchers, hoppers, and dewaterers
   Pulpers and finishers
   Floors, walls, ceilings, and gutters (or drains)
   Other
4. Canning
   Conveyors
   Packaging or filling equipment
   Floors, walls, ceilings, and gutters (or drains)
   Other
5. Cooking
   Exhaust box
   Syrupers
   Steamers
   Floors, walls, ceilings, and gutters (or drains)
   Other
6. Storage
   Tanks and pipes
   Other containers
   Floors, walls, ceilings, and gutters (or drains)
   Other
7. Welfare Facilities
   Lockers
   Wash basins
   Toilets and urinals
   Floors, walls, ceilings
   Other
8. Personnel
   Cleanliness
   Head covering
   Health records
   Other

gram through visual inspection and laboratory tests. Effective sanitation starts with reduced contamination of raw materials, water, air, and supplies. If the facility and equipment are hygienically designed, cleaning is easier and contamination is reduced.

Cleaning labor can be reduced through use of portable or centralized high-pressure or foam cleaning systems, and CIP systems can be used in large operations. Many facilities, if designed of durable material, can be cleaned effectively with acid cleaning compounds and sanitized most adequately and economically by paints and other protective coatings as additional sanitary precautions. The effectiveness of a sanitation program can be evaluated through the establishment of standards as guidelines, visual inspection, and laboratory tests.

STUDY QUESTIONS

1. Where is CIP cleaning used most in fruit and vegetable processing plants?
2. What percentages of raw materials from the citrus juice industry are normally handled as waste products?
3. What is the maximum water temperature that should be used for cleaning fruit and vegetable processing plants?
4. Which sanitizer that can be applied in fruit and vegetable plants is the most stable and acts the longest amount of time?

REFERENCE

Shapton, D.A., and N.F. Shapton, eds. 1991. Aspects, microbiology safety in food preservation technologies. In *Principles and practices for the safe processing of foods*, 305. Oxford: Butterworth-Heinemann.

SUGGESTED READINGS

Jowitt, R. 1980. *Hygienic design and operation of food plant.* Westport, CT: AVI Publishing.

Lopez, A. 1987. *A complete course in canning (Book 1).* 12th ed. Baltimore: Canning Trade.

CHAPTER 18

Beverage Plant Sanitation

Because the soils found in beverage plants are primarily high in sugar content and are water soluble, they are less difficult to remove than those described in other plants. Soil removal and microbial control present more of a problem in breweries and wineries. Therefore, a large percentage of the discussion in this chapter will concentrate on these two areas.

MYCOLOGY OF BEVERAGE MANUFACTURE

Because beverage plants such as breweries must maintain a pure yeast culture, it is important to retain the desirable microbes and to remove those that cause spoilage and unsanitary conditions. Ineffective sanitation can cause product acceptability problems because contaminating microorganisms, although kept under control, are never eliminated from the environment. Breweries differ from most plants in that commonly recognized pathogenic microorganisms are normally of minimal concern, primarily because of the nature of the raw materials, processing techniques, and limiting environmental characteristics of the final product (low pH, high alcohol concentration, and carbon dioxide tension). Stewart (1987) suggested that an exception to this is the unlikely possibility that significant levels of toxic metabolic products from certain fungi may pass from infected raw materials into finished products. Rigid control of raw materials is essential to ensure an acceptable product because there is no satisfactory method to detoxify a finished product that is contaminated.

SANITATION PRINCIPLES

An adequate supply of urinals should be provided, kept in a sanitary condition, and located within a short distance from the bottling area and other production areas. Employees must be required to wash their hands after using the toilet facilities. Drinking fountains should contain guards to prevent contact of the mouth or nose with the metal of the water outlet.

Employee Practices

As with other food operations, sanitation is a team job. It is important in beverage plants that employees clean as they go. Periodic cleaning increases tidiness, reduces contamination, and minimizes cleanup time at the end of the production shift or during a production change from the manufacture of one product to another. Furthermore, one or more employees who operate equipment that fills

303

bottles or cans frequently have time to pick up debris or to hose down spills or other extraneous matter.

Effective housekeeping in a beverage plant depends on training and standards for the development of appropriate employee working habits. Rigid sanitation practices and work habits should be cultivated through effective communication, training programs, educational material, and continuous supervision and instruction. Employees should be instructed how, when, and where to clean to immediately remove soil and debris that can provide nutrition for pests and microorganisms. Leaking equipment should be corrected immediately. If rodents, birds, insects, or molds are detected, employees should either perform the necessary corrective steps or report the problems. Employees should be instructed regarding proper storage practices so that pest harborages are not created and proper cleaning can be accomplished. Further instruction should be related to closing doors and windows, removal of infested and extraneous matter, and proper storage of tools and cleaning and sanitizing equipment.

The following sanitation rules apply for beverage plants:

1. All employees visiting a lavatory must wash their hands before returning to work.
2. Any spilled materials or products must not be returned to the production area.
3. Waste materials must be placed in containers suitable for disposal that have tight-fitting covers.
4. Each employee is required to keep the immediate work area clean and tidy.
5. Tobacco use is forbidden, except in designated areas.
6. Spitting is prohibited anywhere in the plant.

7. A periodic inspection of clothing, lunchrooms, and lockers by management should be conducted to ensure proper cleanliness.
8. Headgear should be worn at all times.

Cleaning Practices

There are five standard steps for cleaning a beverage plant:

1. Rinse to remove large debris and nonadherent soil, to wet the area to be cleaned, and to increase the effectiveness of the cleaning compound.
2. Apply a cleaning compound (usually via foam) to provide intimate contact of water with the soil for removal through effective wetting and penetrating properties.
3. Rinse to remove the dispersed soil and the cleaning compound.
4. Sanitize to destroy residual microorganisms.
5. Rinse quaternary ammonium sanitizers [especially if present in more than 200 parts per million (ppm)] before exposing the cleaned area to any beverage materials.

NONALCOHOLIC BEVERAGE PLANT SANITATION

It is beyond the scope of this text to discuss sanitation principles for all nonalcoholic beverage plants. Although ultra-high temperature as a technique for aseptic packaging is becoming more common and is expected to be the "wave of the future," the ramifications of this technology are too extensive and specific for this general discussion. If further information about sanitation in these specialized operations is desired, a technical publication about aseptic technology should be reviewed.

Proper hygiene in a beverage processing facility includes the use of sanitary water, steam, and air. O'Sullivan (1992) reported that high-quality liquids and gases are required when they are incorporated into finished products or included in the packaging material that contacts the product. The desire to manufacture acceptable products and to meet safety standards has resulted in several beverage processors incorporating various types of filtration to remove microorganisms and other particulate or suspended materials. Filtration for the clarification or microbial control of water, air, and steam is accomplished by absolute filtration to prevent contaminants larger than the filter pore size to pass through and into the filtrate.

Because beverages such as soft drinks, bottled water, beer, and distilled spirits should be manufactured from microbial- and particulate-free water, some form of treatment is necessary. Various treatments include flocculation, filtration (i.e., through a sand bed), chlorination, sterile filtration, reverse osmosis, activated carbon, and deionization. The use of the water determines the type and extent of treatment.

Conditioning of water for use in beverage plants is accomplished primarily through particulate removal and microbial control. Particulate contaminants that may be present in water are most frequently removed by flocculation and sand filtration. The installation of an absolute-rate depth filter behind the sand filter will remove all of the contaminants larger than the rated pore size prior to chlorination and activated carbon treatment.

Activated carbon is incorporated to remove excess chlorine, trihalomethanes, and other compounds associated with chlorine disinfection. However, activated carbon sheds carbon fines and provides sites for microbial growth. According to O'Sullivan (1992), carbon beds are potential microbial contamination sources and are difficult to disinfect. Therefore, the use of filtration before and after carbon beds will reduce the loading of microorganisms and particles.

Resin beds for deionization of water are potential sites for microbial growth and can unload or shed resin beads into the treated or conditioned water. An absolute rated filter will ensure that particles or microorganisms larger than the removal rating of the filter do not enter the treated water. As a final treatment, the incorporation of a sterilizing nylon 66- to 9.2-μm filter will remove microbes present in the water if the unit has been presterilized (O'Sullivan, 1992). Sterile filtration requires no chemicals and is beneficial because of its ease of use and low energy input. A microbially stable product may be produced through use of a combination of flocculation and filtration steps followed by an absolute-rated filter.

Although steam is frequently incorporated in a production operation, it can be a viable contamination source. Steam is normally generated in carbon steel boilers, which are highly susceptible to rusting. A fine impervious film of rust, which acts as a protective barrier against further corrosion, normally deposits as a result of a continual operation of a boiler. Intermittent use allows a continual supply of fresh air containing oxygen into a boiler and promotes the oxidation of iron to iron oxides, or rust. The continual generation of rust causes flaking and steam contamination.

Use of steam permits contamination, and the particles of rust from the boiler transfer lines will damage equipment surfaces, block steam valves, fill orifices and filter pores, and stain equipment surfaces. Processing efficiency is reduced through the alteration of the heat transfer characteristics of heat exchangers. This problem is reduced by the injection of culinary steam with an uninterrupted supply provided through the installation of porous stainless steel filters in parallel to permit the cleaning of one set while the other is in use. Nonculinary-grade steam will add contaminants.

During the past, bottlers have been installing cleaning-in-place (CIP) equipment to clean tanks, processing lines, and filters. Most bottlers that manufacture multiple flavors prefer CIP as a tool to prevent flavor "carry-over" (especially of root beer). Remus (1991b) advocates the TACT (time, action, concentration, and temperature) approach to cleaning beverage plants. He has suggested that, within reason, the parameters can be varied; for example, a 1% cleaning compound concentration at 43.5°C can be equivalent to a 0.5% concentration at 60°C.

Increased efficiency and superior lubricity may be attained through an automated solid-lubricant dispensing system. This equipment saves labor and lubricant costs, and reduces contamination during lubrication.

The following discussion relates how common soils found in beverage plants can be removed. Although this discussion addresses carbonated beverage plants, these cleaning principles apply to other beverages. Principles of cleaning of floors, walls, and the bottling area, as discussed under winery sanitation later in this chapter, should also be considered for carbonated beverage plants. Cleaning applications and principles not discussed here are normally similar to those discussed for dairy processing plants (see Chapter 13).

Automated Cleaning Equipment

A portion of the carbonated beverage industry has turned to automated equipment to facilitate cleaning. A variety of automated solutions are now offered, such as automated chemical formulation and an allocation and control system to streamline the operation.

A microprocessor-controlled system can be assessed by keying in an identification number or by using custom magnetic swipe cards. The controller contains a detailed application list that indexes sanitation procedures and equipment types with the proper chemicals and usage rates. Then the system dispenses the product into a reusable chemical container for use in plant sanitation. A smaller auxiliary dispensing station may allocate acids and other specialty chemicals to cleaning stations. This equipment can maintain detailed records to help monitor regulatory compliance, perform cost analyses, and create custom reports. Data reporting includes which chemicals have been incorporated into each application, when and in what quantity, and times and dates (Flickinger, 1997). Chemical barrel labels may be color-coded so that workers need replace only the empty drums that correspond to colored spots on the floor.

A computer-controlled CIP unit directs water and solutions to the appropriate location and automatically maintains operating conditions. The four basic parameters to be controlled are time, temperature, chemical concentration, and impingement, which relates to flow velocity through a pipeline. Rinse water may be recycled one time for reuse and cleaning compounds several times. The initial prerinse may use recycled water from the previous final rinse.

Tire Track Soil Accumulation

Tire tracks are a difficult soil to remove. The most effective cleaning compound for this application is one that is solvated and alkaline. To facilitate cleaning ease and effectiveness, a mechanical scrubber should be considered. Floor soils should be removed daily to enhance cleaning ease and to avoid soil being further ground into the floor surface.

Conveyor Track Soil Accumulation

This accumulation is most likely to be spilled product, bearing grease, container and track filings, and precipitated soap. Incorporation of a track lubricant containing a detergent will reduce contamination. An effective way of removing this soil is through foam cleaning with a high-pressure rinse.

Film Deposits

Film deposits most frequently occur inside storage tanks, transfer lines, and filters. Thin films cause a dull surface, but as buildup increases, a bluish hue develops. As the film becomes thicker, a white appearance may occur. Although residues from sugars are relatively easy to remove, films from aspartame and certain gums are difficult to eliminate. Tanks may be cleaned manually, but circulation cleaning is frequently practiced. To remove surface films, a chlorinated cleaning compound (or one specially formulated with surfactants for food soils) should be applied.

Biofilms

Residual beverages or their ingredients provide nutrients for microbial growth and their biofilms. Biofilms can occur inside cooling towers, in and outside of warmers and pasteurizers, and inside carbocoolers. As with film deposits, biofilm removal is enhanced by use of a chlorinated alkaline cleaning compound. A quaternary ammonium sanitizer or another biocide should be applied to reduce biofilm deposition because this formation can occur within 24 hours after use.

Hot Sanitizing

Sanitation of beverage plants differs from that of other food facilities. During the past few years a trend toward hot sanitizing has occurred. Hot sanitizing can be incorporated when cleaning products contact surfaces of production equipment, such as batch tanks, low mix units and fillers, and carbocoolers. Although this sanitizing method is not economical because of the required energy costs and ineffectiveness in bacteria removal, it has some merit because of its penetrating ability. Heat can effectively penetrate equipment and destroy microorganisms behind gaskets or in tiny crevices.

Hot sanitizing is not sterilization. Hot sanitizing involves raising the surface temperature to 85°C for 15 minutes. Sterilization requires 116°C for 20 minutes. Sanitizing only reduces the microbial population to an acceptable level. A few of the more resistant microorganisms (yeasts and spores) remain viable. Chemical sanitizers can accomplish the same microbial kill as hot sanitizing, with a much quicker action.

Specially formulated cleaning compounds can be incorporated in a hot sanitizing procedure to loosen and remove soils and biofilms (Remus, 1991a). These compounds are specifically formulated to handle the soil, to condition the water, and to be free-rinsing in the hot sanitizing procedure. Soil and biofilm removal are essential for effective sanitation. A nonviable but intact biofilm is an easy attachment site and nutrient base for other films to develop.

BREWERY SANITATION

Because breweries have been traditionally production oriented, prophylaxis has normally superseded detailed taxonomic interest in microorganisms associated with these operations. The environment typical of a brewery can restrict pathogen activity and impose limitations on the array of spoilage microbes. Bacteria of greatest significance in this environment are non–spore-formers. However, spore-forming bacteria, such as *Clostridium* species, may be involved in the spoilage of brewery by-products, such as spent grain. Non–spore-forming bacteria that are found in breweries may contribute to a wide variety of problems in wort, including pH elevation, acidification, acetification, incomplete fermentation, ropiness, and slow runoff time. Such infection may also be directly or indirectly responsible for various off-odors and biological hazes in finished beer.

Lactobacillus is usually regarded as the most troublesome genus of bacteria in the brewery because its species represents a potential spoilage hazard at the various stages of production, including finished beer. Other genera are less versatile under brewery conditions; therefore, their spoilage potential is more limiting. However, enterobacteria may have an impact on the fermentation, flavor, and aroma of beer. According to Stewart (1987), the most common techniques used for detecting and differentiating the various brew contaminants are selective and differential culture media (either alone or in combination with centrifugation) or millipore filtration, depending on the expected cell density and various serologic techniques and impedance measurement.

Control of Microbial Infection

Contamination can be controlled by removing excess soil and microorganisms that cause off-flavors. Although Stanton (1971) stated that beer will self-sterilize in 5 to 7 days, undesirable bacteria, yeasts, and molds grow rapidly in freshly cooled wort that is contaminated through poor sanitation. There-

fore, it is necessary to clean and sanitize the brewery equipment that processes the wort. Stanton (1971) reported that clean kettles and coolers can transfer heat faster because 1 mm of soil on the inner cooling surface is equivalent to 150 mm of steel. He has hypothesized that 1 mm of soil could have a similar insulating effect. Furthermore, high-speed equipment, such as fillers, cappers, casers, and keggers, performs more effectively if kept clean.

The most effective means of preventing spoilage of beverage products is to control infection by developing and maintaining a comprehensive cleaning and sanitizing program. A program can be developed by sanitation personnel or with the help of a reliable sanitation consulting group or a dependable cleaning compound and sanitizer supply firm. Discussions in other chapters relating to equipment and facility design, cleaning equipment, cleaning compounds, and sanitizers should be reviewed to determine guidelines for the implementation of a sanitation program. It is especially important to review discussion related to CIP equipment (Chapter 9). These systems are quite adaptable to cleaning beverage equipment, and the trend in the industry is toward automation through this concept.

Fermentation facilities such as breweries require sterile air for the production of starter cultures or the maintenance of sterile conditions within a storage tank. O'Sullivan (1992) identified the optimum practice as coarse-filtering air with a coarse depth or pleated filter to remove the bulk of contaminants, followed by filtering with a 0.2-μm membrane or sterile filter. Thus, the sterile air can blanket the stored product by creating a positive pressure within the storage vessel. An inert gas can be substituted for air to reduce oxidation. Blanketing a storage tank is an easy way to create a sterile environment, especially with large storage tanks.

The control of microorganisms may be enhanced through ultraviolet (UV) light to reduce the airborne microbes, eliminate pests, and treat water. Several breweries have implemented UV light in water treatment as it is the main ingredient of the final product and allows for a residue-free water that will not affect the chemistry of beverage manufacture, as do most sanitizer residues. This treatment does not have a detrimental effect on water as UV light is a nonionizing and nonresidual disinfectant.

This sanitizer functions through irreparably damaging microbial DNA, which absorbs these high-energy wave lengths. The disruption of DNA prevents the microorganism from repair and replication. A visible purplish-colored light may be mistaken for UV light, but this wave length cannot be detected by the human eye. The violet-colored light of the nearby visible wave length region can be generated by the UV lamps, which are beneficial in alerting personnel to the presence of UV light but can ultimately diminish its effectiveness (Rosenthal, 1992). In some applications, UV light is cost effective and can be easily incorporated into an existing sanitation program. The nonselective nature of UV light permits the nonresidual cleaning of air, water, packages, and some foods.

Cleaning Compounds

Efficient cleaning can be attained only if the proper cleaning compound is incorporated. Spray cleaning is most effective with the incorporation of a properly blended cleaner having specific cleaning properties for the soil that exists. The cleaning compound should be low-foaming because foam reduces velocity during circulation and tends to prevent contact of the solution with part of the surface. The appropriate cleaning compound will prevent "beerstone" formations. It should also be formulated to prevent metal attack, and it must be easily rinsed to avoid the uptake of objectionable flavors by the beer. (Other information on cleaning compound selection, application, and safety during use is discussed in Chapter 7.)

Sanitizers

Sanitizers such as chlorine, iodine, or an acid-anionic surfactant should be incorporated with the final rinse in fermenters, cold wort lines, and coolers. Because water can contain viable microorganisms exceeding 100/mL, it is possible to have a sterile surface after cleaning but bacteria or yeasts deposited on the equipment surface after the final rinse. (Additional information related to sanitizers and their application is given in Chapter 8.)

Heat Pasteurization

Heat pasteurization is still the most common method for microbial control in beverage plants, such as those producing packaged beer. Although the energy costs are high, it is, nevertheless, a convenient method. Alternative procedures have been investigated because of high energy costs and the adverse effect of heat on the flavor of drinks such as beer. Such alternative procedures, frequently called cold pasteurization, include the use of chemical compounds, such as propyl gallate, as well as millipore filtration, either followed by aseptic packaging or used in conjunction with other chemical treatments. Official approval of chemical compounds is subject to change as new technology and information related to safety become available. The bacterial count of pitching yeast may be reduced by treatment with dilute acids such as phosphoric, sulfuric, and tartaric acid. Acid treatment can reduce bacterial infection, but it has an adverse effect on the yeast culture, and retarded fermentations can occur in the first few cycles after treatment. Sulfur dioxide (SO$_2$) has been used in the past for control of wort bacteria.

Aseptic Filling

Aseptic filling is considered to be a nonpasteurization process that involves ultra-filtration techniques to remove the spoilage organisms from beer before packaging. Because ultrafiltration occurs before packaging, spoilage microorganisms can enter the product. The comments that follow were provided by Remus (1991b) to ensure deliverance of a high level of sanitation in aseptic packaging.

Hygiene Practices

It is important to have closed filling rooms with a positive pressure of filtered air. The workers' apparel should always be clean, and before workers enter a room, their hands should be washed with a sanitizing hand soap. A conveyor lubricant system that reduces microorganisms should also be utilized.

The interior of the filler should be cleaned and sanitized daily, utilizing recirculating CIP equipment. The exterior of the filler, conveyors, associated equipment, floors, and walls should be foamed or gelled, then sanitized daily. For added protection, this process should provide a residual antimicrobial activity because a detergent or sanitizer remains on a surface after its application and subsequent drying, preventing recontamination of the sanitized surfaces.

There should be a regular program of surface and air monitoring for bacteria, yeasts, and molds in the filling area. HACCP (Hazard Analysis Critical Control Point) utilizes chemical and microbial monitoring to guarantee safe food production. These monitorings are always compared against reference standards. Microbial monitoring in aseptic beer filling needs to be developed within that aspect of the beer industry. A base line of data should be gathered and statistically evaluated against finished product quality.

Bottle Cleaning

Centralized high-pressure, low-volume cleaning equipment has improved the efficacy of bottle cleaning. (Chapter 9 describes the principle and capabilities of this equipment.) Tenacious soil can be removed from very difficult-to-reach areas such as conveyors, bottle fillers, cappers, and casers.

Sanitation in Storage Areas

In addition to suggestions provided for storage areas of other food facilities discussed in this book, it is appropriate to recognize the need for proper storage of materials such as grain, sugar, and other edible dry products. Screw conveyors should be cleaned on a schedule basis. This is especially true for the dead ends of conveyors where dormant residues can accumulate. The ends and junctions of conveyors should be cleaned at least once a week. The free-flowing section of a conveyor should be equipped with hinged covers for easy cleaning and inspection. After conveyors have been cleaned thoroughly, they should be fumigated with a nonresidual fumigant. Empty bins should be thoroughly swept (and preferably vacuumed) prior to fumigation. Regular checks should be made of material cleaned out for possible infestation. (Chapter 11 provides a detailed discussion of recommended pest control measures that may apply to storage areas at beverage plants.)

Brewing Area Sanitation

Spray cleaning is faster and more dependable than manual cleaning and can reduce downtime. Although unheated water can be used, a water temperature of up to 45°C can increase the chemical reaction of the cleaning compound with the soil. If glass-lined tanks are used, the maximum water temperature should be 28.5°C to reduce damage due to sudden temperature fluctuation. Temperatures above 45°C should be discouraged because of condensation problems and increased refrigeration requirements. In fact, it is advisable to lock in specific temperature or high-temperature cutoff switches to control water temperature. Caustic soda cleaning compounds should not be used because they attack soldered ends. Scale formation in aluminum vessels can be removed with 10% nitric acid, applied as a paste mixed with kaolin.

Initial and maintenance costs of hoses and fittings suggest the viability of stainless steel lines (even though stainless steel is quite expensive). Circulation cleaning of product-in and product-out lines can be accomplished by the use of U-type fittings to connect the tank valve to both lines. Industrial spray nozzles for equipment cleaning can be positioned to clean areas such as vapor stacks on kettles and strainer troughs in hop strainers and to provide continuous cleaning for conveyor belts. The brewing area should be cleaned at least once per week, and debris and other soils should be removed daily.

Beer stone, which is primarily organic matter in a matrix of calcium oxalate, is one of the most difficult beverage soils to remove. This deposit is most effectively removed with extensive scrubbing and use of a strong chelating agent and alkaline cleaning compound.

Bottle Washing

All empty bottles must be carefully examined on return by both the retailer and the bottler. New bottles should also be inspected at the bottling plant to detect any obvious contamination. All new and used bottles should be mechanically washed immediately before filling with a washer that applies a heavy spray of caustic solution, both internally and externally, with subsequent rinsing. The spray and rinse temperature should be 60° to 70°C. Chlorination of the final rinse with up to 0.5-ppm concentration can be incorporated without affecting the flavor of beer. Chlorination is not necessary unless the water characteristics dictate this purification technique.

Beer Pasteurization

Most brewers pasteurize their beer to maintain a stable condition, flavor, and smoothness. Certain brewers have incorporated sterile filtration as a substitute for sterilization. If filtration is incorporated, the filters should be replaced every other week to reduce the risk of microbes penetrating the series of filters. In a sanitary operation, sterile filtration can be effective.

Pasteurization during containerizing is practiced by much of the brewing industry because it can protect the beer against contamination after packaging. Overheating during pasteurization, however, can have an adverse effect on flavor and can cause haze. Therefore, it is essential to subject the beer to the minimum time and temperature for effective microbial destruction. Most of the brewing industry now has conveyor systems for a pasteurization cycle of approximately 45 minutes. During pasteurization, the temperature of the beer is gradually raised from 1° or 2°C to 61° up to 63° C, with subsequent cooling to ambient temperature at the end of the cycle. The moving belt speed can determine the length of exposure time in the pasteurization environment. Pasteurization is known to speed up the reactions that give oxidation haze, so the effects of an excess of air may be accentuated with pasteurized beers. The total air content of packaged beer should not exceed 1 mL/220 mL of beer.

Haze may develop in beer. A nonbiological haze may form from the slow precipitation of products with unstable solubility—a condition caused or accelerated by oxidation. A biological haze may be caused by the growth of bacteria or yeasts. A sufficient period in the cold conditioning tank and fine filtration will minimize the chances of nonbiological haze. The exclusion of air in the beer container, as well as the selection of suitable container materials, will also minimize the chances of nonbiological haze occurrence. Other hazes have been traced to metallic influences, especially that of tin. Haze of beer in brightly colored bottles due to bacteria or yeast growth suggests either imperfect filtration or subsequent infection. A bacterial or yeast haze can be attributable to lack of proper sanitation in the plant or unclean storage containers or filters.

Cleaning of Air-Conditioning Units

The following procedure for cleaning air-conditioning units is suggested:

1. Clean air-conditioning units every 6 months. Insert a ball spray through a special opening above the coils on top of the air-conditioning units.
2. Run water for 10 minutes to flush the unit.
3. Run a hypochlorite solution (200 ppm) at 40°C for 5 minutes.
4. Let the unit soak for 5 minutes.
5. Rinse the unit with warm water for 10 minutes.
6. Check the unit and clean the pan bottom.

Water Conservation in Brewery Sanitation

Water usage during cleaning can be reduced with a wash-rinse cycle sometimes called the *slop cycle*. This cycle includes a prerinse, in which a cleaning solution is pumped through a spray device for 20 seconds, with 1 minute permitted for chemical action and a subsequent burst rinse with wa-

ter—the same procedure as used in most home dishwashers. Reuse of cleaning solutions is practical and economical. The length of reuse can be increased if the solution tank has a top overflow to skim off floating soil and a drain valve to permit bottom-draining of the heavy soil. Furthermore, the final rinse water can be salvaged for the prerinse on the next tank to be cleaned. This technique can reduce water and sewage treatment costs in areas where both water supply and sewage are metered.

WINERY SANITATION

It is essential to remove soil contaminants that affect the taste, appearance, and perishability of wine. Included among the contaminants are the reddish tartrate deposits that form or build up on tank interiors as a result of fermentation. Other tenacious soils should be cleaned from the surfaces of processing equipment to reduce microbial growth throughout the winery. In general, the more sanitary a winery is, the smaller the quantities of SO_2 that must be added to the wine at the end of the winemaking process. Although SO_2 has been used to control microbial growth, use of this compound has been discouraged and may be discontinued in the future. As a complement to SO_2, sorbic acid is effective in the prevention of fermentation of sweet wines if there is a low initial count of yeast and free SO_2 is still active to prevent bacteria from destroying the sorbic acid. Zoecklein et al. (1995) have suggested that rigid sanitation is a viable alternative strategy for microbial destruction.

Because rigid sanitation will not destroy all microorganisms, as does sterilization, the reduction of viable cell number to an acceptable level may be attained. As stated by Zoecklein et al. (1995), effective sanitation accomplishes another important goal—the elimination of hospitable environments for growth.

Although the requirements for sanitation increase during the winemaking process and peak at bottling time, it is important to recognize that the vineyard tools and harvesting equipment must be washed to remove dirt, pomace, soil, and leaves. Destemmers, crushers, and grape processing and bulk storage areas require a brush detergent and water. Harvester heads, pipes, hoses, pumps, faucets, spigots, and anything else coming in direct contact with the juice or wine will require the five cleaning steps discussed early in this chapter. The same steps apply to the bottling line, but additional control and checking are necessary to reduce the microbial load of the wine.

Water Quality

The water used in a winery must have certain chemical and microbial properties. A low pH is inimical to steel and other surfaces, and a high pH will favor calcium precipitation. The biochemical oxygen demand (BOD) should be less than 3 mg/L. Because water can be a potential carrier of molds, yeasts, and acetic or lactic acid bacteria, pure water should be used.

Winery design and layout should incorporate hygienic principles. Floors must be easy to clean and have nonslip, sloped surfaces. Walls and ceilings should be impervious and easily cleaned. Sanitation in a winery can be enhanced through proper location of equipment to reduce the creation of corners and crevices that are difficult to clean and to facilitate the cleaning of floors. As with other food manufacturing facilities, equipment should be constructed with sanitary features that enhance effective cleaning.

Cleaning Floors and Walls

Although a winery may be somewhat seasonal in operation, year-around sanitation is necessary. A combination of wet and dry cleaning is usually most appropriate. The heavy-duty, wet-dry vacuum cleaner can be effectively incorporated in cleaning. Floors should be cleaned at least once a week by dry or wet methods, depending on the nature of the soil. To facilitate cleaning, floors should be constructed of concrete, sloped, and should contain trench drains. Spilled wine, especially any that has spoiled, should be washed away immediately.

It is necessary to remove as much of the visible debris as possible before use of cleaners. This task is accomplished manually or by mechanized cleaning (spray balls, etc.). Spray applications should be directed at an angle to the surface being cleaned.

The area should be mechanically scrubbed, washed with lime or a strong hypochlorite solution, and rinsed with water. Floors and the outside of the wood cooperage (if applicable) should be periodically washed and disinfected with a dilute hypochlorite solution. When dry cleaning is possible, the humidity of the winery can be kept lower than when wet cleaning is practiced, with resultant reduced mold growth in areas where wood is present. Tank tops, overhead platforms, and ramps can be vacuumed, cleaned, or washed, taking precautions that no water gets into the wine. The walls should be washed with a warm alkaline solution, such as a strong solution of a mixture of soda ash and caustic soda, followed by rinsing with water and spraying with a hypochlorite solution containing 500 mg/L of available chlorine. Amerine et al. (1980) recommended that all free chlorine be removed through washing.

Equipment Cleaning

Improper equipment cleaning is one of the most viable sources of contamination. Crushers, must pumps and lines, presses, filters, hoses, pipes, and tank cars are all difficult to clean completely. Less complicated equipment, such as wine thieves, hydrometer cylinders, buckets, and shovels, can also be difficult to clean. This equipment should be dismantled as much as possible, thoroughly washed with water and a phosphate or carbonate cleaner for nonmetallic surfaces and caustic soda or equivalent for cleaning metal equipment, and sanitized with hypochlorite or an iodophor if the material being cleaned is adversely affected by chlorine. Enzymes are useful as cleaning agents because they can hydrolyze proteins, fats, and pectins. They are currently used in enology because their maximum efficiency is at nearly a neutral pH. Where possible, circulating the cleaning solutions is recommended. Hoses, after cleaning and rinsing, should be placed in sloping racks instead of on the floor to facilitate drainage and drying. Thorough cleaning and sanitizing are essential for equipment that has been in contact with spoiled or contaminated wine.

During the harvest season, conveyors, crushers, and must lines should be kept clean. They should not be permitted to stand with must in them for more than 2 hours. After use for 2 days, they should be washed, drained, and thoroughly flushed with water before reuse.

Bottling Area Cleaning

Effective cleaning of the bottling area is essential to reduce bacterial or metallic contamination. This area is usually observed most closely by public health agencies. To facilitate effective sanitation in this area, the room should be well lighted and ventilated and should have glazed-tile walls and epoxy-finished floors. Ample space between equipment is essential to facilitate easy cleaning, and the equipment should be easy to disassemble. All pumps, pipes, and pasteurizers should be constructed of stainless steel be-

cause freshly cut corks contain debris, mold spores, and yeasts. They should be cleaned and sterilized before use by soaking for 2 hours in 1% SO_2 and a little glycerol, then rinsed with water. Corks can be gamma-radiation sterilized to prevent off-odors occurring because of mold growth.

Pomace Disposal

It is essential to dispose of the pomace as rapidly as possible after pressing. This material must not stand in or close to the fermentation room because it rapidly acetifies, and the fruit flies carry acetic acid bacteria from the pomace pile to clean fermenting vats. Pomace should be further processed or scattered as a thin layer on fields, where it dries quickly and does not become a serious breeding ground for fruit flies.

Cleaning of Used Cooperage

According to Amerine et al. (1980), alkaline solutions (soaking with 1% sodium carbonate) are most effective in removing tannins from new barrels. If further treatment is necessary, steam and several rinsings should be applied.

Other viable cleaning compounds are sodium ortho- and metasilicates (Na_2SiO_3) which are less caustic and less corrosive than NaOH, with superior detergent properties. A lighter organic load can permit the application of milder alkalies, such as sodium carbonate (soda ash) or trisodium phosphate. Sodium carbonate (Na_2CO_3) is an inexpensive, frequently used, cleaning compound. However, Na_2CO_3 contributes to precipitate formation in hard water.

Polyphosphates are frequently included in cleaning compound formulations because of their ability to chelate calcium and magnesium and to prevent precipitation. Examples are sodium tetraphosphate (Quadrofos) and sodium hexametaphosphate (Calgon). The amount to be included in the formulation depends on water hardness. Acid cleaners are formulated in specialized detergent formulations (approximately 0.5%) to reduce mineral deposits and to soften water. Phosphoric acid is preferred because of its low corrosiveness and compatibility with nonionic wetting agents (Zoecklein et al. 1995).

Past practices have involved washing empty containers with water and spraying with a hot (50° C) 20% solution of a mixture of 90% soda ash and 10% caustic soda or caustic potash (KOH). Both NaOH and KOH have excellent detergent properties and are strongly antimicrobial against viable cells, spores, and bacteriophage. After subsequent washing with hot (50° C) water, containers should be sprayed with a chlorine sanitizer solution containing 400 ppm of available chlorine. A cold-water rinse should follow, with subsequent drainage and drying using a dry-wet vacuum. If mold is present, it should be scraped off because it cannot be removed by washing. Further precautions include washing with a quaternary ammonium compound. Paints containing copper-8-quinolate can also control mold growth. Burning a sulfur wick in the tank (700 mg/hl) or adding an equivalent amount of SO_2 from a cylinder of gas is also effective. Before use, the tanks should be rewashed, and the cooperage should be inspected visually and by smelling before being filled. A warm 5% soda-ash concentration is too high and, if exposed too long, the wood can deteriorate. The outside surface of wooden containers should be washed with a dilute solution once a year. Propylene glycol can be applied to discourage mold growth on the surface of the tanks. Stainless steel tanks should be cleaned with a 400-ppm or less concentrated solution to prevent mold growth.

Removal of Tartrate Deposits

It is necessary to remove tartrate deposits to smooth the inner surface, which becomes very rough. Scraping is labor-intensive and may injure the wood. Installation of a circular spray head inside the tank can help remove tartrates. Soaking with 1 kg of soda ash and caustic soda in 100 L of water will also aid in the removal of tartrates.

Storing Empty Containers

Concrete tanks should be left open and kept dry when not in use. Before reuse, they should be inspected and cleaned. An example of fermentation tanks is shown in Figure 18–1. Open wooden fermenters are sometimes painted with a lime paste when not in use, but this surface is difficult to remove. A better approach is to clean the fermenters thoroughly with an alkaline solution, followed by a chlorine solution. They can then be filled with water and approximately 1.6 kg of unslaked lime per 1,000 L of water added. Stored empty barrels can be sulfured by a sulfur wick or by introducing SO_2. However, Amerine et al. (1980) have suggested that the use of sulfur wicks can be disadvantageous because sulfur may sublime into the walls of the container, and pieces of elemental sulfur from the wick may fall to the bottom of the cask. If containers with elemental sulfur are used, hydrogen sulfide might be reduced.

Other Cleaning and Sanitary Practices

Fillers, bottling lines, and other packaging equipment can be cleaned with CIP systems. A chlorinated alkaline cleaning compound can clean, sanitize, and deodorize in one operation if the soil is light. However, the presence of organic matter can negate the effect of the chlorine sanitizer because chlorine will

Figure 18–1 Red wine fermentation tanks. Courtesy of Bruce Zoecklein, Virginia Polytechnic Institute and State University, Blacksburg, Virginia.

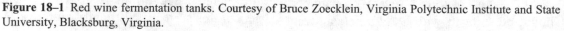

react with the organic matter in the soil. The addition of approximately 7 g of a sanitizer with a 4.25% available chlorine per liter of water should provide approximately 200 ppm of available chlorine for rapid destruction of microorganisms.

Heat has been considered the safest sterilization process available. It can be applied to everyday wines, but it does not yield high-quality premium wines. Heat can potentially transform some taste elements. Bottle sterilization is the best method for sterilizing sparkling wine.

Bottling-line sterilization can be accomplished (albeit expensively) with steam or hot water. Where hot water is employed to sanitize lines, Zoecklein et al. (1995) recommend a minimum temperature of 82°C for at least 20 minutes. The temperature should be monitored at the farthest point from the steam source (i.e., fill spouts, end of the line, etc.) Ultraviolet light is effective against mi-

crobes, but it has low penetrative capabilities, and a thin film provides a barrier between radiation and the microbes. Ozone can be used to sanitize in cold-water recirculation.

Sterile Filtration

Sterilization by filtration can be attained through use of sterile filter pads or, better, with membranes. Diatomaceous earth filtration can reduce the yeast population but will not eliminate bacteria. This filtration can be followed by membrane filtration.

Reinfection

Most problems arise from incorrect use of the sterile filtration process. Any efforts in sterile filtration are ruined by postfiltration infections if the entire bottling line is not sterile. The filter, as well as the bottling line, should be sterilized before admitting the wine.

The most effective method of sterilization is to use a steam generator hooked to the filter, which is hooked to the bottling line. A slow flow of low-pressure steam is run for 30 minutes through the entire system. The steaming is followed by cold water to cool off the machinery before allowing the wine to enter.

In some cases, steam is not available or may injure some equipment, such as plastic filter plates. Sterilization may then be achieved by running a solution of 300 g/hl citric acid and 10 g/hl SO_2 (or 20 g/hl metabisulfite) at a temperature of 60°C.

Some parts of the bottling line, such as the corker, are more difficult to sterilize. The corker jaws or diaphragm should be sterilized with alcohol. Membrane filters may be sterilized with water at 90°C.

Corks

Modern cork suppliers provide sterile corks. In case of doubt, corks should be dipped before use in a 10 g/hl of SO_2 solution.

Bottles

Bottles normally come in cases and can be recontaminated from dust and cardboard. A rinsing and sterilization station must be provided, and a solution of SO_2 at 500 ppm is used. Sterile water (obtained after cooling the sterile filter) is used to wash off excess SO_2 solution from the bottle. A dispenser for SO_2 can be set in line off of the main water supply using a medicator or other similar device. Iodophors are used frequently for bottling sanitation, followed by a cold-water rinse.

Control

Good bottling practices require checking sanitation standards. Special kits are available to evaluate the level of sanitation through a count of the number of receivable yeasts or spoilage bacteria left in the wine after filling.

Pest Control

Fruit flies are especially attracted to fermenting musts. A large proportion of the fly population is brought to the winery from the vineyard. The most effective control measures are the prompt crushing of grapes after picking, removing of all dropped and culled fruit from the winery, disposing of all organic wastes, use of repellent insecticides around the vineyard, washing of all containers and trucks after handling grapes, and use of attractant insecticides on dumps. Maximum fly activity occurs in the range of 23.5° to 27.0°C in low light intensity and low wind velocity. Fans blow air out of winery entrances. Mesh screens and air curtains are also helpful.

Insecticides that kill fruit flies are available, but Amerine et al. (1980) suggested that the heavy fly population in adjacent unsprayed areas makes the effectiveness of this practice questionable. If insecticides are used, U.S. Food and Drug Administration tolerances must be observed. (Chapter 11 provides additional information related to fly, rodent, and bird control.)

Sanitation Monitoring

The most common method of sanitation evaluation is sensory (Zoecklein et al., 1995). Visual appearance is assessed, as are smell and, sometimes, touch—to determine whether the surface feels clean. A slippery surface suggests inadequate cleaning and/or rinsing. In some instances, microbial sampling should be conducted as a means of verification.

Each microbial technique has limitations, such as surface characteristics, definition of area to be sampled, amount of pressure applied to the surface, and time of application (Zoecklein et al., 1995). Furthermore, cotton swabs will not recover all microbes. Thus, standardization of sampling procedures will improve the success of sanitation monitoring.

DISTILLERY SANITATION

As with breweries and wineries, the commonly recognized microorganisms are normally of minimal concern in distilleries because of the nature of the raw materials, processing techniques, and high alcohol concentration. A possible safety exception is the potential for contamination by significant levels of toxic metabolic products. Control of raw materials is essential because a contaminated finished product cannot be effectively detoxified. Both yield and product quality are compromised when sanitary conditions are not maintained (Arnett, 1992).

Reduction of Physical Contamination

To practice effective sanitation, corn and other grains are inspected upon arrival at the plant. Insects are the major concern at this stage, because a contaminated grain shipment can infect the storage silos, as well as the entire plant. The most common insect pests for grain are flour beetles and weevils. Off-odors are also important to detect at this point, because many will persist through the fermentation and become detectable in the final product. Grain storage silos are routinely emptied 2 to 4 times a year, sprayed with high-pressure hoses, and allowed to air dry. The area surrounding the outside of the silos is kept clean from grain dust by washing the area with water and by periodic insecticide spraying.

When the grain enters the plant by auger conveyors, it should be sifted on shaking screens to remove any corn cobs, debris, or insects that may have found their way into the silos. The mill room should be washed down with water to reduce grain dust, and approximately every 2 to 6 months, it should be heated to 50° up to 55°C for 30 minutes to kill any insects that may be present.

Reduction of Microbial Contamination

Bacterial and wild yeast contamination of the fermentation is the most important sanitation aspect to control in whiskey production. The source of most of the contamination is the malted barley. Malt routinely has bacterial counts of 2×10^5 to 5×10^8. The malt is added at 60° to 63°C. Thus, many of these microbes survive to propagate during the fermentation. Bacterial levels of the corn and other grains added before the cooking process or at temperatures greater than 88°C are not given much attention because they are killed at these high temperatures.

The most common bacterial contaminants include *Lactobacilli, Bacilli, Pediococci, Leuconostoc,* and *Acetobacter.* These microbes propagate at the expense of yeasts. Microbial contamination will cause plant yield to decrease because the sugar substrates are being utilized by these bacteria to produce compounds other than alcohol. Many of these bacteria produce acids, mainly lactic and acetic acid, which can alter fermentation conditions, as well as lower product quality. Other compounds that can alter the consistency of the whiskey, such as esters and aldehydes, are also produced.

The fermentation process is a rather hostile environment for many microbes. Initially, sugar concentrations may exceed 16%, which provides for high osmotic tension. Initial pH is between 5.0 and 5.4 for a sour mash whiskey and, by completion, will be between 4.0 and 4.5. Final alcohol concentration is approximately 9%, and little, if any, oxygen will be present in this high carbon dioxide environment. These conditions severely restrict the types of contaminating organisms that will proliferate.

Contamination of the fermentations should be minimized. Because many bacteria are airborne, dust is kept to a minimum by water washing all plant surfaces (walls, floors, etc.). Incoming shipments of malt are probed for the determination of bacterial counts. Most specifications limit total bacterial counts to between 200,000 and 1,000,000/g.

Equipment Cleaning

Large fermentation vessels (120,000 to 180,000 L) should be cleaned by filling with hot water and detergent while steam is spread through a CIP sparger in the center of the tank. This process should continue for 30 minutes, at which time the tank is emptied, rinsed with water, and steamed for 2 to 3 hours for sterilization purposes before new mash is pumped into the vessel.

The cooling coils should keep fermentation temperatures below 32°C. Higher temperatures promote yeast cell death and off-flavor production. These coils tend to sustain a buildup of beer stone, which is a hard, rock-like material composed of calcium carbonates, phosphates, and, sometimes, sulfates. As this material builds on the coils, heat transfer efficiency is reduced. To combat this problem, every 6 months, the vessels should be filled with a 1% caustic solution (NaOH) and water, and allowed to soak for 3 days to remove the buildup.

Both the cookers, where the mash is prepared, and the beer still, where the finished beer is pumped, tend to get residual grain buildup through continuous operation. To remedy this, a 1% caustic solution should be prepared weekly to wash the cookers, the beer still, and all connecting lines. Some distillers use a 1% solution of acetic acid instead. The caustic solu-

tion can be prepared in a large tank and pumped to the beer still, then through the connecting lines, and finally into each of the cookers. These areas should then be rinsed with water to wash away all caustic residue.

The lines and stainless steel tanks that accommodate alcohol that is distilled from the beer and pumped to receiving tanks and into barrels for maturation should be periodically rinsed. The nature of the product, crystal-clear alcohol at 140 proof, alleviates the need for more stringent sanitation.

Water quality for distillery products is important to ensure an acceptable end product. The blending of water for distilled spirits is typically from a chlorinated and carbon-treated well or city water supply that is visually clarified using a depth filter prior to blending with the high-proof spirits. The microbial safety/acceptability is ensured through water chlorination, and polishing only prior to blending is required for visual clarity.

SUMMARY

Most soils found in beverage plants are high in sugar content, water soluble, and relatively easy to remove. Ineffective sanitation in a beverage plant can reduce product acceptability because contaminating microorganisms are difficult to remove from the environment. Rigid control of raw materials is essential to ensure a method of detoxifying a finished product that is contaminated.

Bacteria of greatest significance in breweries are non–spore-formers. The most effective means of preventing spoilage of beverage products is to control infection through a comprehensive cleaning and sanitizing program appropriate to each manufacturing establishment. Spray cleaning is most effective, with the incorporation of a properly blended, low-foaming cleaning compound with specific cleaning properties for the soil that exists. Sanitizers such as chlorine, iodine, or an acidanionic surfactant are recommended for the final rinse in fermenters, cold wort lines, and coolers.

The requirements for sanitation increase during the winemaking process and peak at bottling time. A combination of wet and dry cleaning is usually most appropriate. Wine manufacturing equipment should be dismantled as much as possible, thoroughly washed with water and a phosphate or carbonate cleaner for nonmetallic surfaces and caustic soda or equivalent for cleaning metal equipment, then sanitized with hypochlorite or an iodophor. Installation of a circular spray head inside a tank will help remove tartrates, as will soaking with soda ash and caustic soda. Fillers, bottling lines, and other packaging equipment can be cleaned with a CIP system. Prompt processing of grapes after picking will reduce fly infestation.

The control of raw materials is essential for distilled spirits. Both yield and product acceptability are compromised when sanitary conditions are not maintained.

STUDY QUESTIONS

1. What is the TACT approach to cleaning beverage plants?
2. What temperature is used in hot sanitizing a beverage plant?
3. What is the maximum water temperature for cleaning glass-lined tanks in a brewery?
4. What spray and rinse temperature should be used for bottle washing in a brewery?
5. What are the two major methods of pasteurizing beer?
6. What cleaning solution is recommended for washing empty wine storage containers?
7. How can tartrates be removed in a winery?
8. How are wine fermenters cleaned most effectively?
9. How should the mill room of a distillery be cleaned?
10. How should large fermentation vessels in a distillery be cleaned?

REFERENCES

Amerine, M.A., et al. 1980. *The technology of winemaking.* 4th ed. Westport, CT: AVI Publishing.

Arnett, A.T. 1992. Distillery sanitation. Unpublished information.

Flickinger, B. 1997. Automated cleaning and sanitizing equipment. *Food Qual III*: 26–32.

O'Sullivan, T. 1992. High quality utilities in the food and beverage industry. *Dairy Food Sanit* 12: 216.

Remus, C.A. 1991a. Just what is being sanitized? *Beverage World* (March): 80.

Remus, C.A. 1991b. When a high level of sanitation counts. *Beverage World* (March): 63.

Remus, C.A. 1989. Arrhenius' legacy. *Beverage World* (April 1991): 76.

Rosenthal, I. 1992. *Electromagnetic radiation in food science.* 65–104. New York: Springer-Verlag.

Stanton, J.H. 1971. Sanitation techniques for the brewhouse, cellar and bottleshop. *Tech Q Master Brew Assoc Am* 8: 148.

Stewart, G.G. 1987. Alcoholic beverages. In *Food and beverage mycology.* 2nd ed. New York: Van Nostrand Reinhold.

Zoecklein, B.W., et al. 1995. *Wine analysis and production.* New York: Chapman & Hall.

CHAPTER 19

Food Service Sanitation

In the United States, there are over 710,000 food service establishments, employing approximately 10 million people. These food service operations range from mobile restaurant stands to large industrial cafeterias and multiunit fast food chains to plush restaurants. Consumers now spend approximately 43% of their food budgets on meals outside the home. As the food service industry has grown, methods of food production, processing, distribution, and preparation have changed. The major changes have included increased prepackaged food as partially or fully prepared bulk or preportioned servings and centralized food production.

As food production, handling, and preparation techniques and eating habits change, one fact remains: Food is a source for microorganisms that can cause illness. As handling and modern processing methods increase the journey from the production area to the table, opportunities for food to be contaminated with microorganisms become a public health concern. During the second half of the past decade, 2,397 outbreaks of foodborne illnesses occurred. Over 90% of the cases originated from bacteria that cause foodborne illness. Food was the source of 58% of the foodborne illnesses, and 58% of the foodborne illness outbreaks were from food

served in restaurants. The economic cost of these outbreaks has been estimated at $1 billion to $10 billion. The relative risk of food service-related illness is 1 in 9,000, or 5 million illnesses from the 45 billion meals served in a year. This problem is compounded by the growing number of centralized kitchens. Mass feeding operations increase the number of people who may be affected by any contamination. Thus, the challenge of protecting food from contamination is made more complicated and critical.

The primary goal of a food service sanitation program is to protect the consumer from contamination and to reduce the effect of contamination that does occur. It is difficult to protect food from all contamination because pathogenic microorganisms are found in so many locations and on approximately 50% of the people who handle food.

SANITARY DESIGN

Maintaining proper sanitation standards in an improperly designed food service facility is difficult. Time and energy are wasted, and the task becomes difficult and frustrating. This can cause a worker to settle for something less than the desired result.

Cleanability

The primary requirement for sanitary design of food service facilities and equipment is cleanability. Cleanability of an item or surface suggests that it can be exposed for inspection or cleaning without difficulty and that it is constructed so that soil can be removed effectively by normal cleaning methods. Minimal inaccessible locations for soil, pests, and microorganisms to collect will enhance the maintenance of a clean establishment. A facility that is easier to clean can be maintained with less contamination.

Design Features

Design for sanitary features should begin when the facility is being planned. Although most managers inherit an established food service facility, they can improve the environment every time remodeling or renovation takes place or new equipment is purchased.

In most areas, the sanitary design of a food service unit is governed by law. Public health, building, and zoning departments may all have the power to regulate construction of a facility. Regulatory agencies frequently provide checklists of features considered desirable or necessary for good sanitation.

Floors, walls, and *ceilings* should be constructed of material that is easily cleaned and maintained and is attractive. The materials used should be inert, durable, resistant to soil absorption, and smoothly surfaced. Absorbency or porosity of floor material should be considered. When liquids are absorbed, flooring can be damaged, and microbial growth is enhanced. Nonabsorbent floor covering materials should be used in all food preparation and food storage areas; thus, carpeting, rugs, or similar materials should not be installed.

Although flooring material is a critical aspect of sanitation, the way the floor is constructed is also important. Covering at a floor-wall joint facilitates cleaning by preventing accumulation of bits of food that attract insects and rodents. Concrete and terrazzo floors should be sealed to make the floors nonabsorbent and to reduce possible health hazards from cement dust.

Many of the same factors apply to the selection of wall and ceiling materials. Ceramic is a popular and satisfactory wall covering for application in most areas. Grouting should be smooth, waterproof, and continuous, without holes to collect soil. Stainless steel, although expensive, is a satisfactory finish because it is resistant to moisture and most soil, and is durable. Walls of plaster painted with nontoxic paint or cinder block walls are satisfactory for relatively dry areas if sealed with soil-resistant and glossy paint, epoxy, acrylic enamel, or similar materials. Toxic paints, such as those with a lead base, should never be used in a food service facility because flaking and chipping can result in food contamination. Ceilings should be covered with smooth, nonabsorbent, and easily cleanable materials. Smoothly sealed plaster, plastic panels, or panels of other materials coated in plastic are all good choices.

When purchasing equipment, the food service manager should specify that all acquisitions comply with generally acceptable standards. The following characteristics are examples of sanitary features needed in food service equipment:

- Minimal number of parts necessary to perform effectively
- Easy disassembly features for cleaning
- Smooth surfaces free of pits, crevices, ledges, bolts, and rivet heads
- Rounded edges and internal covers with finished smooth surfaces
- Coating materials resistant to cracking and chipping
- Nontoxic and nonabsorbent materials that impart no significant color, odor, or taste to food

Cutting boards are frequently identified as a means of cross-contamination. Wooden cutting boards absorb juices, and plastic boards can harbor microbes in crevices. Knife cuts on plastic surfaces do not heal and offer crevices in which bacteria can evade removal during manual cleaning and contaminate surface sampling. Used foamed polypropylene cutting boards retain high numbers of bacteria, even after a thorough wash (Park and Cliver, 1997). Continued use of polyethylene boards results in numerous knife marks, holes, cracks, and a furry and shaggy appearance, which contribute to bacterial entrapment.

Equipment Arrangement and Installation

Equipment should be arranged to reduce food contamination and to make all areas accessible and cleanable. For example, the soiled-dish table should not be located next to the vegetable preparation sink. Waste processing and the food preparation areas should be located as far apart as possible, and food preparation equipment should not be placed under an open stairway.

When feasible, mobile equipment should be considered to permit easy cleaning of walls and floors. Equipment that is not mobile should be sealed to the wall or to adjoining equipment. If sealing is not practical, equipment should be located approximately 0.5 m from the wall or adjoining equipment to permit easy cleaning. Equipment that is not mobile should be mounted approximately 0.25 m off of the floor or sealed to a masonry base. If the latter approach is used, a 3- to 12-cm toe space should be allowed.

A nontoxic sealant must be used to seal equipment to the floor or wall. Wide gaps caused by faulty construction should not be covered with a sealant because such buried mistakes will be exposed ultimately, opening new cracks to soil, insects, and rodents.

Hand Washing Facilities

Hands are the most viable source of microbial contamination. Therefore, management should provide hand washing facilities in locations where hands are likely to become contaminated, such as food preparation areas, locker or dressing rooms, and areas adjacent to toilet rooms. Because employees may be reluctant to walk very far to wash their hands, these facilities should be conveniently located. Hand washing facilities should consist of mechanized hand washing equipment or a wash bowl equipped with hot and cold water, liquid or powdered soap, and individual towels or other hand-drying devices, such as air dryers. Foot-operated faucets should be installed. More discussion about hand washing is included in Chapter 4.

Welfare Facilities

Dressing rooms or locker rooms should be provided for employees. Street clothes are a viable source of microbial contamination, so uniforms should be provided for wear during the production shift. Dressing rooms should be located outside of the area where food is prepared, stored, and served, and should be separated physically from the other areas by a wall or other barrier. Hand washing facilities should be provided next to the dressing rooms and toilet rooms, with mirrors hung away from the hand washing equipment. The washing facilities and toilet rooms should be scrubbed at least once a day. Receptacles should be provided for waste materials and should be emptied at least once a day.

Waste Disposal

Disposal of garbage and trash is an important operation in food service sanitation because waste products attract pests that can contaminate food, equipment, and utensils.

Waste containers should be leak-proof, pest-proof, easily cleaned, and durable. Plastic bags and wet-strength paper bags are used effectively to line waste containers. Tight-fitting lids are essential.

Waste should be removed from food preparation areas as soon as possible and disposed of often enough to prevent the formation of odor and the attraction of pests. Accumulation of waste materials should be permitted only in waste containers. Waste storage areas should be easily cleaned and pest-proof. If a long holding time is required, refrigerated indoor storage should be provided. Inside storage areas should be easily cleaned and made pest-proof. Large waste containers, such as dumpsters and compactors located on the outside, should be stored on or above a smooth surface or nonabsorbent material, such as concrete or machine-laid asphalt.

An area equipped with hot and cold water and a drain should be provided for waste containers. They should be located such that food in preparation or storage will not be contaminated when the containers are washed.

The volume of trash found in a food service facility can be reduced through the use of pulpers or mechanical compactors. Pulpers can grind waste material into components small enough to be flushed away with water. The water can then be removed so that the processed solid wastes can be trucked away. Mechanical compacting of dry, bulk waste materials is beneficial for establishments with limited storage space because the process can reduce volume to 20% of its original bulk.

Incineration of burnable trash and garbage is another alternative, provided it is permitted in the area and the incinerator is constructed according to all federal and local clean-air standards. Most waste from a food service establishment is high in moisture content and does not burn well. Incinerators should be used only as temporary collection containers for waste.

CONTAMINATION REDUCTION

The wholesomeness of prepared food should be safeguarded through sanitary practices in kitchen and storage areas. Precautions discussed here should be considered when attempting to minimize contamination.

Preparation Area

The prevention of contamination with food poisoning and food spoilage microorganisms, as well as with filth, is especially important in the food preparation area after food preparation and during service. Contamination during service can permit transfer of disease-causing microorganisms directly to the consumer.

Utensils

Thorough washing and disinfecting of utensils are needed to prevent contamination and to maintain a hygienic condition. Disinfection can be accomplished by subjecting utensils to a 77°C environment for at least 30 seconds after cleaning. If chemical germicides are applied at room temperature, 10 minutes or longer of exposure time is required. Contamination will be reduced if cracked, chipped, creviced, or dented dishes or utensils are thrown away. Food particles and microorganisms can collect in the damaged parts and are more difficult to reach during cleaning and sanitizing.

Contamination can be further reduced by requiring that the server not touch any surface that will come in direct contact with the mouth or with food. Dishes or utensils where surfaces touch the counter or table tops should not be permitted to contact foods. When microorganisms are transferred from hands and surfaces to dishes, utensils, or food, they will be transferred to consumers.

Factors that must be considered in the safe preparation and handling of food to reduce the risk of foodborne disease are:

- *Multiple step preparation*: Increased handling leads to more exposure to contamination.
- *Temperature changes*: Heating and cooling places foods in the "danger zone" (3° to 60°C).
- *Large volume*: Products in large volumes require multistep handling and longer times to heat and cool, giving microorganisms more time for growth.
- *Naturally contaminated foods*: Raw produce can be contaminated by field dirt or pesticides, whereas raw red meats and poultry can be contaminated during slaughter, and raw seafood can carry a variety of viruses, bacteria, or parasites.

Another consideration is to survey the progress of items through the establishment—from delivery at the receiving area to service at the table. Temperatures and times should be recorded at the beginning and end of each handling step. Time/temperature curves will help to determine whether existing procedures are adequate to retard microbial growth. Although several control points exist, only a few will be critical control points.

Critical control points are as follows:

- Prevent microbial growth by holding foods below 2°C or above 60°C.
- Ensure microbial destruction by cooking foods above 74°C.

SANITARY PROCEDURES FOR FOOD PREPARATION

The wholesomeness of prepared food can and should be safeguarded through sanitary practices in preparation and storage. Food contaminated with poisonous substances and certain microorganisms can cause food poisoning. Three sanitary procedures can help reduce contamination:

1. *Wash food*. Processed foods do not necessarily require washing; however, all fruits and vegetables to be eaten raw and cooked should be washed. Dried fruits and raisins should be washed, as should raw poultry, fish, and variety meats. Washing poultry will reduce contamination of the body cavity by *Salmonella* and other microorganisms. Fruits, vegetables, and meats should be washed with cold to lukewarm running water to remove dirt. Washed foods should also be drained. If these items are not cooked immediately, they should be refrigerated until time for cooking. If insect infestation is suspected, fresh foods should be soaked in salted water for 20 minutes; any insects will rise to the top of the water.

2. *Protect food from contamination*. Protection from contamination of all food with poisonous substances and bacteria that cause foodborne illness is part of a sanitation program. Cleaning compounds, polishes, insect powders, and other compounds used in a food service operation can get into food inadvertently. To prevent contamination, all chemicals should be stored separately from food and never in the food preparation area or other locations where food is stored and handled.

3. *Thoroughly heat questionable foods*. All foods that may be expected to harbor illness-producing microorganisms—raw meat, poultry, and any foods that may have been recontaminated after processing—should be thoroughly heated when feasible. Heating to 77°C is necessary to destroy non–spore-forming bacteria, such as staphylococci, streptococci, and salmonellae. Time-temperature exposure for destruction of spore-forming bacteria depends on the genus and species.

SANITATION PRINCIPLES

The food service operator has an arsenal of cleaning and sanitizing procedures and products available for selection. The challenge is to determine the most appropriate procedures and products and to apply them properly.

Cleaning Principles

A clean and sanitary establishment is the result of a planned program that is properly supervised and followed according to schedule. Workers who are rushed trying to meet the needs of customers frequently neglect correct practices. A knowledgeable, alert, and strong manager is needed to prevent a breakdown in sanitation discipline. He or she must be able to recognize and institute proper sanitary conditions. Cleaning and sanitizing form the basis of good housekeeping. All food contact surfaces must be cleaned and sanitized after every use, when there is service interruption during which contamination is possible, or at regularly scheduled intervals if the surfaces are in constant use.

Cleaning has been previously referred to as "a practical application of chemistry." A specific cleaning compound should be selected for its special cleaning properties. Also, a compound that is effective for one application may be ineffective for other uses. In addition to being effective and compatible with its intended use, a cleaning compound should fit the needs of the establishment. The important characteristics of a cleaning compound were discussed in Chapter 7. Because some cleaning agents are more effective than others, the quantities required to achieve desired results should be considered in making cost comparisons.

Certain types of soil are not affected by alkaline cleaners—for example, the lime encrustations in dishwashing machines, rust stains that develop in washrooms, and tarnish that darkens copper and brass. Acid cleaners, usually in a formula that contains a detergent, are used for these purposes. The kind and strength of the acid vary with the purpose of the cleaner.

If soil is attached so firmly to a surface that alkaline or acidic cleaners will not be effective, a cleaner containing a scouring agent, usually finely ground feldspar or silica, is used to attack the soil. Worn and pitted porcelain, rusty metals, or seriously soiled floors can be effectively cleaned with abrasives. Abrasives should be used cautiously in a food service facility. Because they can mar a smooth surface, they should be used sparingly on food contact surfaces. (Chapter 7 provides a more detailed discussion of cleaning compounds.)

Sanitary Principles

It may appear to be unnecessary to sanitize cooking utensils that are subjected to heat during cooking. However, heat from cooking is not always uniform enough to raise the temperature of all parts of the item high enough for a long enough time to ensure effective sanitizing.

Sanitizing may be accomplished through heat or chemicals. Heat sanitizing occurs through a high enough temperature to kill microorganisms. Chemical sanitizing is thought to be accomplished primarily through interference of metabolism of the bacterial cell. Regardless of the method used, it is necessary to clean and rinse thoroughly the area and equipment. Soil not removed by cleaning may protect microorganisms from the sanitizer. (Sanitizing methods and compounds are discussed in detail in Chapter 8.)

For food service establishments, chemical sanitizing is accomplished by immersing the object in the correct concentration of sanitizer for approximately 1 minute or by rinsing, swagging, or spraying twice the normally recommended concentration on the surface to be sanitized. The strength of the sanitizing solution should be tested frequently, because the sanitizing agent is depleted in the bacterial killing process. The sanitizer should be changed when it is no longer effective. Sanitizer manufacturing firms normally provide free test kits for monitoring sanitizer strength. Any sanitizing agents that are toxic to humans should be applied only on non–food-contact surfaces.

Sanitizers may be blended with cleaning compounds to create detergent sanitizers. These products can sanitize, but sanitizing should be a separate step from cleaning. A separate step is necessary because the sanitizing power can be destroyed during cleaning. The chemical sanitizer can react with organic matter in the soil. Generally, detergent sanitizers are more expensive than regular cleaning compounds and are more limited in their applications than are detergents.

Cleaning and sanitizing most portable food contact items require a washing area away from the food preparation area. The work station should be equipped with three or more sinks, separate drain boards for clean and soiled items, and an area for scraping and rinsing food wastes into a garbage container or disposal. If hot water is used to sanitize, the third compartment of the sink must be equipped with a heating unit to maintain water near 77°C and with a thermometer. Requirements for cleaning and sanitizing equipment vary among areas. Therefore, regulations that apply to the area should be checked.

Cleaning Steps

There are eight basic steps for manual cleaning and sanitizing of a typical food service facility:

1. Clean sinks and work surfaces before each use.
2. Scrape heavy soil deposits and presoak to reduce gross deposits that contribute to deactivation of the cleaning compound. Sort items to be cleaned, and presoak silverware and other utensils in a solution designated for that purpose.
3. Wash items in the first sink in a clean detergent solution at approximately 50°C, using a brush or dish mop to remove any residual soil.
4. Rinse items in a second sink. It should contain clear, potable water that is approximately 50°C for removal of all traces of soil and cleaning compound that may interfere with the activity of the sanitizing agent.
5. Sanitize utensils in a third sink by immersing the items in hot water (82°C) for 30 seconds or in a chemical sanitizing solution at 40° to 50°C for 1 minute. The sanitizing solution should be mixed to twice the recommended strength if items are immersed in water. Therefore, water carried from the rinse sink will not dilute the sanitizing solution below the minimum concentration required to be effective. Air bubbles that could shield the interior from the sanitizer should be avoided.
6. Air dry sanitized utensils and equipment. Wiping can recontaminate sanitized utensils and equipment.
7. Store clean utensils and equipment in a clean area more than 20 cm off the floor for protection from splash, dust, and contact with food.
8. Cover the food contact surfaces of fixed equipment when not in use.

Stationary Equipment

Stationary food preparation equipment should be cleaned according to the manufacturer's instructions for disassembly and cleaning. Generally, these procedures are:

1. Unplug all electrically powered equipment.
2. Disassemble, wash, and sanitize all equipment.
3. Wash and rinse the balance of the food contact surfaces with a sanitizing solution mixed to twice the strength required for sanitizing by immersion.
4. Wipe all non–food-contact surfaces. Periodically wring out cloths used for wiping down stationary equipment and surfaces in a sanitizing solution. Keep them separate from other wiping cloths.
5. Air dry all cleaned parts before reassembling.
6. Clean stationary items that are designed to have a detergent and sanitizing solution pumped throughout according to the manufacturer's instructions. High-pressure, low-volume cleaning equipment (as discussed in Chapter 7) can be used for cleaning, and spray devices can be used for sanitizing. For sanitizing, spray for 2 to 3 minutes with double-strength solution of the sanitizer.
7. Any wooden cutting boards should be scrubbed with a nontoxic detergent solution and stiff-bristled nylon brush (or a high-pressure, low-volume cleaning wand). Also, a sanitizing solution should be applied after every use. As wooden cutting boards show wear from cuts and scars, they should be replaced with polyethylene boards. Wooden cutting boards should not be submerged in a sanitizing solution.

Floor Drains

Floor drains should be cleaned daily at the end of the cleaning operation. Sanitation workers should wear heavy-duty rubber gloves to remove the drain cover and the debris with a drain brush. When the cover is replaced, it should be flushed with a hose through the drain. Water should not splatter. A heavy-duty alkaline cleaner should be poured down the drain, following the manufacturer's directions for solution preparation. Then the drain should be washed with a hose or drain brush and rinsed. If a quaternary ammonium plug (quat plug) is not used, a chlorine or quat sanitizing solution should be poured down the drain.

Cleaning Tools

Cleaning tools should be stored separately from those used to sanitize equipment and other areas. Clothes, scrubbing pads, brushes, mops, and sponges should be rinsed, sanitized, and air dried after use. Clothing should be laundered daily. All buckets and mop pails should be emptied, washed, rinsed, and sanitized at least once a day.

Mechanized Cleaning and Sanitizing

If properly operated and maintained, mechanized cleaning can more effectively remove contamination from utensils and equipment than can hand cleaning. A trend toward more emphasis on sanitation, combined with increased volume, has been responsible for extensive use of dishwashing machines. In addition, portable high-pressure, low-volume cleaning can be effectively adapted to larger food service establishments.

There are two basic types of dishwashing machines, as discussed by The National Restaurant Association Educational Foundation (1992). They are high-temperature washers and chemical sanitizing machines.

High-Temperature Washers

The major high-temperature washers will be discussed according to model. The sanitizing temperature for these washers should be a minimum of 82°C and maximum of 90°C.

1. *Single-tank, stationary-rack type with doors*. This washer contains racks that do not move. Utensils are normally washed by a compound and water at 62° to 65°C introduced from beneath, but headers can be installed above the rack. A hot-water final rinse follows the wash cycle.
2. *Conveyor washer*. This equipment features a moving conveyor that takes utensils through the washing (70° to 72°C), rinsing, and sanitizing (82° to 90°C) cycles. Conveyor washers may contain a single tank or multiple tanks.
3. *Flight-type washer*. This washer is a high-capacity, multiple-tank unit with a peg-type conveyor. It may have a build-on dryer. This washer is commonly installed in large food service facilities.
4. *Carousel or circular conveyor washer*. This multiple-tank washer moves a rack of dishes on a peg-type conveyor or in racks. Some models have an automatic stop after the final rinse.

Chemical Sanitizing Washers

A brief description of the major chemical sanitizing dishwashers will follow. Glassware washers are also chemical sanitizing machines.

1. *Batch-type dump washer*. The water temperatures for chemical sanitizing should be 49° to 55°C. This washer combines the wash and rinse cycle in a single tank. Each cycle is timed, and the cleaning compound and sanitizer are automatically dispensed.
2. *Recirculatory door-type, nondump washer*. This washer is not completely drained of water between cycles. The wash is diluted with fresh water and re-used during the next cycle.
3. *Conveyor type washer with or without a power prerinse*. The name of this equipment defines its function.

The following considerations are important for the procurement and operation of dishwashing equipment.

1. Optimal capacity should be provided.
2. A booster heater with sufficient capacity to supply the dishwashing equipment with 82°C water for a sanitizing rinse in hot water should be provided.
3. Proper installation, maintenance, and operation are necessary to ensure that the equipment adequately cleans and sanitizes.
4. Dishwashing equipment should be incorporated in an efficient layout for optimal utilization of the unit and personnel.
5. Accurate thermometers are needed to ensure that appropriate water temperature is used.
6. A prewash cycle should be considered to omit scraping and soaking of soil utensils.
7. If machines have compartments, the rinse water tanks should be protected by a device to prevent wash-water flow into the rinse water.
8. Larger dishwashers should be cleaned at least once a day, according to the manufacturer's instructions.

Table 19–1 provides further information about symptoms, causes, and cures related to dishwashing problems.

Cleaning-in-Place (CIP) Equipment

Equipment such as automatic ice-making machines and soft-serve ice cream and frozen

Table 19–1 Dishwashing Difficulties and Solutions

Symptom	Possible Cause	Suggested Solution
Soiled dishes	Insufficient detergent	Use enough detergent in wash water to ensure complete soil removal and suspension
	Low wash-water temperature	Keep water temperature within recommended ranges to dissolve food residues and to facilitate heat accumulation (for sanitation)
	Inadequate wash and rinse times	Allow sufficient time for wash and rinse operation to be effective (time should be automatically controlled by a timer or by conveyor speed)
	Improper cleaning	Unclog rinse and wash nozzles to maintain proper pressure-spray pattern and flow conditions; overflow should be open; keep wash water as clean as possible by prescraping gross soil from dishes, etc.; change water in the tanks at proper intervals
	Improper racking	Verify that racking or placement is done according to size and type; silverware should always be presoaked and placed in silver holders without sorting or shielding
Films	Water hardness	Use an external softening process; use the proper detergent to provide internal conditioning; check temperature of wash and rinse water (water maintained above recommended temperature ranges may cause a precipitate film)
	Detergent carryover	Maintain adequate pressure and volume of rinse water; worn wash jets of improper angle of spray may cause wash solution to splash over into final rinse spray
	Improperly cleaned or rinsed equipment	Prevent scale buildup in equipment by adopting frequent and adequate cleaning practices; maintain adequate pressures and volume of water
Greasy films	Low pH; insufficient detergent; low water temperature; improperly cleaned equipment	Maintain adequate alkalinity to saponify greases; check cleaning compounds and water temperature; unclog all wash and rinse nozzles to provide proper spray action (clogged rinse nozzles may also interfere with wash tank overflow); change water in tanks at proper intervals
Foaming	Detergent dissolved or suspended solids in water	Change to a low-sudsing product and reduce the solid content of the water
Streaking	Alkalinity in the water; highly dissolved solids in water	Use an external treatment method to reduce alkalinity within reason (up to 300–400 ppm); selection of a proper rinse additive will eliminate streaking; above this range, external treatment is required to reduce solids
	Improperly cleaned or rinsed equipment	Maintain an adequate pressure and volume of rinse water; alkaline cleaners used for washing must be thoroughly rinsed from dishes

continues

Table 19–1 continued

Symptom	Possible Cause	Suggested Solution
Spotting	Rinse-water hardness	Provide external or internal softening; use additional rinse additives
	Rinse-water temperature too high or too low	Check rinse-water temperature; dishes may be flash-drying, or water may be drying on dishes rather than draining off
	Inadequate time between rinsing and storage	Change to a low-sudsing product; use an appropriate treatment method to reduce the solid content of the water
	Food soil	Adequately remove gross soil before washing; the decomposition of carbohydrates, proteins, or fats may cause foam during the wash cycle; change water in the tanks at proper intervals
Coffee, tea, metal staining	Improper detergent	Food dye or metal stains, particularly where plastic dishware is used, normally require a chlorinated machine washing detergent for proper destaining
	Improperly cleaned equipment	Keep all wash sprays and rinse nozzles open; keep equipment free from deposits of films or materials, which could cause foam buildup in future wash cycles

Source: National Sanitation Foundation, 1982.

yogurt dispensers are designed to be cleaned by passing a detergent solution, hot-water rinse, and sanitizing solution through the unit. These machines must be designed and constructed so that the cleaning and chemical sanitizing solution remains within a fixed system of tubes and pipes for a predetermined amount of time, and the cleaning water and solution cannot leak into the remainder of the machine (National Restaurant Association Educational Foundation, 1992). All food contact surfaces of these machines must be reached by the cleaning and sanitizing solutions. The CIP equipment must be self-draining, and the units should be designed for inspection through exposure of the area that has been cleaned.

Cleaning Recommendations for Specific Areas and Equipment[1]

AREA Floors
FREQUENCY Daily and weekly
SUPPLIES AND EQUIPMENT Broom, dustpan, cleaning compound, water, mop, bucket, and powered scrubber (optional)

Daily
1. Stack chairs on table surfaces or remove from area to be cleaned
2. Sweep and remove all trash from floor

Other Requirements
1. Self-explanatory

2. Use push broom

---

[1]The recommendations in this section are adapted from Marriott, 1982.

3. Clean all table surfaces

3. Wipe food particles into a container. Wash table with warm soapy water. Rinse the surfaces with clean water.

4. Post signs warning of wetness
5. Mop floors or use a mechanical scrubber for larger operations

4. Self-explanatory
5. Mix 15 g of detergent per liter of clean water. Rinse with clean water

Weekly
1. Apply steps 1–4 for daily cleaning
2. Scrub floors

Other Requirements
1. See above
2. Use powered scrubber and/or buffer on floor. Rinse with clean 40°–55°C water. Squeegee the floor and dry mop.

AREA Walls
FREQUENCY Daily and weekly
SUPPLIES AND EQUIPMENT Hand brush, sponge, cleaning compound, bucket, water, and scouring powder

Daily
1. Spot clean as necessary

Other Requirements
1. Use 15 g of cleaning compound per liter of water. Hand wipe all dirty areas. Rinse with clean water. Wipe dry. Mop floor areas to remove spillage.

Weekly
1. Remove all debris away from walls
2. Assemble cleaning equipment

3. Scrape walls
4. Rinse wall surfaces
5. Wipe dry
6. Scrub floor area to remove any spillage

Other Requirements
1. Self-explanatory
2. Mix 15 g of cleaning compound per liter of water
3. Use hand brush. Scrub tiles and grout
4. Use clean, warm water
5. Use clean cloths or paper towels
6. Self-explanatory

AREA Shelves
FREQUENCY Weekly
SUPPLIES AND EQUIPMENT Hand brush, detergent, sponge, water, and bucket

Weekly
1. Remove items from the shelves
2. Brush off all debris
3. Clean shelves in sections

4. Replace items on the shelves

5. Mop floor to remove soil

Other Requirements
1. Store items on a pallet or other shelves
2. Brush debris into a pan or container
3. Mix detergent and warm water and scrub shelves
4. Check for damaged cans and discard as appropriate
5. Use a clean, damp mop

EQUIPMENT   Stack oven
FREQUENCY   Clean once a week thoroughly; wipe daily
SUPPLIES AND EQUIPMENT   Salt, metal scraper with long handle, metal sponges, cleaning compound in warm water, 4-L bucket, sponges, stainless steel polish, ammonia, vinegar, or oven cleaner, as appropriate

Weekly	*Other Requirements*
1. Turn off the heat and scrape the interior	1. Sprinkle salt on hardened spillage of oven floor. Turn thermostat to 260°C. When the spillage has carbonized completely, turn off the oven. Cool thoroughly. Scrape the floor with a long-handled metal scraper. Use a metal sponge or hand scraper on the inside of doors, including handles and edges.
2. Brush out scraped carbon and other debris	2. Begin with top deck of stack oven. Brush out with stiff-bristle brush and use dustpan to collect
3. Wash doors	3. Usc a hot detergent solution on enameled surfaces only; rinse; wipe dry
4. Brush interior chamber	4. Use a small broom or brush for daily cleaning
5. Clean and polish exterior	5. Wash the top, back, hinges, and feet with warm cleaning compound solutions; rinse; wipe dry. Polish all stainless steel

Note: Never pour water on or use a wet cloth or sponge to soak. Do not squeegee, drip, or pour water inside oven to clean.

EQUIPMENT   Hoods
FREQUENCY   Once a week (minimum)
SUPPLIES AND EQUIPMENT   Rags, warm soapy water, stainless steel polish, degreaser for filters

Weekly	*Other Requirements*
1. Remove filter	1. After removal, carry filter outside and rinse with a degreaser and run it through the dishwasher after all dishes and eating utensils have been cleaned
2. Wash hood inside and outside	2. Use warm, soapy water and a rag to wash hoods completely on the inside and outside to remove grease. Clean drip trough in an area below filters
3. Shine hood with polish	3. Spray polish on hood and wipe off. Use a clean rag on inside and outside

| 4. Replace filters | 4. Put filters back into proper place after they have completely drained |

EQUIPMENT   Range surface unit
FREQUENCY   Thoroughly once a week
SUPPLIES AND EQUIPMENT   Putty knife, wire brush, damp cloth, hot detergent water, 4-L container, vinegar or ammonia, as appropriate

Weekly	*Other Requirements*
1. Clean back apron and warming oven (or shelf)	1. Let surfaces cool before cleaning. Use a hot, damp cloth, wrung almost dry. Wipe back apron and warming oven. Remove hardened substances with a putty knife; scrape edge of plates. Scrape burned material from top, flat surfaces with a wire brush
2. Remove top sections, scrape edges and flat surfaces	2. Lift plates. Remove burned particles with a putty knife; scrape edge of plates. Scrape burned material from top, flat surfaces with a wire brush
3. Wipe heating element	3. Wipe heating elements with a damp cloth
4. Clean base and exterior	4. Wipe with a cloth dampened with hot detergent water
5. Clean grease receptacles and drip pans	5. Soak grease receptacles and drip pans in a detergent solution for 20–30 min; scrub, rinse, and dry

Note: Do not immerse heating elements in water.

EQUIPMENT   Griddles
FREQUENCY   Daily
SUPPLIES AND EQUIPMENT   Spatula, pumice stone, paper towels, hot detergent solution

Daily	*Other Requirements*
1. Turn off heat. Remove grease (after each use)	1. Scrape surface with a spatula or pancake turner after surface has cooled. Wipe clean with dry paper towels. Use pumice stone block to clean hard-to-remove burned areas on plates after use. Avoid daily use of pumice stone where possible
2. Clean grease and/or drain troughs	2. Pour a hot detergent solution into a small drain and brush. Rinse with hot water
3. Empty grease receptacles	3. Remove grease from scrapings and supporting pans with hot detergent solution. Rinse and dry

4. Scrub guards, front, and sides of the griddle

4. Using a hot detergent solution, wash off grease, splatter, and film. Rinse and dry

EQUIPMENT   Rotary toaster
FREQUENCY   Daily
SUPPLIES AND EQUIPMENT   Warm hand-detergent solution, brush, rags, stainless steel polish, nonabrasive cleaner

Daily

1. Disconnect and disassemble

2. Clean surface and underneath

3. Clean frame and interior as far as is accessible

Other Requirements

1. After the toaster is cooled, remove pan, slide, and baskets. Move basket midway up front. Press to left carrier chain to permit pins to slip out of holes in the basket

2. Use a soft brush to remove crumbs from the front surface and behind bread racks

3. Wipe clean with a warm hand-detergent solution. Rinse and dry. Polish if necessary with a nonabrasive cleaning powder. The exterior casing should not be allowed to collect excessive grease or dirt. Prevent water and cleansing compounds from touching the conveyor chains. If the frame is stainless steel, polish all of this material

EQUIPMENT   Coffee urns
FREQUENCY   Daily
SUPPLIES AND EQUIPMENT   Outside cleaning compound (stainless steel polish), inside cleaning compound (baking soda), urn brush, faucet, and glass brush.

Daily

1. Rinse urns
2. Heat water and half fill the urn tank

3. Brush the liners, faucet, gauge glass, and draw-off pipe

Other Requirements

1. Flush with cold water after use
2. Be certain that outer jacket is three-quarters full of water. Turn on heat. Open water inlet valve and fill coffee tank with hot water to the coffee line. Add recommended quantity of cleaning compound (15 g/L). Allow solution to remain in the liner for approximately 30 min, with the heat on full

3. Scrub inside of tank, top rim, and lid. Draw-off pipe 2 L of solution and pour it back to the fill valve and sight gauge. Insert the brush in the gauge glass and coffee draw-off pipe, and brush briskly

4. Drain

4. Open the coffee faucet and completely drain the solution. Close the faucet

5. Rinse

5. Open the water inlet valve into the coffee tank. Use 4 L of hot water. Open the faucet for 1 min to allow water to flow and sterilize the dispensing route

6. Disassemble faucet and thoroughly clean

6. Scrub with brush. Rinse spigot thoroughly. Clean

7. Refill (twice weekly)

7. Make a solution (1 cup of baking soda in 4 L of hot water) and hold in the urn for approximately 15 min. Drain. Flush thoroughly with hot water before use.

Note: Place a tag on the faucet while the urn is soaking with the cleaning compound.

Biweekly

1. Fill urn with a destaining compound solution

Other Requirements

1. Fill urn with 80°C water. Add destaining compound in the ratio of 2 tablespoons to 20 L of water (or as directed by manufacturer)

2. Draw off mixture and repour

2. Open spigot and draw off 4 L; thoroughly remix to allow the mixture to come into faucet. Allow the solution to stand for 1 hr at 75°–80°C

3. Scrub liner, gauge glass
4. Clean faucet

3. Use a long-handled brush to loosen scale
4. Take faucet valve apart and clean all components. Soak in hot water until reassembled

5. Rinse and reassemble faucet

5. Rinse urn liner three or four times with hot water. Repeat until all traces of the compound are removed

6. Refill urn

6. Place enough hot water in the urn to fill and allow to remain until next use. Drain and replace with fresh water when ready to make coffee

Note: To destain vacuum-type coffee makers, use a solution of 1 teaspoon of compound per liter of warm water. Fill the lower bowl up to within 5 cm of the top and assemble the unit.

EQUIPMENT   Iced tea dispensers
FREQUENCY   Daily
SUPPLIES AND EQUIPMENT   Rags and warm, soapy water

Daily
1. Clean exterior
2. Wash drip pan

3. Wash trough

4. Inspect parts

5. Wash plastic parts

Other Requirements
1. Wipe exterior parts with a damp cloth
2. Empty drip pan and wash drip pan and grill with a mild detergent and warm water
3. Open front jacket, remove mix trough, and wash in a mild detergent and warm water
4. When inspecting parts, remember their order of removal, so they will be replaced properly
5. Do not soak plastic parts in hot water or wash in dishwashing machines

EQUIPMENT   Steam tables
FREQUENCY   Daily
SUPPLIES AND EQUIPMENT   Dishwashing detergent, spatula, scrub brush, and rags

Daily
1. Turn off heating unit

2. Remove insert pans and transport them to the dishwashing area

3. Drain water from the steam

4. Prepare the cleaning solution; assemble supplies
5. Scrape out food particles from the steam table
6. Scrub interior and clean exterior
7. Rinse exterior

Other Requirements
1. Turn steam valve counterclockwise (steam heated). Turn dial to OFF position (electrically heated)
2. Lift one end up until clear; then pull forward, grasping the other with the free hand, and remove. Clean inserts thoroughly after each use by hand cleaning and sanitizing processes. Air dry. Store in a clean area until needed
3. Remove the overflow pipe, using a cloth to prevent injury
4. Dissolve 30 mL of dishwashing compound in a suitable container
5. Use a spatula or dough scraper

6. Use a scrub brush and cleaning solution
7. Use enough clear water to remove all traces of detergent

Note: Hot-food tables, electric, mobile: Clean corrosion-resistant steel after each use. Ordinary deposits of grease and dirt can be removed with mild detergent and water. Whenever possible, thoroughly rinse and dry after washing.

EQUIPMENT   Refrigerated salad bars (with ice beds or electrically refrigerated)
FREQUENCY   After each use
SUPPLIES AND EQUIPMENT   Detergent, plastic brush, and sanitizing agent

Daily
1. Transfer shallow pans or trays and to preparation areas following meal service
2. Clean and sanitize table counter

3. Periodically descale to prevent rust, lime, or hard-water scale formation (nonrefrigerated types)

4. Defrost electrically refrigerated units

Other Requirements
1. Run insert pans and/or trays through dishwashing machine

2. Wash off and/or scrub table surfaces with detergent and plastic brush. Rinse. Sanitize by swabbing with a solution containing a sanitizing agent

3. Fill table bed with boiling water. Add a descaling compound in proportions recommended by the manufacturer. Allow to stand for several hours. Scrub with a plastic brush. Drain. Rinse thoroughly. Sanitize by spraying on solution

4. Turn off electric current and defrost ice formation from the coils as often as required. Follow up with a cleaning procedure, as described above

EQUIPMENT   Milk dispenser
FREQUENCY   Daily
SUPPLIES AND EQUIPMENT   Sanitized cloth or sponge, mild detergent, sanitizing agent

Daily
1. Remove empty cans from the dispenser

2. Wipe up spillage as it occurs

3. Clean interior when units are empty

4. Clean exterior

Other Requirements
1. Place a container under the valve. Open the valve and tip can forward in dispenser to drain out the remaining milk. Extract tube. Lift out the oar

2. Use a sanitized cloth or sponge to prevent possible contamination

3. Wash entire inner surface with a milk cleaning solution. Rinse

4. Follow normal procedures for cleaning stainless steel. If steel shows discoloration or stains, swab with a standard chemical to stand 15–20 min before rinsing with clean water and polishing with a soft cloth

5. Disassemble and clean valves daily or as frequently as empty cans can be removed to keep valves clean and sanitary

5. To remove, lift valves; swing valve upward and slide pins free of recesses to disengage from the plastic well upward to remove. Wash in detergent water. Rinse and sanitize

6. Place full cans in the dispenser

6. Wipe the bottom of milk cans with a sanitizing solution before placing in dispenser. Clamp-type dispensing valves should be thoroughly cleaned and sanitized before reuse

EQUIPMENT   Deep fat fryer
FREQUENCY   Daily
SUPPLIES AND EQUIPMENT   Knife, spatula, wire brush, detergent, long-handled brush, vinegar, nylon brush, dishwashing compound

Daily
1. Turn off the heating element
2. Drain and filter the fat (after each use)

3. Remove baskets

4. Remove strained container or cup as often as necessary for cleaning

5. Close the drain. Fill the tank with water

6. Turn on heating element

7. Turn off heat
8. Scrub interior

Other Requirements
1. Allow fat to cool to 65°C
2. Open drain valve and catch drained fat in a container. Drain entire kettle contents and filter into a container. Place a clean fat container into the well or wash and replace the original container

3. Scrape off the oxidized fat with a knife. Remove loose food particles from the heating units with a spatula or a wire brush. Flush down sides of the kettle with a scoop of hot fat. Soak basket and cover in a deep sink with hot detergent

4. Clean off sediment and place container in the kettle. Stir hot fat and whirl sediment to permit settling in the sediment container

5. Add water up to fat level. Add 60 mL of dishwashing compound

6. Set control at 121°C and boil 10–20 min, depending on need

7. Open drain. Draw off cleaning solution
8. Using a long-handled brush, scrub the interior. Flush out with water. Clean the basket with a nylon brush and place it back in the kettle

9. Rinse and sanitize

9. Fill the kettle with water. Add one-half cup of vinegar to neutralize the remaining detergent. Turn on the heating element. Boil 5 min and *turn off heat*. Drain. Rinse with clear water

10. Air-dry parts

10. Expose baskets and strainer to air and dry

11. Clean exterior

11. When kettle is cool, wipe off exterior with a grease solvent, or a detergent solution. Rinse

Weekly

1. Fill kettle to fat level with water. Heat to at least 80°C or allow to boil for 5–10 min. Turn off the heat.
2. Add one-half tablespoon of destaining compound (stain remover, tableware) per liter of water. Agitate solution and loosen particles remaining on sides of the kettle.
3. Place screens and strainers in 80°C water containing one-half tablespoon of destaining compound per liter. Allow to stand overnight. Rinse thoroughly and air-dry.
4. Drain kettle and rinse thoroughly before replacing cleaned screen and strainer.

EQUIPMENT   Vegetable chopper
FREQUENCY   Daily
SUPPLIES AND EQUIPMENT   Brush, sponge, cloth, bucket, detergent, sanitizer solution

Daily

1. Disassemble parts after each use

2. Clean knives, bowl guard, and bowl

3. Clean parts and under chopper surface

4. Reassemble detachable parts

Other Requirements

1. Turn off power. Wait until knives have stopped revolving

2. Remove blades from the motor shaft and clean them cautiously and carefully. Wash with a hand detergent solution. Rinse and air-dry. Remove all food particles from the bowl guard. If the bowl is removable, wash it with other parts; if the bowl is fixed, wipe out food particles from table or base. Clean with a hand detergent solution; rinse and air-dry

3. Immerse small parts in a hot hand detergent solution; wash, rinse, and air-dry

4. Replace comb in guard. Attach bowl to the base and knife blades to the shaft. Drop guard into position

Note: Choppers vary considerably in mechanical operational details.

EQUIPMENT   Meat slicer
FREQUENCY   Daily
SUPPLIES AND EQUIPMENT   Bucket, sponge, cloth, brush, detergent, sanitizer solution

Daily	*Other Requirements*
1. Prepare equipment for cleaning	1. Disconnect. Remove meat holder and chute by loosening screw. Remove scrap tray by pulling it away from the knife. Remove the knife guard. Loosen bolt at the top of knife guard in front of the sharpening device. Remove bolt at the bottom of the knife guard behind chute. Remove guard
2. Clean slicer parts	2. Scrub parts in a sink filled with hot detergent solution. Rinse with hot water. Immerse in a sanitizer solution. Air-dry
3. Clean the knife blade	3. Use a hot detergent solution to wipe off knife blade. Wipe from center to edge. Air-dry
4. Clean receiving tray and underneath tray	4. Wipe the receiving tray with a hot detergent solution. Rinse in hot clear water. Air-dry

Note: Do not pour water on or immerse this equipment in water.

AREA   Welfare facilities (see Chapter 15).

FOOD SERVICE SANITATION REQUIREMENTS

Effective cleaning does not occur without effective management. An effective cleaning program results from deliberate efforts to understand the cleaning needs of the food service establishment and to provide the necessary equipment and cleaning agents to fill the requirements. Sanitation managers must fully ensure that available sanitation tasks are not omitted and must plan ahead to maximize the use of resources, familiarize new employees with cleaning routines, establish a logical basis for such supervisory tasks as inspections, and save employees time that might be spent in deciding which tasks to perform. Table 19–2 provides a partial sample cleaning schedule for a food preparation area. A full cleaning schedule can incorporate the same format. The schedule adopted should constitute a detailed and comprehensive list arranged logically so that nothing will be overlooked.

Major cleanup functions should be scheduled when contamination of foods is least likely to occur and interference with service is minimized. Vacuuming and mopping should not occur during preparation and serving of food. However, cleaning should be accomplished as soon as possible after these operations to prevent soil from drying and hardening and to reduce bacterial multiplication. Food service sanitation management should also include scheduling of cleaning operations for even spacing of periodic cleaning and arrangement of jobs in the proper order.

Table 19–2 Sample Cleaning Schedule (Partial), Food Preparation Area

Item	When	What	Use	Who
Floors	As soon as possible	Wipe up spills	Cloth, mop and bucket, broom, and dustpan	_____
	Once per shift between rushes	Damp mop	Mop, bucket, or mechanical scrubber	_____
	Weekly, Thursday evening	Scrub	Brushes, bucket detergent (brand)	_____
	January, June	Strip, reseal	See procedure	_____
Walls and ceilings	As soon as possible	Wipe up splashes	Clean cloth; portable high-pressure, low-volume cleaner; or portable foam cleaner	_____
	February, August	Wash walls		Contacted specialists
Work tables	Between uses and at end of day	Empty, clean, and sanitize drawers; clean frame, shelf	See cleaning procedure for each table	_____
	Weekly Saturday p.m.		See cleaning procedure for each table	_____
Hoods and filters	When necessary	Empty grease traps	Container for grease	_____
	Daily, closing	Clean inside and out	See cleaning procedure	_____
	Weekly, Wednesday evening	Clean filters	Dishwashing machine	_____
Broiler	When necessary	Empty drip pan, wipe down	Container for grease; clean cloth	_____
	After each use	Clean grid tray, inside, outside, top	See cleaning procedure for each broiler	_____

Source: Adapted from *Applied Foodservice Sanitation,* 4th Edition. Copyright 1992 by the Educational Foundation of the National Restaurant Association.

A new cleaning program should be discussed with employees at a meeting, which also can serve as an opportunity to demonstrate the use of new equipment and procedures that relate to the program. It is important to explain the need for the program and its anticipated benefits, and to emphasize the importance of following the procedures exactly as written. Communication with em-

ployees can be responsible for less deviation from specified procedures.

The sanitation program should be evaluated for effectiveness after it has been implemented. Evaluation can be accomplished during continuous supervision and self-inspection. Monitoring is necessary to verify that all of the procedures are followed. Evaluations should be documented to verify that the program is being followed and that expected results have occurred. Documentation can be in the form of periodic inspection reports.

Employee Training

Training requires time away from the job for both workers and management and should involve the employment of training specialists. Printed material, posters, demonstration, slides, and films should be used as training devices.

It is difficult to measure the return on the investment in sanitation training. In fact, the benefits are not always measurable. The savings can sometimes be realized through prevention of an outbreak of foodborne illness or of establishments being closed until local health standards are met. It is difficult to measure the improved image attained through a sanitary operation, even though increased sales will result.

Employee training is important because it is difficult to recruit competent and motivated workers. Periodic training is essential because the industry has a higher rate of employee turnover than do most organizations.

On-the-job training can be effective for certain tasks but is not comprehensive enough for sanitation training. Each employee involved in food service sanitation should become familiar with the sanitation concept and sanitary practices required for job performance.

An ideal method for training employees of a large firm is to set up a training department and hire a training director. This approach is being adopted in many large and medium-sized food service operations. In fact, food service trainers have established their own professional association, the Council of Hotel and Restaurant Trainees. Furthermore, the Association of Food and Drug Officials develops and publishes food sanitation codes and encourages food protec-

tion through the adoption of uniform legislation and enforcement procedures.

In most food service operations, the supervisors, rather than professional trainers, normally conduct the sanitation training. Therefore, a previously trained employee or one certified in food service sanitation should personally conduct the training.

The effectiveness of a training program can be evaluated by the ability of employees to perform their assigned tasks. If standards of achievement have been set before training and are understood, progress can be determined by measuring individual achievement against those standards. Employee turnover data, absenteeism and tardiness reports, and performance data determine the value of the training program. The quality of training is also reflected in the amount of guest complaint reports and customer return rates.

The National Restaurant Association Educational Foundation (1992) recognizes two methods that are most frequently used to evaluate the effectiveness of training. An objective method involves the use of tests or quizzes to determine employee comprehension. The other method is job performance by employees, as evaluated by management. Training effectiveness may be enhanced through praise of employees, wall charts that recognize superior performance, pins, and certificates. Organizations such as the National Restaurant Educational Foundation and some regulatory branches provide certification courses that provide both training and recognition.

SUMMARY

Food is a source for microorganisms that cause food spoilage and illness. Increased handling of food is responsible for a more complicated and critical challenge of protecting food from contamination. To improve sanitation in food service establishments, the facility and equipment should be designed for cleanability. Choosing equipment with sanitary features has been simplified by a number of equipment standards provided by organizations and manufacturers.

Food should be safeguarded through sanitary practices in the receiving, storage, preparation, and serving areas. It should be handled with equipment and utensils and in a physical facility that has been thoroughly cleaned and sanitized. If properly operated and maintained, mechanized cleaning by means such as a dishwasher can effectively remove contamination from utensils and equipment. To manage the sanitation operation of a food service facility properly, a cleaning and sanitizing program should be written, supervised, and evaluated, with subsequent documentation of results.

STUDY QUESTIONS

1. What construction materials should be used for the (a) floors, (b) walls, and (c) ceilings of food service facilities?
2. What kind of faucets should be installed for hand washing in food service facilities?
3. What temperature is needed for the disinfection of utensils?
4. What end-point cooking temperature is recommended to ensure microbial destruction?
5. What water temperature is needed for the third compartment sink?
6. What water temperature is needed for the first and second compartment sinks?
7. What water temperature is needed for a dishwasher?

REFERENCES

National Restaurant Association Educational Foundation. 1992. *Applied foodservice sanitation*. 4th ed. Chicago: Education Foundation of the National Restaurant Association.

Park, P.K., and D.O. Cliver. 1997. Cutting boards up close. *Food Qual III* 22, 57.

SUGGESTED READINGS

Guthrie, R.K. 1988. *Food sanitation*. 3rd ed. New York: Van Nostrand Reinhold.

Harrington, N.G. 1987. How to implement a SAFE program. *Dairy Food Sanit* 7: 357.

Longrée, K., and G. Armbruster. 1996. *Quantity food sanitation*. 5th ed. New York: John Wiley & Sons.

National Sanitation Foundation. 1982. Procedures for spray-type dishwashing machines. *International recommended field circulation*. Ann Arbor, MI.

CHAPTER 20

Management and Sanitation

Because many jobs in the food processing and food service industry (including cleanup) do not require previous formal training or education, many unskilled workers select the food industry as the area of their first employment. High school and college students frequently work in the food industry. The age and multiple interests of these employees have been blamed for the high employee turnover in the food service industry.

Most management personnel in the food industry will agree that the rapid turnover rate of sanitation employees can be attributable to a lack of training and education. This condition has apparently contributed to a lower salary scale, especially among food service employees. Therefore, management has a challenge in recruiting and training employees for the sanitation operation. Another challenge, sometimes difficult to accept, is the need to give sanitation a professional and exciting image so that employees will proudly and enthusiastically accept their responsibilities related to maintenance of a hygienic operation. Management plays a key role in the effectiveness of a sanitation program.

Sanitation employees are paramount to food safety. Turnover of sanitation employees should be minimized. A complete, stable and well-trained team can cross-train new employees to attain maximal efficiency and reduce plant downtime.

The sanitation team is a valuable asset. Through its efforts, future production problems can be prevented. The team should be recognized for its efforts to inspire more productivity. If the sanitation team is perceived as a necessary expense, that is how it will perform. Employees tend to perform at the level of their employer expectations. Carsberg (1998) stated, "If you give enough people what they want, then you will be assured to get everything you want." The sanitation team needs to know that through its work, the manufactured food will be clean and safe.

MANAGEMENT REQUIREMENTS

The success or failure of a sanitation program is attributable to the extent that management supports it. Management can affect the success of a sanitation program. The discussion that follows will suggest the key role that management plays in the organization and implementation of an effective sanitation program.

Management Philosophy

Unfortunately, too many managers in the food industry are not convinced that an organized sanitation program is necessary for the success of their operations. Yet, their philosophy or attitude toward sanitation "sets the state" for the entire organization. Holland (1980) suggested that managers lack interest in sanitation and indicated that companies operating with such an outlook start up and vanish long before the benefits of a sound sanitation program can be realized. Sanitation programs have not been consistently supported by management because they reflect a cost where dividends cannot always be accurately measured in terms of increased sales and profits. Frequently, lower and middle management have difficulty selling the sanitation concept when top management does not fully comprehend it.

However, some progressive management teams have been more enthusiastic about a sanitation program. They have recognized that it can be used in promotion and can improve sales and product stability. Other managers have been able to improve the image of their organizations through sanitary practices and quality assurance laboratories. Sophisticated cleaning equipment can also add to the impressiveness of an operation. Progressive firms have realized that an effective sanitation program will ultimately save money.

Management Knowledge of Sanitation

If management is not educated on the value of a sanitation program, progress in this area will be slow. Without understanding and support, the effectiveness of a sanitation program is reduced.

Management must support and promote sanitation because of its direct impact on corporate planning, marketing, and the company's relationship with the law. Sanitation programs have a direct impact on the industry-regulatory interface. The U.S. Food and Drug Administra-

tion (FDA) can prohibit the preservation, production, packaging, storage, or sale of any food under unsanitary conditions. Management should recognize that an effective sanitation program reduces cleaning expenses through increased efficiency.

Management Commitment

Before a successful sanitation program can be implemented, management must accept the fact that rigid sanitary practices must be incorporated. After recognition of the need for effective sanitation, the commitment should be communicated to all employees, followed by the adoption of a workable program.

Program Development

A successful sanitation program should be tailored to the operation. An example is that a meat processing operation will mostly operate conveyors, mixers, and other open equipment and containers, requiring more handheld hoses and wands than foam and high-pressure units. In contrast, a milk plant can incorporate more automated cleaning equipment, such as cleaning-in-place (CIP) technology.

According to Graham (1992), program planning should involve:

1. Preventive measures to reduce food spoilage and growth of potential microbial contaminants that can be extremely costly to a firm
2. Employee input, especially that of production supervisors and line workers
3. Quality assurance (QA) personnel to identify areas that require attention or improvement and have knowledge of current technical developments in sanitation and control of microbial growth

4. Plant engineering evaluation of equipment and layouts, as required equipment must be maintained for effective sanitation
5. Purchasing department input to reduce expenditures for equipment and supplies
6. Delegation of sanitation responsibility to a trained sanitation manager who reports directly to the plant manager and has authority and accountability to make the program work

A complete sanitation program should incorporate a Hazard Analysis Critical Control Point (HACCP) program. Good Manufacturing Practices (GMPs) devised by the FDA are a driving force for sanitary program design and hygienic operations because the primary objective of these practices is the prevention of adulteration (contamination).

Program Follow-Through

Effective management means that everyone involved with sanitation works as a team to share problems, solutions, and knowledge. A successful sanitation program that has been developed and implemented must be regularly checked through monitoring and recording results. Another effective check is through an outside sanitation audit. Trained auditors with valuable experience provide a fresh perspective (Graham, 1992) and new ideas. An inside audit by the sanitation manager or general manager should also be conducted periodically. Detailed deficiency lists should be maintained, and action should be taken to correct the problems noted on the list.

EMPLOYEE SELECTION

Employees who handle food should be carefully selected to be free of infectious diseases. They should have a personal hygienic level above the average of the population (Jowitt 1980), and they should maintain appropriate hygiene.

The level of expertise of the sanitation team is changing rapidly. In the past, it was a standard practice to hire inexperienced employees and assign them to the sanitation team without any training. Today, educational courses are being developed and updated to meet the demands of effective sanitation, and many sanitation employees are provided with some form of training. The American Institute of Baking is a leader in providing sanitation technology courses through classes conducted at its headquarters in Manhattan, Kansas, or by correspondence courses.

Employee Training

The importance of adequate training of employees has been suggested several times throughout this book. It is especially important to train sanitation employees in the basics of sanitation because *nothing happens in a food establishment until the facility is clean*. Sanitation employees should be serious, dedicated professionals who clearly comprehend the company's policy and their role in the organization. Although the sanitarian reports directly to someone on the management team, indirect responsibilities and allegiance belong to management, labor, regulatory agencies, and consumers. A finely tuned sanitation program consists of effective interaction between a QA department and a research and development laboratory—within the organization or in a private laboratory—for an accurate assessment of sanitary practices.

Management must ensure that the sanitarian is well qualified. The sanitarian should be educated in the operations of the food facility, the role of cleaning compounds and sanitizers, and food microbiology. Additional expertise that the sanitarian should have as a result of experience and/or training includes knowledge of specific surface design and hardness, porosity, vulnerability to oxidation, and corrosion of surfaces to be cleaned, so that the appropriate cleaning equipment, cleaning compounds, and sanitizers may be determined.

An effective management team should ensure that the sanitarian is educated in the safety and efficacy of cleaning compounds, the functions of detergent auxiliaries and sanitizers, and the most effective cleaning equipment. A sanitarian who understands the characteristics of cleaning equipment, cleaning compounds, and sanitizers can reduce waste and employee injury, and simultaneously optimize cleaning efficiency. Further benefits include reduced water consumption, sewage load, and sanitation labor.

Management must disseminate information to sanitation workers in a form for easy comprehension. Information should be presented in a clear, easily accessible instruction manual that provides facts related to cleaning all areas and equipment, including the selection and application of cleaning compounds and sanitizers for all cleaning applications. The instruction manual should also include a sanitation plan and material on operational methods, pest control, hygienic practices, and preventive maintenance. The adoption and application of these principles will affect operational appearance, practices, and performance, and will positively reflect on the company image.

Some companies conduct intensive, formal in-house training programs for sanitation employees. These firms can provide sanitation technology based on their QA program needs. Those needs that are most frequently addressed include determination of required manpower and effective communication to nontechnical personnel.

Management should realize that consumers desire and deserve wholesome products. Responsible managers should acknowledge the importance of well-trained employees and should conduct sound employee training programs as an integral part of their processing or food service operations. Therefore, sanitarians should attend training courses and should seek the assistance of regulatory personnel for discussion of sanitation standards and public health needs in order to discharge their responsibilities in the training of employees. As a necessary adjunct to training, management has the responsibility to provide the materials and facilities necessary for employees to practice what they have been taught. Managers have found that they benefit from such employee training activities and that the success of the activity depends partially on their own leadership.

A number of industry activities has produced meritorious results through a coordinated industry/regulatory authority approach to problems in the field of food protection:

- Minimum sanitation standards for membership eligibility in food associations have been established.
- An achievement citation to food sanitarians for outstanding accomplishments has been devised.
- Encouragement and support of uniform and impartial interpretation and application of food sanitation ordinances have been accomplished.
- Promotion of an understanding of the need for and the benefits to be derived from industry self-inspection has been accomplished.
- Management's leadership role has been asserted by providing individualized on-the-job training for employees.

If these recommended practices are followed, it is possible to optimize human resources and available technology in attaining good hygiene throughout the food operations and especially in raw materials, production, and the finished product.

A well-trained sanitation team will reduce production downtime, reduce product recalls, and improve employee morale. A clean plant is a more productive plant. Because motivation comes from within the individual, it cannot be forced upon employees. However, they can be given reasons to be positive about their work. Carsberg (1998) indicated that sanitation team members should be given reasons to justify why they must do quality work. Because most employees want to do a good job, the importance of their assignments should be emphasized.

Other Sources for Sanitation Training and Education

Trade associations and regulatory agencies provide information beneficial in educating and training employees. Examples are the FDA, Food Safety and Inspection Service, American Association of Meat Processors, American Meat Institute, National Food Processors Institute, and National Restaurant Association Educational Foundation. These organizations periodically send their membership educational information related to sanitation and conduct short courses in this area.

Professional associations such as the International Association of Milk, Food and Environmental Sanitarians and Affiliates are committed to improving the professional status of the sanitarian and to educating the food industry regarding the need for effective sanitation programs. Universities serve in a similar capacity. Professional associations and universities both contribute to the education of food organizations by offering courses related to sanitation.

MANAGEMENT OF A SANITATION OPERATION

Management experts define *management* as "getting things done through people." Sanitation management has three basic responsibilities:

1. Delegation of responsibilities or telling employees what must be done
2. Training employees by showing them how responsibilities should be executed
3. Supervision to ensure that all responsibilities are properly executed

Managers should continue to make certain by regular inspections that assignments are being properly performed. Although employees are properly trained, they must be supervised to ensure proper conduct of responsibilities.

The competence of food sanitation personnel and the effectiveness of the program administration are major factors in achieving the objectives of a food sanitation program, regardless of the type of enforcement methods employed. Managers cannot afford to be mistaken in their judgment or unreasonable in their decisions, because such actions are concerned with the health of consumers. Success in food sanitation and consumer protection programs also depends on understanding, interest, and support within the top levels of the regulatory authority and other branches of government.

Management and Supervision

The key to success of any sanitation program is supervision. The role of management in supervision involves the audit of the sanitation program to ensure that the rules are being followed. Program requirements may be considered the cement that holds the building blocks of sanitation together. Supervisors should always be on the alert to identify unsafe practices that may creep into an operation. Thorough supervision should be reinforced, with a continuous training program to keep employees informed of their responsibilities.

As in other management positions requiring continuous surveillance of the operation, monitoring a food production facility involves an organized supervision routine. Food handler supervision should incorporate the same health standards to regulate workers as are adopted in screening prospective employees—for example, daily checks of employees for infections that can be transmitted through food. In fact, many local health ordinances require that the proprietor who knows or suspects that an employee has a contagious disease or is a carrier must notify health authorities immediately.

The burden of managers is reduced and supervision is made easier if employees are motivated to do a good job. Effective training can be a motivating force. Professional treatment of employees can improve morale and can be a positive motivating force.

Management should convey an attitude that sanitation is an important job. Sanitation employee efforts should be recognized, not ignored. Instead of ignoring their efforts and criticizing failures, they should be commended for maintaining a hygienic environment and for their contribution to safe food products. This approach provides positive reinforcement and motivates employees to perform to a higher level. The sanitation team needs to know that its work is valued and critical to food safety.

Public Relations Considerations

Managers must understand the principles of public relations and constantly practice them in the course of executing a food sanitation program. It must employ every legitimate practice and technique to interpret the program's needs and objectives and to motivate people to cooperate. Food sanitarians should acquire and apply basic public relations skills.

The mass media can be the sanitarian's most important tool in communicating and marketing the hygienic concept. A relationship with news reporters should permit a free exchange of information and create an atmosphere of mutual understanding. It is highly desirable to stress improvements, achievements, new programs, appointments, promotions, and similar developments. These practices promote better understanding and an appreciation of the program by everyone concerned: the public, the food industry, and food sanitation personnel.

Sanitarians should have a practical understanding of the fundamentals of human motivation. It is more productive to work with groups than with individuals. Food sanitarians should recognize that there is more to their duties and responsibilities than making inspections. They can find that other types of activities are also productive and rewarding—for example, taking the necessary time to talk to a class or civic group, preparing news announcements, participating in radio or TV programs, and designing educational material. The promotion and interpretation of food sanitation needs and goals can often be more easily accomplished when understood and supported by community leaders or civic groups.

Whenever a food sanitarian makes recommendations, the operating costs are frequently increased. Selling the need for and benefits of such recommendations is a public relations challenge.

Cooperation with Other Agencies

Joint regulatory/industry advisory committees have frequently provided valuable assistance in the evaluation of new developments, techniques, and procedures. Consumers can also be represented on these advisory committees, which may be helpful in counseling on broad policy matters and in establishing and maintaining wholesome industry and regulatory agency relationships. Benefits accrue to all when program development and improvement are brought about through co-operative efforts.

Job Enrichment

Many employees, including managers and supervisors, consider the sanitation operation to be a second-rate job. Therefore, it is important that sanitation workers be made aware of the importance of their responsibilities. Through effective management, sanitation can be glamorized and made more exciting. An effective job enrichment program can create more interesting and rewarding work for employees. This program can also make them feel more a part of the operation and can actually be more demanding of employees by assigning them more responsibilities and emphasizing self-inspection.

Self-Supervision

A major challenge of the supervisor is to set a good example for other employees. A supervisor who does not follow the rules will not be effective. The supervisor is frequently the most experienced employee in the operation—and may be the most immune to learning and to eradicating bad habits. This problem can be solved if supervisors recognize that their major challenge is to provide their customers with a wholesome product.

Self-Inspection

Self-inspection should be considered a regular task performed by trained personnel who are familiar with the establishment's operation. These inspections may be conducted by the owner/operator or by managers, supervisors, or sanitation consultants. These inspections are more beneficial if they are conducted with the aid of a checklist.

TOTAL QUALITY MANAGEMENT

Gould (1992) has described *total quality management* (TQM) as the modern term to describe how firms are becoming more successful today. He considers TQM to be a new philosophy that "sets goals for employees working with management, employees having a voice in the operation, and employees who feel that they are a part of the ownership of the firm."

This author agrees with Gould (1992) that TQM is more than a buzz word. This innovative concept involves management and employees working together for improved productivity, cost reduction, and product uniformity and acceptability.

In the TQM approach, management contributes resources and direction but does not dominate. Small groups are encouraged to identify opportunities and to discuss them with a cross-functional steering committee for prioritization. Projects are implemented unless costs are increased. Employees are empowered and work within acceptable guidelines.

Relationship of TQM to Sanitation

The TQM philosophy is applicable to the management aspect of sanitation. To more effectively maintain a sanitary environment, sanitation must become more exciting, and everyone must accept responsibility for the maintenance of a hygienic operation.

In the past, sanitation operations have been primarily a policing program instead of a direct responsibility of the individual employee. However, TQM stresses the involvement of all employees in decisions and accountability. The term itself denotes *providing the customer with a uniform and acceptable product through the training, instruction, and efforts of all employees.*

A *hygienic product* implies product safety, shelf stability, and compliance of the item with the latest regulations. It appears at the date of this writing that additional emphasis will be placed on the use of TQM principles for sanitation programs of the future. TQM can assist those involved with sanitation as it has guided firms in the manufacturing and service industries. It is a management philosophy that has arrived and when incorporated will be a valuable tool for sanitarians and firms that adopt the fundamentals of TQM and practice the principles.

SUMMARY

A major challenge of management in the food industry is to recruit and train employees for an effective sanitation operation. The success or failure of a sanitation program depends on the extent to which management supports the program.

An effective sanitation program includes provisions for constant training and education of employees. Educational information can be disseminated through sanitation training manuals and short courses given by trade associations, professional organizations, or regulatory agencies.

The major functions of sanitation management are to delegate responsibilities and to train and supervise employees. Self-supervision and self-inspection are two tools that contribute to a more effective sanitation program.

STUDY QUESTIONS

1. What is management?
2. What health requirements should be considered when selecting employees?
3. What sources exist for sanitation training and education?
4. What are three basic responsibilities of sanitation management?
5. What is the major key to success in a sanitation program?

REFERENCES

Carsburg, H. 1998. Motivating sanitation employees. *Food Qual* 5, no. 1: 68.

Gould, W.A. 1992. *Total quality management for the food industries*. Baltimore: CTI Publications.

Graham, D. 1992. Five keys to a complete sanitation system.

Prepared Foods 101, no. 5: 50.

Holland, G.C. 1980. Education is the key to solving sanitation problems. *J Food Prot* 43: 401.

Jowitt, R. 1980. *Hygienic design and operation of food plant*. Westport, CT: AVI Publishing.

SUGGESTED READING

Guthrie, R.K. 1988. *Food sanitation*. 3rd ed. New York: Van Nostrand Reinhold.

APPENDIX A

Glossary

Acid: A substance with a pH of less than 7.0.

Acids, strong: Substances that release high concentrations of hydrogen ions in a solution giving a low pH; examples are muriatic and sulfuric acids.

Acids, weak: Substances with a moderately low pH; examples are organic acids, such as acetic and hydroxyacetic acids.

Adulteration: The addition of an improper, foreign substance.

Aerobic: The ability to live and reproduce only in the presence of oxygen.

Air screen: A unit that provides a strong downward air movement of air at doors to prevent refrigeration loss and insect entry.

Alkali: A substance with a pH of more than 7.0.

Alkalies, strong: Substances that release high concentrations of hydroxyl ions in a solution giving a high pH; examples are sodium hydroxide and potassium hydroxide.

Alkalies, weak: Substances that release moderate to low concentrations of hydroxyl ions in a solution giving a moderately high pH; examples are sodium bicarbonate and sodium tetraphosphate.

Anaerobic: The ability to live and reproduce in the absence of oxygen.

Antibiotic: A compound produced by a microorganism that interferes with the growth of another microbe.

Antiseptic: A chemical substance used to interfere with or inhibit the growth of certain microorganisms.

A_w: The unit of measurement for water requirement of microorganisms.

Bacilli: Rod-shaped bacteria.

Bacteria: Single-celled microorganisms that decompose matter, resulting in subsequent product spoilage and/or foodborne illness.

Bactericide: A chemical substance that will kill certain bacterial cells.

Bacteriostat: An agent that inhibits the growth of bacteria but does not necessarily kill them.

Botulism: Intoxication resulting from consumption of a toxin produced by *Clostridium botulinum*.

Buffer: A material that moderates the intensity of an acid or alkali in solution without reducing the quantity of acidity or alkalinity.

Builder(s):   An adjunct added to cleaning compounds to control properties that tend to reduce the surfactant's effectiveness.

Celsius: Temperature scale related to the Fahrenheit scale by the formula $5/9$ ($^{\circ}$Fahrenheit -32°) = $^{\circ}$Celsius (centigrade).

Clean: Free of visible soil.

Cleaning: The physical removal of soil from a surface.

Cocci: Spherically shaped bacteria.

Contaminate: To add foreign and unwanted matter to an object or environment.

Control point: Any step or procedure where biological, physical, or chemical factors can be controlled.

Coving: A curved sealed edge between a floor and wall to facilitate cleaning and retard insect harborage.

Critical control points: A step or procedure at which control can be applied and a food safety hazard can be prevented, eliminated, or reduced to an acceptable level.

Critical limits: Tolerances prescribed to ensure that critical control points effectively control a hazard.

Cross-contamination: The transfer of microorganisms from one food to another through a nonfood surface, such as equipment, utensils, or human hands.

Deflocculation (dispersion):   The action of breaking up aggregates into individual parts.

Detergent: A chemical cleanser similar to soaps but of differing chemical nature.

Disinfect: To remove potentially pathogenic microorganisms from an object or environment.

Disinfectant: A chemical used to destroy the growing forms, but not necessarily the spores, of potentially pathogenic microorganisms.

Dispersion: Deflocculation; breaking up of a mass into fine particles that are suspended in solution.

Endotoxin: A toxin produced within a microorganism and liberated when the microorganism disintegrates.

Exotoxin: A toxin excreted by a microbe into the surrounding medium.

Fahrenheit: A temperature scale related to Celsius (centigrade) by the formula $9/5$ ($^{\circ}$Celsius $+ 32^{\circ}$) = $^{\circ}$Fahrenheit.

Flocculation: Agglomeration or building of a macrofloc resulting from coagulation into larger particles until the sheer force of water movement prevents further building or until it settles out.

Germicide: A chemical that will kill certain microbial cells.

GRAS substances: Food additives that are designated as "Generally Regarded As Safe" for use.

Host: A plant or animal harboring another as a parasite or as an infectious agent.

Hygiene: Practices necessary for establishing and maintaining good health.

Immunocompromised: An individual who is susceptible to becoming ill from a foodborne illness due to an existing disease or weakened physical condition.

Infection: A condition caused by the invasion of the tissues of a host by living pathogenic microorganisms.

Infestation: Occupation or invasion by parasites other than bacteria.

Intoxication: A disease caused by consumption of poisons naturally occurring in food or produced by pathogenic microorganisms.

Listeriosis: The foodborne infection caused by *Listeria monocytogenes* and having a high mortality rate among immunocompromised individuals.

Mycotoxins: Compounds or metabolites produced by a wide range of fungi that have toxic or other adverse effects on humans and animals.

Nonionic: Lacking an electrical charge through a balance of negatively and positively charged compounds.

Organism: An individual living thing.

Parasite: An organism that derives its nourishment from a living plant or animal host and does not contribute to the host's well-being but does not necessarily cause disease.

Pathogen: A microorganism capable of producing disease when it enters the human or animal body.

pH: A logarithmic measurement, on a scale from 0 to 14, of acidity and alkalinity due to hydrogen and hydroxyl ion concentration.

Pollution: The accumulation of foreign, unwanted matter in an environment in which it becomes a nuisance or a danger to the health of the environment.

Potable: Suitable or safe for drinking.

Precipitate: A deposit of an insoluble substance resulting from chemical or physical changes in a solution.

Precision: Representative of how closely replicate values approximate each other.

Sanitary: Free of disease-causing microorganisms and other harmful substances.

Sanitation: The creation and maintenance of conditions favorable to good health.

Sanitize: Treatment by heat or chemicals to reduce the number of microorganisms present.

Soap: A compound of fatty acids and alkalies that has cleaning properties.

Spore: An inactive, resistant, resting, or reproductive body that can produce another vegetative individual under favorable conditions.

Sterile: Free from all living microorganisms.

Taint: To contaminate with undesirable organisms.

Toxin: A chemical produced by living organisms that is poisonous to humans and animals.

Virus: Any of a large group of infectious agents that require a living host for reproduction.

Index

A

Accelerated death phase. *See*
  Microorganisms, growth pattern of
Acid cleaning compounds. *See* Cleaning
  compounds
Acquired immune deficiency syndrome
  (AIDS), 30
Activated carbon absorption, 204
Activated sludge. *See* Wastewater
  treatment
Aeromonas, hydrophila, 27, 36
Alkaline cleaning compounds. *See* Cleaning
  compounds
Arcobacter butzleri, 31–32
Assay procedures for evaluation of
  sanitation effectiveness, 98–99
  direct contact contamination removal,
    98–99
  direct surface agar plating technique
    (DSAP), 99
  surface rinse method, 99
  vacuum method, 99
A_w, 17–18

B

Bacillus cereus, 26, 36
Bacteria. *See* Microorganisms
Bactericides, 19, 353
Bacteriostats, 19, 354

Biochemical oxygen demand, 187, 189
Biofilms, 19–20
Bird infestation, 220
Bottle cleaning, 310, 317
Botulism. *See Clostridium botulinum*
Brewery sanitation, 307–312
  air conditioning unit cleaning, 312
  aseptic filling, 310
  beer pasteurization, 311–312
  bottle cleaning, 310
  cleaning compounds, 309
  control of microbial infection, 308–309
  heat pasteurization, 309–310
  hygienic practices in, 310

C

Campylobacter, 25, 27–29, 56, 261
Campylobacteriosis, 25, 28–29
Cetylpyridinum chloride, 154–155
Chain of infection, 54
Chloramines, 144
Chlorine dioxide. *See* Sanitizing
Chelating agents. *See* Cleaning compounds
Cleaning compounds, 119–137
  builders for, 128
  characteristics, 119
  chemical burns from, 136–137
  classification of
    acid, 122–124
    alkaline, 121–122, 266

auxiliaries, 127–130
    cleaning auxiliaries, 128–130
    neutral, 130
    solvent, 124, 127
  detergent auxiliaries, 127
  enzyme based, 126
  factors affecting effectiveness of, 120
  function of, 119–121
  handling, precautions for, 131–137
  properties of, 119
  selection of, 131
  sequestrants, 120–121, 128–129
  surfactants, 121, 129–130
  terminology, 120–121
Cleaning equipment, 160–184
  centralized high pressure and foam, 168, 266
  cleaning-in-place (CIP), 169–181, 251–253, 266, 297
    multiuse, 176–179
    reuse, 174–176
    single use, 171, 173–174
  cleaning out-of-place, 181–182, 253–254
  foam
    centralized, 167, 265–266, 296
    portable, 167, 265–266, 296
  gel, 167–168, 266, 296–297
  high pressure units, 162–163
  high-pressure, low-volume, 163–165, 265, 296
    centralized, 164–165
    implementation, 159–160
      portable, 163–164
  hot water wash, 163, 265, 296
  low pressure units, 162
  mechanical abrasives, 160–161
  microprocessor control unit, 179–181
  selection, 159, 165–167
  slurry, 168, 297
  steam guns, 162–163
Cleaning media
  air, 118
  water, 118–119
Clostridium botulinum, 24–26, 29
Clostridium perfringens, 24, 26, 53, 261

Cockroaches, 207–210
  common species of, 208–209
  control of, 209–210
  detection of, 209
Contamination
  human, 65–66
  protection against, 57–58
  sources of, 55–57
Control point(s), 83, 354
Critical control points, 83–86, 354
Critical control points decision tree. *See* HACCP
Critical limit(s), 84–85
Cumulative sum (CUSUM) control charts, 111–112
Current good manufacturing practices (CGMPs), 80–81

D

Dairy processing plant
  CIP equipment for, 251–253
  cleaning compounds for, 248, 249
  cleaning steps for, 250–251
  construction considerations for, 246–247
  COP equipment for, 253
  pathogens in, 244–246
  sanitation for cheesemaking, 255–256
  sanitation principles for, 248
  sanitizers for, 249–250
  soil characteristics of, 247
  storage equipment cleaning in, 254–255
Decimal reduction time (D value), 37
Detergent(s). *See* Cleaning compounds
Detergent auxiliaries. *See* Cleaning compounds
Diagnostic tests
  assay for *E. coli*, 49
  CAMP test, 48
  crystal violet test, 48
  diagnostic identification kits, 48
  DNA hybridization and colorimetric detection, 50
  enzyme-linked immunoassay tests, 46–47

Fraser enrichment broth/modified oxford agar, 48
IDEXX bind, 47
immunomagnetic separation and flow cytometry, 48
methyl umbelliferyl glucuronide test (MUG), 48–49
micro ID and minitek, 49–50
polymerase chain reaction, 50
random amplified polymorphic DNA, 47–48
Salmonella 1-2 Test, 47
Disease transmission, 67, 70
Disolved air flotation, 196–197
Distillery sanitation, 318–319
  contamination reduction, 318–319
  equipment cleaning, 319

E

Electrodialysis, 204
Electronic pasteurization, 38–39
Environmental regulations, 6–7
Escherichia coli, 32, 36, 66
Escherichia coli 0157:H7, 32–34, 37, 152, 154, 245–246
  growth requirements, 34
  illness from, 32–34
Ethylenediaminetetraacetic acid (EDTA), 129

F

Flies. *See* Housefly
Food and Drug Administration (FDA), 5–6, 346, 347, 349
Foodborne disease. *See* Foodborne illness
Foodborne illness, 22–37
  foodborne disease outbreak, 22
  impact from, 1, 22
  psychosomatic food illness, 23
Food infection, 22
Food intoxication, 22
Food toxicoinfection, 32

Food Safety and Inspection Service (FSIS), 6, 349
Food service sanitation, 321–344
  CIP equipment for, 329
  cleaning steps for,
    coffee urns, 335–336
    deep fat fryer, 339–340
    floors, 331–332
    floor drains, 328
    griddles, 334–335
    hoods, 333–334
    iced tea dispensers, 336–337
    meat slicers, 341
    milk dispensers, 338–339
    range surface units, 334
    refrigerated salad bars, 338
    shelves, 332
    stack ovens, 333
    stationary equipment, 327–328
    steam tables, 337
    toasters (rotary), 335
    vegetable choppers, 340
    walls, 332
  contaminated reduction, 324–325
  employee training, 343
  requirements, 341–343
  sanitary design for, 321–323
  sanitary principles, 326–327
  sanitary procedures, 325
  washers for, 328–329
  waste disposal, 323–324
  welfare facilities, 323
Fruit fly, 211, 215
Fruit and vegetable plant sanitation, 291–302
  cleaning considerations, 294–295
  cleaning equipment for, 296–297
  cleaning procedures for, 298–300
  construction considerations for, 292–294
  contamination sources, 291–292

G

Generation interval, 20
Glutaraldehyde. *See* Sanitizing
Good manufacturing practices (GMPs), 6

H

Hand washing, 66–67
Hazard
    categories, 80–83
    definition of, 76
    risk categories, 83
Hazard Analysis Critical Control Points
    (HACCP) 8, 75–88, 96, 98, 160, 263, 269,
    289
    decision tree, 84
    development of, 77–79, 86–88
    implementation, 79–80, 86–88
    maintenance of, 86–88
    principles of, 78, 81–86
    validation, 88
    verification, 86
Helicobacter pylori, 32
Hepatitis, 28
Housefly, 210–215
    control of, 211–215
    effects of, 210–211
Hygiene
    definition of, 60
    employee, 60–65
    requirements, 71–73
Hygiene design. *See* Sanitary design of food
    facilities

I

Inorganic chloramines. *See* Sanitizing
Insecticide(s). *See* Pesticides
Integrated pest management, 222–224
Iodophor(s). *See* Sanitizing
Ion exchange, 204

J

Job enrichment, 93, 351

L

Lag phase. *See* Microorganisms, growth
    pattern of
Legionellosis, 25
Listeria monocytogenes, 20, 27, 29–31, 77,
    148, 244–245, 261–262, 263
Listeriosis, 29–31, 355
Logarithmic growth phase. *See*
    Microorganisms, growth pattern of

M

Meat contamination, 260–261
Meat discoloration, 259–260
Meat and poultry plant sanitation, 258–282
    cleaning compounds for, 266–267
    cleaning equipment for, 265–266
    effects on production discoloration,
        259–260
    equipment design for, 263
    pathogen control in, 261–262
    plant design, 262–263
    practices, 268–269
    principles, 264
    procedures for
        brine curing and packaging areas,
            275–276
        dry curing areas, 276
        fresh product processing area, 273–274
        garments, 281
        livestock pens, 271
        livestock trucks, 271
        locker and rest rooms, 280–281
        low-temperature rendering area,
            279–280
        offices, 280
        packaged meats storage area, 279
        poultry mechanical eviscerators,
            272–273
        poultry pickers, 272–273

processed products, offal, and storage cooler, 273

processed products area, 274–275

receiving and shipping area, 273

slaughter area, 271–272

smokehouses, 276–278

spiral freezers, 278–279

trolley washing, 280

wash area, 279

wire containers, 280

sanitizers for, 267–268

troubleshooting tips for, 281

Mesophiles, 17

Mice. *See* Rodent infestation

Microorganisms

bacteria, 13

definition of, 11

destruction of, 37–39

deteriorative effects of, 20–22

diagnostic tests for, 46–50

growth pattern of, 14–16

inhibition of, 39–40

molds, 12

proliferation of, 16–20

quantitative determination of, 40–46, 98–99

viruses, 13–14

yeasts, 13

Molds. *See* Microorganisms

Monitoring program. *See* Quality assurance monitoring program

Most probable number (MPN), 42

Mycotoxins, 35–36, 355

N

Nonalcoholic beverage plant sanitation, 303–307

cleaning practices for, 304

sanitation principles for, 303–304

O

Organic chloramines. *See* Sanitizing

Oxidation-reduction potential, 18

P

Pesticides, 211–215

nonresidual, 212–213

precautions related to, 213–214

residual, 211–212

use of, 211–215, 220–222

pH, 18

Pheromone traps, 215

Polishing ponds, 204

Protection auxiliaries, 127–128

Psychosomatic food illness, 23

Psychrotrophs, 17

Pulsed light, 39

Q

Quality

characteristics, 91

definition of, 91

Quality assurance (QA)

basic tools, 102

components of, 91

description of, 90

elements of, 95–96

functions of, 92

management role in, 93

organization for, 92, 94–96

product recall, 100–101

sampling for, 101–102

sanitation programs, administration of

assay procedures for, 98

data interpretation for, 99

establishment of, 95

evaluation of, 98

major responsibilities, 92–93
monitoring program, 99–100
structure of, 94
total quality assurance, 95–96

R

Radiation, 38, 140
Randomization, 101
Rapid methods of microbial load
    determination. *See* Microorganisms,
    quantitative determination of
Rats. *See* Rodent infestation
Rotating biological contactor (RBC). *See*
    Wastewater treatment
Reduced death phase. *See* Microorganisms,
    growth pattern of
Rodent infestation, 215–220
    control of, 217–218
    determination of, 216–217
    eradication of, 218–220

S

Salmonella, 23–24, 26, 53, 56, 66, 154, 235,
    261
Salmonellosis, 23–24, 26
Sampling. *See* Quality assurance, sampling
    for a QA program
Sanitary design of food facilities, 226–230,
    232–236, 246–247, 262–263, 283–285,
    292–294, 321–323
    construction considerations for, 227–229,
        232–236
    construction materials for, 230
    dust removal, 235
    equipment for, 235
    pest control considerations, 230
    processing area, 229

site preparation for, 226, 232
ventilation for, 235
Sanitation
    costs, 158
    definition of, 2, 355
    employee selection, 347
    employee training, 347–349
    importance of, 1
    laws and regulations, 4–8
    low-moisture food manufacturing and
        storage, 236–241
    management requirements, 345–351
    practices, 8–9
    program, 1
    program development, 346–349
    standard operating procedures, 81
    voluntary programs, 8
Sanitizing
    acid, 149–151, 153, 154
    acid-quat, 151, 153
    bromine, 147
    chemical, 140–155, 250, 267–268, 297–298
    chlorine, 141–145, 153, 154, 267–268,
        297
    equipment, 154, 182–184
    gluteraldehyde, 151, 153
    hydrogen peroxide, 151
    iodine, 145–147, 153, 154, 155
    microbiocides, 151–152
    oxine, 144–145
    ozone, 151
    phenols, 298
    quaternary ammonium compounds, 147–
        149, 153, 154, 297–298
    radiation, 140
    strength determination, 155
    thermal, 139, 250, 267
Scouring compounds, 130
Seafood plant sanitation, 283–290
    cleaning schedule, 287–288
    contamination sources, 285–286

construction considerations for, 283–285
HACCP models for, 289
recovery of by-products for, 288
sanitation principles, 286–287
sanitizers for, 288
site requirements for, 283–284
voluntary inspection for, 289
Shigella, 26, 32, 53, 66, 67
Soap. *See* Cleaning compounds
Soil
  attachment characteristics, 117–118
  chemical characteristics of, 114, 115–116
  characteristics in dairy plants, 247
  definition of, 114
  deposit classification, 116
  physical characteristics of, 114–115
  removal of, 117–118
  solubility characteristics of, 116
  surface characteristics of, 116
Staphylococci, 66
Staphylococcus aureus, 23, 26, 61, 261
Stationary growth phase. *See*
  Microorganisms, growth pattern of
Statistical quality control
  charts, 104–111
  program standards
    definition of, 111
    lower control limits, 107–108
    upper control limits, 107–108
  rating scales
    exact measurement, 111
    subjective evaluation, 111
  role of, 102–103
Streptococci, 36–37, 64–66
Surfactants. *See* Cleaning compounds

T

Thermal death time, 37
Thermophiles, 17

Total quality management, 91, 351–352
  relationship to sanitation, 351–352
Trichinella spiralis, 27

U

United States Department of Agriculture
  (USDA) regulations, 6

W

Waste disposal, 187–205
  solid, 192
  strategy, 188
  survey, 188
Wastewater
  residue in, 189–192
  fats, oil, and grease (FOG), 191
  nitrogen, 191
  phosphorus, 191
  sampling of
    biochemical oxygen demand (BOD),
     189
    chemical oxygen demand (COD),
     189–190
    dissolved oxygen (DO), 190
    total organic carbon (TOC), 190
  settleable solids (SS), 191
  sulfur, 191
  total dissolved solids (TDS), 191
  total suspended solids (TSS), 191
Wastewater treatment, 192–205
  disinfection, 205
  pretreatment, 193–195
    flow equalization, 195
    screening, 195
    skimming, 195
  primary treatment, 195–197
    flotation, 196–197
    sedimentation, 195–196

secondary treatment, 197–203
   activated sludge, 200–201
   aerobic lagoons, 200
   anaerobic lagoons, 198–199
   contact stabilization process, 201
   extended aeration process, 201
   land application, 202
   magnetic separation, 203
   oxidation ditch, 201–202
   rotating biological contactor, 202–203
tertiary treatment, 203–205
   chemical oxidation, 205
   filtration, 204
   microstraner separation, 203–204
   physical separation, 203–204
   tertiary lagoons, 204
Water activity (A_w). *See* A_w

Web of causation, 54
Winery sanitation, 312–318
   bottling area cleaning, 314
   cleaning floors and walls, 313
   cleaning used cooperage, 314–315
   equipment cleaning, 313–314
   pest control, 317
   sanitation monitoring, 317–318
   sanitary practices, 315–316
   sterile filtration, 316
   tartrate deposit removal, 315

Y

Yeasts, 13
Yersinia enterocolitica, 28, 32
Yersiniosis. *See Yersinia enterocolitica*